SCIENCE

A Four Thousand Year History

SCIENCE

A Four Thousand Year History

PATRICIA FARA

OXFORD
UNIVERSITY PRESS

OXFORD
UNIVERSITY PRESS

Great Clarendon Street, Oxford OX2 6DP

Oxford University Press is a department of the University of Oxford.
It furthers the University's objective of excellence in research, scholarship,
and education by publishing worldwide in

Oxford New York

Auckland Cape Town Dar es Salaam Hong Kong Karachi
Kuala Lumpur Madrid Melbourne Mexico City Nairobi
New Delhi Shanghai Taipei Toronto

With offices in

Argentina Austria Brazil Chile Czech Republic France Greece
Guatemala Hungary Italy Japan Poland Portugal Singapore
South Korea Switzerland Thailand Turkey Ukraine Vietnam

Oxford is a registered trade mark of Oxford University Press
in the UK and in certain other countries

Published in the United States
by Oxford University Press Inc., New York

British Library Cataloguing in Publication Data
Data available

Library of Congress Cataloging in Publication Data
Fara, Patricia.
Science: a four thousand year history/Patricia Fara.
p. cm

Includes bibliographical references and index.
ISBN 978-0-19-922689-4
1. Science-History.   2. Science and civilization.   I. Title.
Q125.F27 2009

509—dc22

2008050975

Typeset by SPI Publisher Services, Pondicherry, India
Printed in Great Britain
on acid-free paper by
CPI Antony Rowe, Chippenham, Wiltshire
ISBN 978-0-19-922689-4

3 5 7 9 10 8 6 4

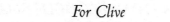

For Clive

Acknowledgements

Ilong ago lost count of the people who started laughing when I outlined this project to them. One person who took the proposal seriously was my former supervisor, Jim Secord, who in 1991 organized 'The Big Picture' conference in London, which I attended as his Ph.D. student. The discussions I listened to that day sparked off my ambition to write a new type of scientific history, although I had no idea then that it would eventually appear as *Science: A Four Thousand Year History*. So my greatest debt of gratitude is to Jim, for continually encouraging me ever since I first met him over twenty years ago, and for providing many helpful suggestions while I was planning and writing this particular book. I also thank all my friends and colleagues who contributed ideas and corrections, especially Jon Agar, Christopher Cullen, David Edgerton, Michael Fara, Lauren Kassell, Nick Hopwood, Eleanor Robson, Liba Taub, Clive Wilmer, and two anonymous referees who gave me invaluable comments on various drafts. And of course, this book would never have appeared without the help and commitment of my agents Tracy Bohan and Andrew Wylie, my editor at Oxford University Press, Matthew Cotton, and my copy editor, Charles Lauder, Jr. More generally, in the 'Special Sources' section I have listed for each chapter the specific papers and books on which I relied for analyses as well as factual information. Although I have only footnoted quotations, I hope that these references will make it clear how much I owe to the work of many other historians. The opinions and mistakes in this book are my own, but I like to think of it as a collective enterprise aiming to bring recent scholarship to a wide audience.

Contents

IV INSTITUTIONS

V LAWS

VI INVISIBLES

VII DECISIONS

List of Illustrations

Introduction

Seeing may be believing, but what you see depends on how you look. Figure 1 instinctively appears wrong, even though there is no intrinsic reason why the world should not be shown this way up. Putting north at the top is a convention established by early European cartographers who viewed the globe from their own vantage point and did not even know that Australia and Antarctica existed. Created by an Australian, this image is not so much a map as a political statement. It also provides a visual metaphor for *Science: A Four Thousand Year History*.

Writing history is not just about getting the facts right and putting the events in their proper order: it also involves reinterpreting the past—redrawing the world—by making choices about what to put in and whom to leave out. In traditional books about science's past, scientists are celebrated as geniuses elevated above common mortals. Like Olympic racers, they pass on the baton of abstract truth from one great intellect to another, uncorrupted by mundane concerns and dominated by their insatiable thirst for absolute knowledge. Through meticulous experiment, logical reasoning and the occasional leap of inspired imagination, they unlock nature's secrets to reveal absolute truth.

In contrast, *Science: A Four Thousand Year History* is not about idealized heroes but about real people—men (and some women) who needed to earn their living, who made mistakes, who trampled down their rivals, or even sometimes got bored and did something else. This book explores scientific power, arguing that being right is not always enough: if an idea is to prevail, people must say that it is right. This new version of scientific history challenges the notion of European superiority by showing how science has been built up from knowledge and skills developed in other parts of the world. Rather than concentrating on esoteric experiments and abstract theories, *Science: A Four Thousand Year History* explains how science belongs to the real world of war, politics, and business.

Compared with areas on a map, it's hard to draw lines around the edge of science. Greek philosophy, Chinese astronomy, and Renaissance anatomy bear little obvious resemblance either to each other or to modern high-tech research projects, but they do somehow seem linked. Pinning science down is difficult. One obvious if irritating definition is to say 'Science is what scientists do,' but

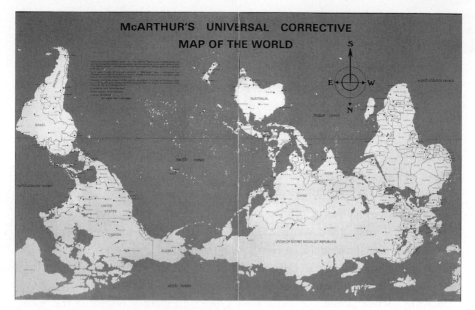

FIG. I McArthur's Universal Corrective Map of the World (1979).
The last part of the caption reads:
Finally South emerges on top.
So spread the word! Spread the map!
South is superior. South dominates!
Long live **AUSTRALIA—RULER OF THE UNIVERSE!!**

even that circular description limps, as the word 'scientist' wasn't invented until 1833. Writing about science's long-term history involves tracing the origins of something that didn't exist until relatively recently, and so it means considering people who weren't doing whatever it is that scientists do now. Many of this book's characters are included not because they were scientists, but because they developed a variety of skills—navigating by the stars, smelting ores, preparing herbal medicines, building ships, designing cannon—that contributed to the global scientific enterprise of today.

When looking at the past from a fresh angle, deciding which questions to ask is as important as ferreting out new information. Instead of worrying about what science is or isn't, there are more interesting problems to think about. Does religion—of any kind—inhibit or encourage science? Are alchemy and magic completely divorced from science? Were there really so few women, or have historians distorted the picture by telling too many exciting adventure stories about intrepid men exploring the female world of nature? Is it possible to have different types of science that are all valid? And if there were indeed different sciences in Patna, Persia, and Pisa, then how are they related to each other and to modern science?

Such questions have no definitive answers, but *Science: A Four Thousand Year History* explains why they are important and suggests ways of tackling them. Then there's the most fundamental question of all: How has science become so important? Men like Kepler, Galileo, and Newton were certainly brilliant, but they are celebrated worldwide because science itself has become so powerful. They seem more significant nowadays than they did to their own contemporaries, who gave greater acclaim to classical and biblical experts. Isaac Newton declared that he was standing on the shoulders of giants, but when he published his big book on gravity in 1687, very few people thought it was worth reading. By the beginning of the twenty-first century, science ruled the globe and Newton had become one of the most famous people ever to have lived. This book examines how that happened by looking at how science and society have changed together—it investigates the financial interests, imperial ambitions, and academic enterprises that made science global.

In black-and-white versions of the world, science is set apart, as though it were a unique type of intellectual activity yielding unassailable truth. Yet what counts as a scientific fact depends not only on the natural world, but also on who is doing the research—and where and when. Scientific knowledge has never travelled neutrally from one environment to another, but is constantly adapted and absorbed in different ways: it has geographies as well as histories. These processes of perpetual transformation are still continuing, so that the significance of science will alter still further.

Paradoxically, as science becomes ever more successful, non experts are becoming increasingly sceptical. Now that governments are preoccupied with fears about global warming, genetic manipulation, and nuclear power, it is clear that scientific, commercial, and political interests are inseparably intertwined. In a sense, the history of science is the history of everything: modern science, technology, and medicine are interwoven, intimately bound up in a giant knotted web with every other human activity all over the globe. *Science: A Four Thousand Year History* is, like the Australian map, committed to challenging assumptions that appear natural yet have been created artificially—it aims to provoke thought and argument, not just provide information. It looks at the past in order to find out how we've arrived at the present. And the whole point of doing that is to improve the future.

Origins

When and where did science begin? This is not a trivial question, but gets right to the heart of what science might be. Looking back at the past, it is possible to pick out ideas and discoveries that later became incorporated within a global scientific enterprise. But at the time, they belonged to other projects—finding an auspicious time for religious festivals, winning wars, vindicating biblical prophecies, and (above all) surviving. This book starts with the ancient Mesopotamian civilizations, whose great store of practical knowledge was inherited and passed on to modern science. Babylonian court advisers developed mathematical, astronomical, and medical expertise not because they were interested in theoretical physics, but because they wanted to divine the future. In contrast, Greek philosophers preferred to build up grandiose systems that aimed to explain the cosmos. Although many of their theories now seem bizarre, they were continuously modified and assimilated, dominating first Islamic and then European thought well into the eighteenth century. Science's very foundations lie in techniques and concepts now often denigrated as magical or pseudo-scientific.

I Sevens

I loved you, so I drew these tides of men into my hands
  and wrote my will across the sky in stars
To earn you Freedom, the seven pillared worthy house,
  that your eyes might be shining for me
    When we came.

—T. E. Lawrence, *The Seven Pillars of Wisdom* (1935)

Seven has always been a very special number. Sanskrit's most ancient holy book, the Rig Veda, describes seven stars, seven concentric continents, and seven streams of soma, the drink of the gods. According to the Jewish and Christian Old Testament, the world was created in seven days and Noah's dove returned seven days after the Flood. Similarly, the Egyptians mapped seven paths to heaven, Allah created a seven-layered Islamic heaven and earth, and the newborn Buddha took seven strides. Seven also has unusual mathematical properties. Many of them sound rather esoteric to the uninitiated, but amongst the more straightforward ones, you only need seven colours to fill in a map on a torus (a ring, or a doughnut with a hole in the middle) so that no two adjacent areas have the same colour.

It's only a short step from being a special number to becoming a magical one. For numerologists, seven signifies creation, because it is the sum of the spiritual three and the material four; for alchemists, there are clear parallels between the seven steps leading up to King Solomon's temple and the seven successive stages of chemical and spiritual purification. Iranian cats have seven lives, seven deities bring good luck in Japan, and a traditional Jewish cure for fever entailed taking seven prickles from seven palm trees and seven nails from seven doors.

Science or superstition? It's not always easy to separate the two. When early astronomical observers looked up into the heavens, they saw seven planets circling the Earth. The Sun and the Moon were the most obvious, but five others were also discovered—Mercury, Venus, Mars, Jupiter, and Saturn (the next one, Uranus, was not identified until the end of the eighteenth century). Finding planets, and working out how they move across the sky, demands skills important for modern science. On the other hand, the first sky-watchers were not aiming to fathom how the Universe operates, but were trying to relate the patterns of the stars to major events on earth, such as famines, floods, or the death of a king.

So it seems wrong to call them scientists. But does it make sense to call them magicians or astrologers? Some of their pronouncements do sound as woolly as those of modern newspaper horoscopes. These two Assyrian examples give a flavour: 'If Venus rises early, the king will have a long life; if this planet rises late, the king of the land will die soon.' Or, 'If the Moon is surrounded by a halo, and the Pleiades [a constellation with seven stars visible to the naked eye] stand in it: in that year women will give birth to male children.'[1]

Easy to laugh—but these were no tea-leaf interpreters or crystal ball gazers: they were skilled astronomers who made detailed calculations based on meticulous observation. Astrology has now been made to look ridiculous, but many civilizations—including Western Europe up to the seventeenth century—have viewed people as being integrated within the Universe, so that strange occurrences in the skies are linked with unusual happenings down on the Earth's surface. Just as one goal of science is to find patterns of relationships, so early diviners were trying to make sense of their lives by examining the world around them. They believed in an interlinked, harmonious Universe, one in which the gods, the stars, and human beings were bonded together and acted in concert.

Modern astronomy rests on a foundation of data collected by expert star-gazers who were also astrologers. Their observations were generally sound, even if their theories have been rejected. Many scientists find it hard to accept that their own expertise is rooted in beliefs that they dismiss as magic. For those who pledge their faith in progress, magical mumbo-jumbo has been eliminated by scientific reason: magic and science are clearly polar opposites, and any notion that they might share common origins is sacrilegious. But this comforting view is not always easy to reconcile with the historical facts.

Consider Pythagoras, the Greek who gave his name to one of the world's most famous geometrical theorems (even though he didn't invent it). This celebrated mathematician was swayed by mystical visions of cosmic harmony and the number seven. According to the traditional anecdote, one day Pythagoras was passing by a blacksmith's forge when he noticed how tuneful the banging sounded. After some investigations and much inspired thought, he realized that a hammer's weight affects the note it produces on an anvil, and he derived some suspiciously neat and straightforward numerical relationships between weights, tones, and string lengths. Pythagoras had been seduced by the seven intervals of the musical scale: like many Greek philosophers, he believed that it was more important to unify the cosmos mathematically than to make detailed observations. Pythagoras imposed regular sevenfold patterns onto the Universe, maintaining that the orbits of the planets are governed by the same arithmetical rules as musical instruments.

Isaac Newton's rainbow is an even more dramatic example of how science and magic are tied together through the power of seven. Over two thousand years after Pythagoras, Newton was the world's leading campaigner for accurate experimentation. Nevertheless, he believed so firmly in the Greek harmonic universe that he divided the rainbow into seven colours to correspond with the musical scale. Before then, although opinions varied, artists mostly showed rainbows with four colours. It is, of course, impossible to make any objective decision about the correct number, because the spectrum of visible light varies continuously: there is no sharp cut-off between bands of different colours, so that how you think about a rainbow affects how you see it. Nowadays, Newton's experiments with prisms are commemorated as the foundation of modern optics, and the magic number seven has become part of scientific colour theory. But be honest—can you tell the difference between blue, indigo and violet?

Since Newton has become an iconic scientific genius, it would seem strange to say that he did not practise science. On the other hand, modern scientists denigrate many of his activities as ridiculous, or even antithetical to science. In addition to his preoccupation with numbers and biblical interpretation, Newton carried out alchemical experiments, poring over ancient texts and carefully recording his own thoughts and discoveries. This was no mere hobby: Newton regarded alchemy as a vital route to knowledge and self-improvement, and he incorporated his results within his astronomical theories. Newton's example illustrates how hard it is to pin down when science began.

Lewis Carroll knew how difficult it can be to decide when a story should start. 'Where shall I begin, please your Majesty?', asked the White Rabbit. Alice listened for the answer. '"Begin at the beginning," the King said, gravely, "and go on till you come to the end: then stop."' Science has no definite beginning, and all historians must—like the White Rabbit—choose their own starting point. But none of them seem ideal.

One possibility is to settle for 1687, when Newton published his great book on mechanics and gravity. But that would mean leaving out many world-famous names such as Galileo Galilei, William Harvey, and Johannes Kepler. The most popular option date is 1543, when Nicolas Copernicus suggested that the Sun and not the Earth lies at the centre of our planetary system. However, there are several objections to this choice, not least that it excludes the Greeks, whose ideas remained enormously influential well into the eighteenth century. So another possibility is to start in Greece itself. Thales of Miletus, who lived on the Turkish coast around 2500 years ago, is often said to be the first true scientist. He was a superb geometer and successfully predicted an eclipse—but choosing him would result in leaving out all *his* important predecessors, such as the Egyptians and the Babylonians.

Everybody has predecessors. When Greek astronomers were tackling this problem of setting a zero-line, they looked back a millennium to Babylon and the reign of King Nabonassar, who sponsored accurate observation projects. So perhaps a better solution would be to go as far back into the past as possible, and examine the earliest surviving evidence indicating any activity that could be labelled 'scientific'. Scattered throughout Europe, there are ancient ruins which show that long-vanished peoples once tracked the movements of the sun and the stars. Unfortunately, they cast little light on the origins of science.

The most famous is Stonehenge, the dramatic stone circle in southern England where Druids still gather to celebrate the midsummer sunrise. Many archaeologists have claimed that Stonehenge was a mammoth astronomical observatory accurately aligned with the sun's passage across the sky. Using complex statistical techniques, they have attached huge significance to the location of holes and stones, even ones that have been moved and altered over the last five millennia. But if you study random patterns long enough, you can always impose some sort of order, and most experts now agree that although Stonehenge and similar monuments did symbolically refer to the heavens, this was a ritual significance rather than a search for precise astronomical knowledge. Deciphering ancient mysteries is fascinating, but does not necessarily help to explain the origins of science.

Another problem is the survival of expertise. Several ancient civilizations—in Latin America, for instance—did have a sophisticated knowledge of the stars, but it was not passed on to future generations around the globe. In order to trace a continuous story of science from the past to the present, the search for an origin must focus on North Africa and the eastern Mediterranean. Roughly five thousand years ago, about a millennium before the Stonehenge monument was a place of worship, Egyptian pharaohs were commissioning equally impressive feats of engineering—the pyramids. These ancient Egyptians generally oriented their pyramids towards the Sun, although, like the builders of Stonehenge, they were not particularly interested in making detailed observations of the heavens. For them, it was far more important to understand the behaviour of the Nile, vital for irrigating their crops. In their calendar, the year was measured out not by the phases of the Moon or the passage of the Sun, but was divided into three seasons dictated by the Nile's flooding patterns.

This book begins at a similar time, but further east in Mesopotamia, then a fertile region lying between two rivers in what is now Iraq. By influencing their followers, the Babylonians bequeathed an indelible legacy to modern scientific culture. And it is literally indelible—instead of writing on fragile papyrus, the Babylonians used durable clay tablets, many thousands of which still survive. So although the Babylonians preceded the Greek philosophers, far more material evidence survives of what they wrote.

The Babylonian way of thinking about the Universe still profoundly affects people today. They developed complex mathematical techniques, recorded the stars, and made predictions. Because their knowledge of the skies was inherited by later observers, it forms the basis of astronomical science and also structures everyday modern life. Thanks to the Babylonians, weeks have seven days corresponding to the interval between phases of the moon, and there are sixty seconds in a minute and sixty minutes in an hour. This ancient way of recording the passage of time is not perhaps the most convenient, but it has become deeply entrenched: during the French Revolution, a more rational system of ten-hour days and ten-day weeks was introduced, but it was soon abandoned.

Eurocentric calendars have another major irrational feature. The years start at the birth of Christ, even though human history extends far, far back beyond that conventional year zero. Imagine travelling backwards in time through that artificial divide to a symmetrical era in the past—the twenty-first century BCE. That is where this book begins. A personal choice, to be sure: but there can be no other kind, because—whatever the King might have told Alice—science has no definite beginning.

2 Babylon

This is the excellent foppery of the world, that, when we are sick in fortune, often the surfeits of our own behaviour, we make guilty of our disasters the sun, the moon, and stars...My father compounded with my mother under the dragon's tail, and my nativity was under Ursa major; so that it follows I am rough and lecherous. Fut! I should have been that I am had the maidenliest star in the firmament twinkled on my bastardising.

—William Shakespeare, *King Lear* (1605–6)

Around four millennia ago, there was a power shift in the Mesopotamian basin. Instead of small independent city-states, a new single kingdom centred on Babylon, a town lying on the river Euphrates about seventy miles south of modern Baghdad, emerged. The Babylonians inherited one especially valuable invention—cuneiform writing, which had already been in use for around two thousand years. Wood and stone were scarce, so people used tablets made from clay, their standard building material, for storing information by marking wedge-shaped characters with a stylus. These early texts reveal the origins of modern mathematics.

Deducing detailed information from clay tablets is, of course, extremely difficult. Historians are faced not only with the task of decoding an unfamiliar language in an arcane script, but also with unearthing and piecing together damaged pieces of tablet from piles of rubble. Although hundreds of thousands of tablets have been discovered, many more are still buried or have been lost for ever, so the work can seem like reconstructing a great library from the evidence of a few torn pages. The situation was made still worse by European archaeologists who pillaged the ruins for trophies rather than for information. Wrenched out of the Mesopotamian soil that had preserved them for millennia, finds were despatched for display in distant museum showcases. Fortunately, some unwanted tablets were consigned to basements, wrapped in newspapers whose dates helped to establish the sites from which they had been taken.

For Europeans, Babylon's origins were shrouded in fables. Until three hundred years ago, the city's very location was uncertain, and it now appears that even the famed Hanging Gardens never existed (although there may have been smaller ones far to the north). Systematic excavations only started in the mid-nineteenth century. At that time, Babylon was still imbued with such a mythological aura that the composer Giuseppe Verdi adopted it as a symbolic location for *Nabucco*, his political opera attacking Austrian domination of his native Italy. First staged in Milan in 1842, *Nabucco* tells the dramatic tale of a highly fictionalized King Nebuchadnezzar, who converted to Judaism when the repressed Israelites cast off Babylonian tyranny. Although Verdi and his contemporaries knew little about Babylonian realities, this legendary ancient city evoked an appropriately mysterious backdrop for his modern allegory about foreign rule over Italy.

Gradually, as archaeologists removed some of the mystique by deciphering concrete evidence, the scientific achievements of this ancient civilization became apparent. Foreign teams competed to retrieve valuable items, shipping huge quantities overseas to be hoarded by private collectors or displayed in museums (one consignment sank to the bottom of the Tigris). Cuneiform experts collected, interpreted, and classified countless tablets carrying information about weights, areas, and star locations. By around 1950, although these code-breakers were still accumulating material evidence, it seemed as though they could do little else but copy out yet more multiplication tables and argue about the best way of translating them into modern algebraic equations.

In the 1980s, historians decided to abandon this unrewarding quest and start asking different types of question. Instead of searching for more objects and more details, scholars began to interpret old evidence in new ways, trying to understand how Babylonians lived and thought. Approached thoughtfully, clay tablets reveal far more information than just the numbers and words trapped in their surfaces. By recreating the ways in which they were used, Mesopotamian specialists can draw important conclusions about ordinary life and how people's daily activities affected future science.

Clay tablets may seem very different from red tape, but Babylonians knew about bureaucracy. As well as keeping written records, their predecessors had developed the mathematical techniques essential for administering an organized settled society—keeping accounts, constructing irrigation systems, parcelling out plots of land. Control lay in the hands of bureaucratic elites who were grouped together around local rulers, and these separate centres of power were bonded together by sharing the same writing system and the same mathematics. Students were expected not only to excel at arithmetic, but also to be skilled with practical equipment such as measuring rods and surveying lines. To capture the wandering attention of their pupils, teachers devised fictionalized scenarios that related abstract arithmetic to the real world of trade, agriculture, and war. For instance, in an imaginary exchange of letters, an emissary reports back to his king how he is trying (unsuccessfully) to alleviate a famine by importing grain. The suspiciously easy sums involved make it clear that this was a teaching text and not a real-life one:

> With grain reaching the exchange rate of 1 shekel of silver per gur [about 300 litres], 20 talents of silver have been invested for the purchase of grain. I heard news that the hostile Martu have entered inside your territories. I entered with 72,000 gur of grain [the author has helpfully provided the right answer to the sum in the first sentence]...Because of the Martu, I am unable to hand over this grain for threshing. They are stronger than me, while I am condemned to sitting around.

House F is the uninspiring name for the site of some inspired research. After World War II, American archaeologists revisited Nippur, an ancient town now in southern Iraq, to continue excavations that had been abandoned in the nineteenth century. When they reached House F, they knew that they had made a unique find—discarded writing tablets had been used to make repairs and build benches. Beneath their plaster coats, the very floors and walls were inscribed with literary texts and arithmetical calculations. Operating far more systematically than their predecessors, the team carefully noted the precise location of their discoveries before distributing them to museums in Baghdad and their home universities. And there many of them remained, sparsely catalogued until scholars painstakingly pieced together their secrets almost fifty years after their recovery.

In around the eighteenth century BCE, House F was a school. Since clay disintegrates in rain, the house had to be rebuilt every twenty five years or so. Experts know that it was a school because many of the tablets were copies of lists and tables: children were learning how to read, write, and do arithmetic. Inside the rooms, archaeologists found bowls and a bread oven, but realized that schooling took place outside in the courtyards, which had recycling bins for soaking old tablets with water and fresh clay so that they could be used again. Scribes made their own tablets, and the presence of clumsy shapes with awkward writing shows that children were being taught this craft. When you hold one of these tablets in your own hand, you feel the contours of the thumb and palm of someone living thousands of years ago.

The mathematical training at House F was geared towards producing educated scribes who could help resolve legal and financial disputes. Like Victorian children condemned to struggling with rods, poles, and perches before the metric system was introduced, these Mesopotamian pupils spent many hours transforming measurements from one unit to another. To help them cope with practical problems of trade, law, and agriculture, they had to memorize long tables for multiplication and division. The teachers valued practical expertise as well as abstract knowledge. 'You wrote a tablet, but you cannot grasp its meaning,' stormed one teacher in a (supposedly) humorous educational text designed to motivate lazy students. 'Go to apportion a field, and you cannot even hold the tape and rod properly,' he complained; 'You cannot figure out its shape, so that when wronged men have a quarrel you are not able to bring peace but you allow brother to attack brother.'[2]

The trainee scribes were taught how to mix the clay with water and skim off all the twigs, leaves, and other natural debris that floated to the surface. They cut reeds from the river to make styluses, slim quarter-cylinders the size of a chopstick, and pounded piles of wet clay with their feet to make it malleable and uniform in consistency. During their lessons, they moulded lumps of clay into flat ovals, and then practised pushing in the pointed corner of their stylus to represent different numbers by vertical and horizontal marks. Like writing with a pen, these are skills that need to be learnt—the final imprint is very sensitive to the angle at which you hold the stylus, and even making a smooth, symmetrical tablet is harder than it sounds. As they indented symbols into the soft clay, the students constantly sprinkled water over the surface to prevent it from drying out too quickly in the hot sun; from time to time, they tossed their old tablets into the recycling bins, and started over again by shaping a fresh one.

The raw materials that Mesopotamians used—the clay, the reeds—affected the numbering systems that they developed: they counted in blocks of sixty, which seems strange for people brought up with tens and hundreds. However, if you try to write with a stylus (a diagonally cut drinking straw works well), you soon realize that sixty was a more sensible choice than it might seem. The Babylonians used two basic symbols, vertical for single numbers and diagonal for tens. They grouped the first nine digits in threes, one row beneath the other, because the human eye can immediately distinguish between one, two, and three adjacent vertical marks—but not four. Reading the horizontal wedge-marks is more tricky, and scribes developed a system enabling them to recognize a set of up to five instantaneously. So after 59 (five horizontals, and three groups of three verticals), they moved everything over one place to the left and started again, rather like 100 is distinguished from 10.

The nearest modern equivalent is a digital clock, in which the number of hours—groups of sixty minutes—is shown to the left of the display. Microelectronic devices operate very differently from clay tablets, but they have inherited a way of counting developed thousands of years ago and compatible with the raw materials then available. And the numerical conventions of modern geometry, which dictate that there are 360° in a circle, originated not with Euclid and the Greeks, but with Mesopotamian surveyors and accountants writing on clay tablets.

The passage of time takes on new meaning when you realize that some of these ancient tablets can only be dated to within five hundred years. For a future historian of European culture, that would be similar to wondering if Copernicus lived now. Although 'the Babylonians' now seem like a single civilization, a whole millennium separates the children learning to count in House F from the eighth century BCE, when Babylonian observational astronomy began. Throughout those forgotten thousand years, sky-watchers had been recording events in the heavens. Because they stored the information on tablets, it provided an enormous inherited reservoir of astronomical knowledge. By encoding their work in clay, Babylonian scholars left tangible evidence of their work and ideas, influencing not only their immediate successors, but also people—you and me—living thousands of years later.

Through deciphering remnants, archaeologists have pieced together vast tracts of information about Babylonian beliefs. Frustratingly, clay tablets do not reveal everything. Experts still know virtually nothing about the daily lives led by the majority of people outside these learned elites. Moreover, although tablet decipherers can interpret the star catalogues that Babylonian astronomers compiled, they can only guess at what instruments were used for measuring stellar positions. No surviving devices have yet been discovered, but it seems likely that they used some sort of alignment rod resembling the pointer of a sundial.

Because some scribes signed their work, archaeologists have been able to compile their own lists, not of stars but of skilled astronomers who clustered like constellations around the temples and the kings. There is no point in rating these courtly advisers on any scientific scale of achievement, because Babylonians had no separate categories of science and religion, of rationality and spirituality, of astronomy and astrology. For early sky-watchers, the stars represented a holy heavenly text that could, if properly read, reveal portents of prosperity or famine, of peace or warfare. They corroborated these omens by examining other evidence, such as the livers of sacrificial animals, and performed the appropriate rituals for warding off impending disaster. These were influential men, whose prognoses could prompt kings to relinquish power or set up a new court in another part of the country.

Gradually, the Babylonians shifted from recording celestial events to predicting when they would happen. They could do this because over many years, a restricted dynasty of scholarly families accumulated massive stores of data—around a third of a million observations were compiled between the eighth century BCE and the first century AD, the longest recording programme in history. By looking back over their readings, Babylonian astronomers worked out cycles of repetition, and so could predict future positions of the Sun, Moon, and planets. Some of their analyses were sophisticated. For instance, when compiling tables of data, mathematicians took account of the Sun's different speeds during its year-long journey across the sky, and compensated for the variable motions of the planets.

Some features of this ancient astronomy seem alien. For one thing, unlike modern astronomers, Babylonians used their calculations not to map planetary orbits, but to work out how the heavens affected individuals—their investigations were politically motivated, directed towards picking up portents of major events, such as Alexander's invasion of Babylon. As another example, Babylonians held a different opinion from nowadays about where the Earth ends and where the heavens begin, so they bracketed the atmosphere with the stars rather than with the globe. Clouds are now considered meteorological effects, but they used to be grouped together with eclipses, planets, and meteors (hence the name of modern weather science). This way of classifying natural phenomena was inherited by the Greeks and remained important in Europe up to the end of the seventeenth century.

The Greeks did not, however, share the Babylonians' mathematical approach to the Universe. Greek philosophers and astronomers thought geometrically, representing the Universe with three-dimensional visions of stars that orbit around the earth as though voyaging across the surface of an imaginary celestial sphere. In contrast, Babylonian mathematicians thought arithmetically and algebraically, showing off their virtuosity to achieve new results rather than developing techniques designed to solve pre-existing problems. They compiled long tables of observations and star positions, but instead of drawing three-dimensional geometrical diagrams, they relied on complicated, repetitive multiplication and division Babylonian astronomers applied to the skies the same techniques of calculation that the school children had learnt in House F for working out the areas of fields, the profiles of ditches, and the structure of dams.

Although so different from modern scientists, these Babylonian star-gazers bequeathed important legacies. The sheer volume of their observations and calculations made them invaluable for Greek geometrical astronomers who encountered their work in Egypt: Babylonian data lay the foundation for modern astronomical catalogues. Another important survival into the present is their zodiac system with its twelve signs. Twelve is a more versatile number than ten, because it can easily be divided into quarters and thirds. It also fitted well with the Babylonian number base of sixty, so that circles could conveniently be divided into 360°—just as they are now. The Babylonians split the heavens into twelve equal sections, one for each lunar month and carrying the name of a prominent constellation. Translated into Latin, these now exist as the twelve signs of the zodiac familiar from newspaper horoscopes, such as Aries the Ram and Taurus the Bull. Yet although twelve was a rational number to choose, other aspects of this system have no place in modern science.

Babylonian experts also established some aspects of modern time-keeping. As well as dividing time into sets of sixty (influencing our seconds, minutes, and hours) and seven (days in a week), they established a sophisticated calendar based on the movement of the Sun and the Moon. Like many subsequent astronomers, they struggled to reconcile the solar year, which is a little over 365 days long, with

FIG. 2 *Les très riches heures du Duc de Berry*: September. Painted by the Limbourg brothers *c.*1412–16.

the lunar month of about 29½ days. Nowadays, this difficulty is resolved by having months of different lengths and introducing leap years, but the Babylonians devised a technique based on adding a thirteenth month every three years.

This Babylonian approach of relating human time to the Moon became the foundation of the Jewish and Christian religious calendars. Figure 2 shows the September page from a particularly resplendent Book of Hours, commissioned in the fifteenth century by a rich nobleman, and designed to display the appropriate liturgical text for each hour of the day. Painted on vellum in strong blue, red, and gold, this striking autumnal scene shows peasants bending, stretching, and indulging in some illicit tasting as they labour to harvest the grapes for the Château de Saumur, here delicately painted in architectural detail. For devout Christians, one particularly important feature was the semi-circular calendar at the top, which enabled them to work out the dates of the religious festivals in any particular year.

This calendar illustrates how ancient influences survived and mingled as they were absorbed into European culture. Its outermost and innermost rings are written in the Arabic number system (1, 2, 3, etc.) that reached Europe in the twelfth century and is still used today. Between them, two bands of Latin show a double Roman inheritance—Latin numbers as well as months such as November and December (from the Latin words for ninth and tenth). Babylonian astronomy features in the broad arch showing pictures of the zodiac signs Virgo (the virgin on the left) and Libra (the scales on the right). The two central bands of the inner arch come directly from the Babylonian nineteen year calendar. Symbols of the moon are associated with different letters of the alphabet arranged in sequence like a code. By using the number nineteen to decipher their message, priests could work out the date of the new moon for any month and year.

Like their distant predecessors, many of the sky-watchers whose theories incorporated Babylonian ideas were interested in predicting the stars either for religious reasons or to gain political power. That is why it makes sense to include a devotional image in a book about science's history—although astronomy is now a scientific discipline, its knowledge was built up over many centuries of close association with prophecy, rites, and music. There are no straightforward trajectories linking modern science with its Mesopotamian past. Nevertheless, if Greek astronomers in Egypt had not inherited and developed Babylonian astronomy, our current stellar charts and measurements would look very different.

3 Heroes

In science the credit goes to the man who convinces the world, not to
the man to whom the idea first occurs.

—Francis Darwin, *Eugenics Review* (1914)

In English, the word 'history' carries two distinct meanings: it refers both to
the past and to the way that that past is described. Perhaps disconcertingly,
historians can tell different histories about the same historical events or periods,
because although they must all stick to facts, they also write creatively. To make
sense of what happened, they construct narratives with plot-lines; they give their
stories beginnings and ends, and they pay special attention to climactic moments
such as winning a battle, discovering a new chemical, or formulating a revolutionary theory. Like novelists portraying an imaginary world, historians impose a
structure on the continuous historical past. And to carry their readers along with
them through their version of events—their particular story—they focus on key
individuals, who may acquire the status of heroes.

This focus on celebrities stems from the ancient Greeks, who knew that heroes
attract readers and make stories memorable. They invented Achilles, Aeneas,
and other mythological champions whose exploits were humanly unachievable.
Conversely, since the Greeks rated intellectual prowess extremely highly, they
converted real-life philosophers into legendary figureheads who did actually live,
but whose scholarly feats surpassed those of normal mortals. Although the members of this intellectual pantheon varied, they always totalled seven, that especially
significant number. This is one common modern version:

> The seven most important Greek scientists were Archimedes, Aristotle,
> Democritus, Plato, Ptolemy, Pythagoras, and Thales.

There are some obvious objections to that claim. For one thing, some of
those Seven Wise Men are no longer well known, while others seem to be
missing. What happened to Hippocrates, celebrated as the father of Western
medicine? Or Euclid, founder of modern geometry, and one of Newton's
favourite authors? Even if you accept the compilers' selection, the word 'scientist'

is extremely problematic—because there weren't any. Although many ideas that originated in Greece were adapted over the centuries and absorbed into science, these men shared neither the goals nor the experimental techniques of modern research teams.

Another criticism of this Greek Super Seven is its alphabetical arrangement. Rewritten chronologically, these intellectual heroes are Thales, Pythagoras, Democritus, Plato, Aristotle, Archimedes, and Ptolemy. However, over seven hundred years separate the last person from the first: an English equivalent of this time compression would be to bracket together Stephen Hawking and the thirteenth-century monk Roger Bacon. At least Hawking and Bacon both studied at England's two traditional universities, Oxford and Cambridge. In contrast, these Greeks were widely scattered in place as well as in time—Thales (of Miletus) lived on what is now the Turkish coast, Plato taught in Athens, and Ptolemy worked in Egypt.

When viewed from the present, Greeks who in reality lived in different centuries and different regions are often imagined as one homogeneous group distinguished by a few famous names. The most common way of subdividing the Greek past is into three time periods: the first is the pre-Socratic era running from around 600 to 400 BCE, and the second is the following century, the high point of Athenian power when Socrates' student Plato founded his Academy and taught Aristotle. Finally, the half millennium between about 300 BCE and 200 AD is the Hellenistic age. By then, Aristotle's most famous pupil, Alexander the Great, had built up a massive empire, and Greek civilization extended along the northern coast of Africa, round the eastern Mediterranean, and overland as far as India and China. Greece's philosophical heroes feature in all three of these time periods.

Plato visualized the search for truth as an intellectual Olympic race run by scholarly athletes who pass on the torch of genius from one to the next. This appealing model was adopted by later generations. Plato's student Aristotle promoted his own status by claiming that he was inheriting the flame of knowledge from Thales, who two centuries previously had launched a new way of thinking about the Universe. A couple of millennia later, Aristotle was himself being revered by European scholars as the founder of Greek science.

This romantic image of a scholarly relay race was still popular in Victorian times, and it does have several advantages. Above all, it encourages historians to portray science as a series of exciting adventures featuring intrepid discoverers, and interspersed with fallow periods when nothing much happened. Indeed, for the Greeks themselves, it is difficult to tell any other sort of story because there are great gaps in the historical record. The original documents of only a few Greek thinkers have survived, and the surviving evidence often comes from far later interpretations. But during the intervening centuries, ideas had become distorted and information about the lives of even the most eminent early Greeks, let alone all the rest, had vanished. It is often hard to sort myth from fact in these biased and incomplete accounts.

Plato's heroic way of telling of the past elevates a few brilliant men (and the occasional woman) into geniuses, but consigns many others to oblivion. It converts people into intellectual equivalents of mythical gods: endowed with superhuman brains, they float above worldly affairs as they think great thoughts. Yet science and philosophy are not divorced from reality, and everyday concerns—politics, finance, personal relationships—affected how theories were developed in Greece just as much as they influence modern academic activities. Plato claimed that Thales was so absorbed in analysing the stars and predicting their behaviour that he tumbled down a well, but according to Aristotle, Thales was an astute business man who made a fortune by forecasting a bumper harvest and buying up all the olive presses. Aristotle's anecdote about his iconic ancestor may be exaggerated, but as a caricature of human behaviour, it does sound more believable than Plato's absent-minded genius.

During their own lifetimes, scientific heroes often appeared less important than they do in retrospect, when they are admired for leading presciently towards a future that their contemporaries could not possibly have known about. For instance, some historians single out Aristarchus as a precursor of Copernicus because, in the third century BCE, he maintained that the Earth revolves around the Sun. But celebrating Aristarchus for happening to hold this modern idea seems pointless, since his theory was rejected at the time and had little subsequent impact—astronomers went on believing for almost two thousand years that the Sun goes around the Earth.

This question of priority crops up again and again in the history of science. Leonardo da Vinci drew something resembling a helicopter, but there's an enormous difference between a rough sketch and getting a manned machine into the air. However brilliant he may have been, Leonardo was not the world's first aeronautical engineer. Similarly, some specialists boast that in the first century BCE, Hero of Alexandria (Heron in Greek) built a small rotating sphere powered by steam. But whatever they might claim, that hardly makes him responsible for the Industrial Revolution, which started in eighteenth-century Britain.

Heroic stories about science's past limp because, unlike Achilles and Aeneas, the Seven Wise Men of ancient Greece were real people living at particular times in particular places. How they thought, wrote, and behaved depended not only on the opinions of their teachers and their friends, but also on their own material and emotional needs. These included earning money, being careful to avoid offending patrons, placating the gods, trying to gain political advantage, and even staving off boredom or recovering from a love affair. Just as importantly, their ideas did not travel through time and space in some sort of intellectual vacuum, but were constantly adapted and modified. In different places, and in different centuries, some aspects of their thought received more attention than others; large chunks were rejected, or even amalgamated with somebody else's. By examining heroic thinkers within their cultural context, it becomes clear that great geniuses are made, not born.

So how can the Greeks be included in a brief story about science's past? Although they approached the world very differently from modern researchers, their philosophical, cosmological, and theological ideas did profoundly affect later science, through being read directly as well as being transformed and transmitted by Christian and Islamic scholars. By modern standards, theories that held sway for centuries were wrong, while ones that now seem right were discarded: the path of scientific change is far from straight. When thinking about how Greek philosophers influenced the future, the concepts that count are the ones that their successors adopted. Whether modern scientists now judge them to be good or bad is irrelevant.

Focusing on key ideas means leaving out great swathes of the past, but there are two ways of doing it. The conventional approach is to pick out an intellectual route that leads unerringly from the (supposedly) ignorant past towards the superior truth of the present: by omitting what now seem like mistakes, historians can tell stories about the triumphant rise of science and its victory over superstition, magic, and religion. In contrast, this book concentrates on what people thought at the time: it explores—without judgment—how beliefs were passed on from one generation to the next. Ancient ideas can nowadays seem very weird, but they deserve to be treated seriously because they were sincerely held by some extremely intelligent men and women.

Over eight centuries, Greek scholars borrowed observations from elsewhere, accumulated a massive body of data, and developed theories about the Universe and its inhabitants. To examine how their collective activities affected future science, the next four sections focus on four broad areas—the structure of the cosmos; life and medicine; the nature of matter; and practical knowledge. For each of these, the Greeks bequeathed rich legacies that may be regarded as bizarre, but were plundered, absorbed, and transformed by the civilizations that succeeded them.

4 Cosmos

Scientists carry out experiments to test their theories—at least, that's the ideo-logical version of what happens. In practice, theoretical preconceptions of how the Universe *ought* to function have often overridden the evidence provided by observation. There are many Greek examples, including Plato, who insisted that the Universe is characterized by cosmic order and mathematical harmony even though he knew that there were seven obstacles to this ideal model—the seven planets, whose irregular movement across the skies contravened common sense as well as philosophy. Right through to the time of Newton, this problem dominated cosmology, as astronomers tried to 'save the appearances' by reconciling the planets' apparently erratic behaviour with theoretical visions of celestial perfection.

Plato shared this quantitative approach inherited from the Pythagoreans living in Italy a couple of centuries earlier. Although Pythagoras is now celebrated for a theorem about right-angled triangles, he didn't invent it—the properties of the square on the hypotenuse had long been known to the Babylonians. (The sim-plest example is as follows: if the lines on either side of the square (90°) angle of a triangle are 3 and 4 units long, then the opposite long side, the hypotenuse, will measure 5 units, because $3^2 + 4^2 = 5^2$.) The word 'geometry' means 'measuring the land', and Greek mathematicians helped convert practical surveying problems into abstract diagrams. Relying at first on techniques familiar to the Babylonian chil-dren in House F, they gradually developed theoretical mathematical knowledge that was fascinating for its own sake, rather than for its utilitarian value.

Like modern scientists, Pythagoras and his followers believed that mathemat-ics is the key to understanding the Universe. However, they also belonged to a secret brotherhood that searched for numbers everywhere and gave them hidden meanings. Triangles with sides of 3, 4, 5 were especially appealing because of their numerical simplicity, which they felt resonated with cosmic beauty. For the Pythagoreans, this quantitative approach to the Universe belonged to their spiritual quest for self-improvement, yet it also characterizes rational science and preoccupied many famous theoreticians such as Newton and Galileo, who influentially imagined the cosmos as a large book that God had written in a mathematical language of triangles, circles, and other geometrical shapes. Having faith in the power of mathematical science does not preclude being religious.

Pythagoras affected the course of science, yet his own research lay in music. He supposedly made careful measurements to demonstrate that there are simple numerical ratios between musical intervals, so that—for instance—a particular string on a musical instrument will produce a note one octave higher than a string twice as long. Nevertheless, theory and perfection were more important for him than everyday reality, and it seems unlikely that Pythagoras could have obtained many of the experimental results he claimed. As the Pythagoreans searched for mystical number relationships, they extended this earthly mathematics of music into the Universe, trying to establish harmonic ratios for the distances between the planets. This Greek alliance of astronomy and arithmetic, of music and magic, still prevailed throughout Europe in the seventeenth century.

As far as specific cosmological models are concerned, the two most important Greek writers were Plato's student Aristotle, who lived at the height of Athenian power, and Ptolemy, who worked in Hellenistic (Greek-ruled) Alexandria almost half a millennium later. Unlike many other Greek philosophers, Aristotle and Ptolemy left substantial written texts that became familiar to mediaeval scholars throughout Europe. Very little is known about these two men's lives, but their cosmological ideas were enormously influential.

Aristotle had little patience for special numbers and the cosmic mathematics of his Pythagorean predecessors, yet he was a theoretical astronomer who relied on the power of thought rather than accurate observations. In any case, Aristotle had no access to the Babylonians' extensive sets of measurements. Rejecting the unifying mathematical approaches of Pythagoras and Plato, Aristotle divided the universe into two clearly demarcated zones with strikingly different properties—the heavenly region and the terrestrial sphere itself (also called sublunar, from the Latin for 'below the Moon'). Aristotle's celestial realm is stable and orderly, composed of a special mysterious aetherial substance through which the heavenly bodies revolve eternally in perfect circles, kept in steady motion (somehow) by an external Unmoved Mover. In contrast, the globe of the Earth is characterized by corruption and mortality; objects naturally move either up or down—think of rising smoke or falling stones—unless unnaturally forced to change direction.

Aristotle's cosmology was scattered through his books rather than being presented as a coherent whole, yet his distinction between the terrestrial and the

I thought it mete also to put here this figure, shewing the placing, compassing, and distances of eche of the tofore sayd *Planetes* in the heauen: whiche distances at my last publyshing were thought impossible. This figure wittily shaped may confirme a possibilitie to agree vnto the true *Quantities* immediatly before put fourth, therfore not omitted here to be placed.

FIG. 3 A Christianized version of the Aristotelian cosmos. Leonard Digges, *A prognostication everlasting*...(1556).

celestial realms dominated scientific ideas well into the seventeenth century, long after Copernicus had placed the Sun at the centre of the Universe. That Aristotle's model survived so long suggests that it made sense and was useful. That our world is stationary seems obvious: if you fire an arrow straight up into the air, it comes down and hits your body because you have stayed in the same place rather than rotating onwards. In addition, Aristotle's cosmos appealed to European Christians because it was so easy to visualize his Unmoved Mover as God. Figure 3 shows a sixteenth-century modification, in which the central zone of the Earth is surrounded by the circular orbits of the seven planets, each identified by name and by their symbol. Beyond the fixed stars and the crystalline heaven (a later theological addition), the outermost ring is labelled 'The fyrst Mover', a common term for God.

This intuitive version of the universe was marred by seven celestial transgressors—the seven planets, which move at variable speeds across the sky and whose brightness fluctuates, as though they were at altering distances from the Earth's surface. Still worse: apart from the Sun and the Moon, they periodically appear to stop and then move backwards before resuming their normal move-ment. Astronomers found this retrograde motion totally baffling, because they were convinced that the Sun and the planets revolve around the Earth in perfect circles. Once you imagine the Sun to be at the centre of the Universe and adopt elliptical orbits, you can easily work out why these strange effects occur. But what now seems clear used to be unthinkable. For centuries, any such suggestions were immediately rejected because they contravened the ideal of perfect circular rota-tion around a fixed Earth.

Aristotle's solution to this planetary problem was exceptionally cumbersome, because of his underlying conviction that the planets move at a uniform speed. His complete system involved fifty-five invisible concentric spheres, all rotating in various ways about the Earth. The Unmoved Mover makes the outermost sphere revolve perpetually, and its motion is transmitted inwards. Each of the seven plan-ets is carried by one of these spheres and—apart from the Moon—is accompanied by several other spheres to compensate for the pull of the rest. Ironically, one of his own pupils contributed to the partial abandonment of Aristotle's complex model. During a visit to Macedonia, Aristotle taught the prince who later became Alexander the Great. When Alexander's empire spread eastwards, Greek geomet-rical astronomers encountered the huge legacy of Babylonian observations. They were forced to recognize that however appealing Aristotle's spheres might be, they had to be modified. This Mesopotamian influence transformed Greek cosmology, because for the first time, elegant geometry could draw on meticulous data to provide accurate quantitative schemes.

Nevertheless, circular motion was too ingrained a concept to relinquish. Instead, Hellenistic mathematicians started tinkering. The next major book to survive was by Ptolemy, an enigmatic figure with a meagre, uncertain biography. He prob-ably spent most of his life in Alexandria, the Egyptian city founded by Alexander the Great, and he probably died around 170 AD. Mediaeval artists often showed Ptolemy wearing a crown, because they confused him with the Ptolemies who had ruled Egypt several hundred years earlier. An expert self-promoter, Ptolemy relied on his predecessors but relegated them to the ranks of the superceded, suc-cessfully establishing his own identity as the hero who had transformed Aristotle's cumbersome model.

Ptolemy dominated later astronomy because his massive compendium of knowledge was transmitted first to the Islamic empire and then to Europe. Usually called by its Arabic name, the *Almagest* (*The Greatest Compilation*), it includes a detailed catalogue of over a thousand stars, as well as numerical tables and geomet-rical diagrams for calculating the future movements of all seven planets. Building on centuries of Greek theory and Babylonian observation, Ptolemy constructed geometrical models to predict the behaviour of the planets. To achieve this, he

FIG. 4 An armillary sphere, probably from the fourteenth century; the wooden stand is more recent.

sacrificed one of Aristotle's most treasured tenets—uniform motion: although Ptolemy's planets move in circles, they travel with variable speeds.

Influentially, Ptolemy described the instruments that he used to observe the skies. He was proudest of his armillary sphere, whose basic structure remained the same for centuries—Figure 4 shows a European hand-held version mounted on a wooden stand. Like some of his theories, Ptolemy claimed to have invented the armillary sphere, but had probably inherited it. The large calibrated rings (the *armillae*) represent imaginary celestial coordinates surrounding the central Earth, so this instrument could function as a model of the cosmos as well as a device for measuring it (this particular example was too crude and small for accurate measurements).

FIG. 5 Volvelle illustrating
Ptolemy's theory of epicycles
for the planet Jupiter.
Petrus Apianus, *Astronomicon
Cæsareum* (1540).

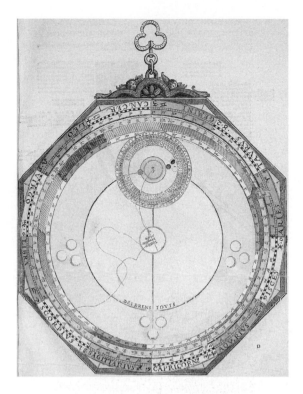

For Ptolemy, its main advantage was that he could directly measure a star's heavenly
coordinates (its celestial latitude and longitude) without undertaking lengthy
calculations. Long after everyone believed that the Sun is at the centre of the planetary system, navigators continued to use Ptolemaic astronomy, because whatever
science may say is the case, when making calculations out in the middle of an
ocean, it is more straightforward to imagine the Sun going round the Earth.

Ptolemy was determined to provide reliable predictions that would correspond
to his measurements and explain how some planets appear to move backwards.
He managed to retain circles, but he did have to renounce simplicity, and diagrams
of his model bristle with geometrical complexities. His major innovation was to
suggest that each planet moves in a small circle whose imaginary centre rotates
in a larger circle about the Earth. Although Ptolemy's scheme may now sound
abstruse, it remained important because it tried to reconcile actual observations
with philosophical and theological commitments to circular motion. Figure 5
shows a teaching device from a famous sixteenth-century book about astronomy.
Rather than being a diagram, it is a paper model in which strings turn coloured

discs to explain the movement of the planet Jupiter. As Jupiter travels round the small circle at the top (called an epicycle), it produces a looping motion as that circle moves along the perimeter of the larger circle (here labelled *Deferens Jovis*, deferent of Jupiter). With some judicious management, the looped pattern could be made to correspond with the actual backwards and forwards motion of Jupiter in the heavens.

The order in which Ptolemy placed the planets also survived for many centuries, even though it was to some extent arbitrary (Figure 3). Outside the planets lay the fixed stars, here divided into bands by later theologians. Next to them Ptolemy placed Saturn, Jupiter, and Mars, the three planets whose behaviour seemed closest to that of the stars. Since Venus, Mercury, and the Moon seemed somehow linked to the Earth, he gave them the inner orbits. Creating a pleasingly symmetrical universe, Ptolemy placed the Sun—the only planet without an epicycle—between these two groups: mediaeval scholars likened the Sun to a king attended on each side by three planetary courtiers.

Ptolemy was a Janus-faced figure who looked backwards as well as forwards. He bequeathed to the future the astrological influences and celestial spheres he had inherited from the past, yet—like modern astronomers—he also insisted on precise geometrical calculations. Like some of his Babylonian and Greek predecessors, Ptolemy believed in a holistic cosmos that integrates human beings with the heavens. Astronomers wanted to trace out the planets' movements not just as an intellectual exercise, but also to discover how they influence people. After all, since the changing position of the Sun clearly affects life on Earth, then why should this not also be true of the other six planets? In Ptolemaic astrology, different parts of the body are related to particular planets and signs of the zodiac, and studying the stars remained important for Islamic and European physicians. In this cosmological medicine, the seven ages of man correspond to the seven planets—or as William Shakespeare explained in *As You Like It*, the Moon represents the 'mewling and puking' infant, while Saturn stands for 'second childishness and mere oblivion'.

5 Life

Vast chain of being! which from God began,
Natures ethereal, human, angel, man,
Beast, bird, fish, insect, what no eye can see,
No glass can reach; from infinite to thee,
From thee to nothing.

—Alexander Pope, *Essay on Man* (1733–4)

'I swear by Apollo the Healer, by Aesculapius, by Health and all the powers of healing.' More than two millennia after his death, Hippocrates is famous for his oath of good medical practice. Nevertheless, he has become a mythological hero as much as a real one. Although his name is still invoked in debates about euthanasia and abortion, many of the sayings attributed to him—including his oath—were written by his followers. Far from being a lone therapeutic voice, Hippocrates was just one of many Greek physicians who between them recommended an immense variety of treatments. And like so many other supposed founders, Hippocrates inherited pre-existing knowledge.

Hippocrates founded his medical school on the Greek island of Cos at around the same time that Socrates was attracting philosophical disciples in Athens. There were no formal qualifications for doctors, so to attract paying students, the Hippocratics engaged in some self-promotional tactics, boasting that they alone were the true medical experts and disparaging their predecessors as mere magicians. By continuing this practice of concealing debts to prior expertise, the successors of Hippocrates converted him into the symbolic father of medicine.

The Hippocratics are justly celebrated for insisting on detailed case reports. They constructed a huge reservoir of practical experience, and this enabled them to predict the course an illness would take, even if they did not understand the reasons. This sensible step gave them the appearance of being in control when in reality, they could do little else but help their patients die comfortably. But although they had few effective cures, Hippocratic physicians emphasized the importance of staying healthy. Unlike modern doctors, their theories focused on the particular constitutions of individuals rather than on universal diseases.

They provided customized prescriptions for maintaining bodies and psyches in good trim by advising people how to keep their essential fluids—their internal humours—in their natural state of balance.

This focus on a patient's personal well being still prevailed in eighteenth-century Europe. In the absence of effective drugs, Hippocratic medicine alleviated feelings of helplessness in the face of illness by putting people in charge of their own well-being as a preventative strategy. The sick (and also the hypochondriac) could monitor their own health by analysing daily fluctuations of their symptoms and trying to restore their normal equilibrium. Patients liked being treated as special individuals, while experienced physicians could charge high fees for wealthy clients who demanded constant personal surveillance. There was also great philosophical appeal in the central Hippocratic tenet that bodies are intrinsically self-healing and searching for their natural balance, because it suggested that the Universe had been intentionally designed rather than arising by chance.

Amongst the Seven Wise Sages of ancient Greece, only one was significant for the life sciences—Aristotle, who lived around a century after Hippocrates. Towards the end of his career, Aristotle rebelled against the conventional view that philosophers should avoid examining the real world. As well as considering environmental topics such as weather patterns and earthquake activity, he literally got his hands dirty by studying plants and animals. Although he made some notorious slips—counting teeth and ribs was not his strong point—Aristotle carried out his own dissections and emphasized the importance of persuading theories to fit the facts, rather than the other way round. Writing in minute detail, Aristotle compiled observations of an immense variety of living beings, including humans.

Unlike modern textbooks, Aristotle's compendium of animal behaviour mingled folklore and medical theory with hard facts. He assured his readers that sheep will produce black lambs if they drink from the wrong stream—yet on the other hand, his counter-intuitive report that dogfish have wombs was finally verified in 1842. Inevitably, Aristotle's theoretical preoccupations affected the types of observation he made, and he tried to unify creation by picking out common features shared by apparently diverse creatures. Ideologically committed to a perfect universe with no gaps, Aristotle searched for continuities rather than differences. He was fascinated by amphibious creatures such as seals, which seemed to form a link between aquatic and terrestrial animals, and also by bats, which have no feathers but fly like birds. He also had a stab at a general law of ageing, relating the growth of fur, hooves, and beaks in different creatures.

Aristotle's catalogue of nature proved extremely popular in Europe because it included detailed descriptions of sexual activities, and later pseudo-versions such as *Aristotle's Master-Piece* provided favourite under-the-counter reading. His biological approach also survived at a more theoretical level, because of his emphasis on small, gradual changes between organisms. In the Christianized version of Aristotle's model, a long continuous chain of being stretches in imperceptible steps from the tiniest organism right up to its living summit on earth—human beings—and then onwards up through the angels to God. At the end of the seventeenth century, the philosopher John Locke explained this Aristotelian concept:

> That in all the visible corporeal World, we see no Chasms or Gaps. All quite down from us, the descent is by easy steps, and a continued series of Things, that in each remove, differ very little one from another...and the Animal and Vegetable kingdoms, are so nearly join'd, that if you will take the lowest of one, and the highest of the other, there will scarce be perceived any great difference between them.[3]

Greek physicians knew far more about the outside than the inside of bodies. Without anaesthetics, internal surgery was prohibitively painful, and dissecting corpses was deemed immoral and not necessarily helpful—why should examining a dead body help you treat a living one? But there were plenty of battle casualties to treat, and victorious armies owed much to their Hippocratic physicians, who learnt from experience how to set fractures and bandage wounds, as well as amputate damaged limbs in record time. During the second century AD, one of these expert surgeons—Galen—treated Roman gladiators as well as soldiers, and his ideas about human anatomy dominated European thought well into the sixteenth century. He also transmitted to Europe his own particular version of Hippocratic theories that had been circulating and shifting during the previous half a millennium.

Galenic physicians learned that every human body is dominated by four special fluids or humours—*blood, yellow bile, phlegm,* and *black bile*—here shown in italics to distinguish them from actual substances with the same name. Each humour has its own function: *blood* is the source of vitality, *yellow bile* aids digestion, *phlegm* is a coolant that increases during fevers, and *black bile* darkens the blood and other bodily secretions. As well as affecting people's physical nature, the humours influence psychological behaviour, so that everybody is characterized by an intrinsic temperament depending on their internal humoral balance. For example, somebody who is thin and sallow is overburdened with *yellow bile* and possesses a mean, acrimonious personality. Conversely, people who are fat, pale, and lazy are afflicted with too much *phlegm*, while Shakespeare's melancholic Malvolio resembles a *black bile* stereotype.

For Galen, understanding anatomy meant studying bodies, not books. He argued that doctors need accurate anatomical knowledge to cope with war wounds and amputations, and so—despite moral protests and practical obstacles—Galen persisted in conducting experiments that would refute older ideas.

Sometimes he got round social taboos against dissecting human corpses through examining battle casualties pecked dry by birds, but he worked mainly with pigs and apes. Nowadays, Galen's research would be banned, since he had no compunction about operating on living animals tied down with ropes. He probed inside pumping hearts, tied off ureters to demonstrate how bladders and kidneys work, and cut through spinal cords to investigate which parts of the body become paralysed. 'Nothing upsets any operation like haemorrhage,'[4] he observed before giving invaluable tips on how to cope with spurting blood. For almost four hundred years, philosophers had maintained that arteries contain air, but Galen proved them wrong by tying an artery at two points and cutting it in between. Obvious—but only if you confront blood on a daily basis and are determined to save a life rather than ponder on its meaning.

Ironically, this surgeon who emphasized the primacy of personal observation perpetuated errors which endured for centuries, enshrined inside received doctrine that nobody dared challenge. In the absence of human corpses, Galen chose the next best option of examining Barbary apes. A sensible strategy—but for well over a thousand years, physicians falsely believed that blood flows through tiny holes in the central wall of human hearts as it does in primate ones. Another striking feature of Galenic physiology is the absence of any circulation system. According to his model, blood is perpetually manufactured in the liver and the veins before being consumed in the body's other organs and limbs. Galen reached this conclusion not only from his common-sense assumption that dark and bright blood must flow in two separate systems, but also because he was conceptually committed to associating the brain, heart, and liver with three different aspects of the soul.

Although a brilliant dissector who believed in wielding his scalpel rather than trusting other people's opinions, Galen was—like so many innovators—constrained by pre-existing notions. The same problem beset Andreas Vesalius, the Renaissance anatomist who adopted Galen's own strategy of first-hand observation and is celebrated for drawing the body as it really is. Although Vesalius exposed many Galenic mistakes, he decided that the holes in the heart must in fact exist—God had made them too small for him to see.

6   *Matter*

I wish men would get back their balance among the elements
And be a bit more fiery, as incapable of telling lies
As fire is.
I wish they'd be true to their own variation, as water is,
Which goes through all the stages of steam and stream and ice
Without losing its head.

—D. H. Lawrence, 'Elemental' (1929)

In seventeenth-century Europe, ancient Greece was still the land of heroes. Many scholars regarded the classical world as the peak of civilization whose achievements would never be surpassed. Greek philosophers had already elaborated the only two views of matter possible at the time— before quantum mechanics made things far more complicated, matter had to be either continuous or else exist in distinct particles separated by gaps. There were, of course, many possible variations on both themes, and none of them were totally satisfactory. And hence the battle between two camps, each with its major classical champion. On one side were ranked the followers of Aristotle, who believed in continuity, and taught that everything on Earth is a blend of four basic elements. Soon to be vanquished, these late Aristotelians were perpetuating academic beliefs that had prevailed in Europe for centuries. Their opponents—young upstarts such as Isaac Newton—had no qualms about lambasting such traditional views. They insisted that matter is built up from discrete atoms, and adopted as their figurehead Epicurus, who had been one of Aristotle's major critics.

Aristotle and Epicurus came to symbolize two fundamentally opposed views of how the universe is put together. The earliest Greeks had gone for continuity, mostly imagining a universe made up from a few essential raw ingredients which change and combine together to form different substances—just as seeds grow into trees, so iron becomes rusty, water solidifies into ice, and people decay into dust. In this type of packed cosmos, light and heat can be thought of either as vibrations in a sort of invisible atmospheric jelly or as superfine fluids that flow like weightless liquids. As critics pointed out, it's hard to explain how such abstract concepts might play out in the real world. For atomists, on the other hand, the basic units are minute, indivisible particles. Remaining unchanged themselves, they bounce around in empty space (in most versions, anyway), colliding to combine in various ways and form different materials. Iron and water corpuscles join up to make rust, water particles are crammed closely together in ice, and light resembles a stream of bullets.

Aristotle was committed to continuity in both the living and the physical realms. His belief in the ladder of being, with its infinitesimal steps between similar beings, fitted ideologically with his conviction that there are no empty spaces anywhere—'Nature abhors a vacuum' was an Aristotelian leitmotif. Although previous Greek philosophers had already outlined various atomic schemes, Aristotle rejected those and deliberately reverted to models developed by the Hippocratics. His model may sound arcane, but it dominated Muslim and Christian thought for centuries.

Aristotle believed that the world is characterized by four idealized, imaginary qualities (italicized in the text)—*hot* and *cold*, *dry* and *wet*—which are possessed by everything in different proportions. For some substances, Aristotelian qualities are clearly linked to physical properties. Milk, for instance, is mostly *cold* and *wet*, whereas the flame of a candle is *hot* and *dry*. Other descriptions are less instinctively obvious. According to the Aristotelian system, women's *cool, damp* bodies make them moody and incapable of the rational thought processes performed by men's *hot, dry* brains. Correspondingly, in the holistic cosmos of correspondences developed by Aristotle's successors, the masculine planets—such as Mars and the Sun—are *hot* and *arid*, while the feminine Venus and Moon are *cool* and *moist*.

Gratifying his love for order, Aristotle complemented these four qualities with four idealized terrestrial elements—*earth, water, air,* and *fire*—which combine together to form all the different materials found on the Earth. These qualities and elements fit together in a neat scheme, illustrated chematically in Figure 6. Symmetry reigns: unlike elements are placed opposite to one another, and each element comprises two contrasting qualities. So at the top, *fire* is flanked by *hot* and *dry*; it faces *water*, with its *cold* and *wet* qualities. Correspondingly, *earth* is *cold* and *dry*, whereas *air* is *hot* and *wet*.

Although Aristotle's ideal elements do not exist in their pure form, they did provide useful hypotheses for thinking about actual matter in the real world. Aristotelian elements can be transformed into each other by changing their qualities. If you heat *cold, wet* water, you drive off the *cold* to produce *hot, wet* air, which provides a reasonable template for what happens when water boils to produce steam. Similarly, it does seem intuitively sensible to think that metals are very *earthy*, or that burning wood is full of *fire*. The elemental constitution of a substance helps to describe its behaviour. Aristotelian *air* and *fire* have an inherent

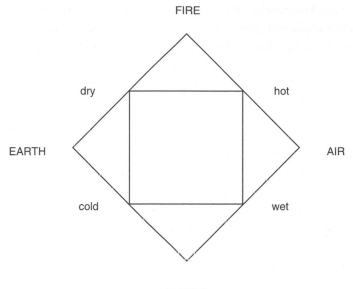

FIRE

dry

hot

EARTH

AIR

cold

wet

WATER

FIG. 6 A Christianized version of Aristotle's elements and qualities.

tendency to move upwards, whereas *earth* and *water* naturally fall downwards. In the Christianized cosmos of Figure 3, these sublunar elements are indicated by the inner land and sea of the central globe, surrounded by outer rings of clouds and flames.

When I look at this diagram, my first thought is: 'So where's the evidence?' Greek scholars were asking different sorts of questions. As philosophers, they were less concerned with empirical justification than in answering fundamental problems about creation—'Why is the Universe stable?' or 'How did this Universe emerge from its initial state of chaos?' For Aristotle, glossing over a few inconsistencies was inconsequential compared with establishing a basic rationale for why a coherent world should exist at all. Above all, Aristotle insisted that there must be some good underlying reason to explain why the world is as it is. In order to make sense of the Universe and of his own life, he adopted a teleological approach based on believing that creation must have a purpose or goal (the Greek *telos*). Eyes provide a simplified example: if you're a teleologist, you believe that animals have eyes because they need to be able to see; if you're not, you argue that animals can see because they happen to have eyes.

An end-based outlook permeates Aristotle's philosophy. For him, it must be an intrinsic property of nature to develop order. That is why his four elements naturally move towards their natural places, as part of the general tendency to establish a steady, systematic cosmos. This purposeful aspect of Aristotelianism made it especially appealing to Christians, whose God is responsible for a similarly goal-directed universe. Teleology has remained central to scientific debates ever since, especially over evolution, where it appears as the argument from design. If you presuppose an intelligent creator, then you are in the very comforting position of being able to explain everything in the Universe as part of a grand plan (although suffering is somewhat tricky). On the other hand, if you carry that argument too far, you run the risk of heading off towards fatalism: personal effort and initiative seem pointless if God has already worked everything out for the best.

A particularly famous anti-teleologist is Epicurus, who disagreed with Aristotle on every fundamental point—although they did share a lack of precision on how their theoretical visions might relate to visible, tangible reality. Arriving in Athens fifteen years after Aristotle had died, Epicurus founded a totally different school of philosophical thought, which flourished around 300 BCE. Not for him the security offered by Aristotelian design and stability: for Epicurus, chance is the key to the cosmos. He argued that our Universe is just one amongst many, and that it arose from the random collisions of atoms streaming through a vast void and occasionally swerving to bang into one another. These indivisible atoms combine in various ways to form chunks of matter with different characteristics, such as heat or colour.

Like many Greek philosophers, Epicurus tried to eclipse his predecessors by denying their significance. Epicurus's ideas are based on those of Democritus, who had lived in the previous century and is now celebrated as the founding father of atomism. Hardly any of Democritus's writing survives, and so his thoughts about atoms must be inferred from later interpreters (Karl Marx chose this challenge for his Ph.D. thesis). Since Greek commentators were inspired by their own agendas, their reports were by no means impartial. They included biased critics such as Aristotle as well as successors who, like Epicurus, wanted to establish their originality. But a few fragments do remain, and these are Democritus's own words:

> By convention there is colour, by convention sweetness, by convention bitterness, but in reality there are atoms and space.[5]

By this, Democritus meant that the Universe is constructed from an infinite number of small indivisible particles which constantly move through infinite, empty space. When Democritan atoms collide, some of them rebound, whereas others attach themselves together to form compounds. These atoms never change, although they do have different shapes, sizes, and properties: for instance, thin angular atoms produce an acid taste, while round ones impart sweetness.

A nice theory—until you must prove it. Even if you did manage to isolate an atom, how could you be sure that it was itself indivisible? Can individual atoms ever be big enough to see? And is it not rather arbitrary to think that sharp atoms yield a sharp flavour? Epicurus did modify Democritus's earlier theories to cope with some of these obvious objections, but he skimmed over other difficulties because he was more interested in ethics than in physics. His major credo was that individuals should free themselves from anxiety—after all, if everything is down to chance, then there's not much point in striving for perfection. With that outlook on life, it's hardly surprising that Epicurus did not spend too much time refining an unprovable theory.

Because physical models based on atomism and continuity were closely linked to moral stances, choosing between them was not based solely on reason or evidence. Many Greeks found Epicurus's cosmology a threatening one, because it lacks any comforting vision of a single world that has been designed with an underlying rationale, such as accommodating human beings. Epicureanism also undermined the exhortations of Plato and Aristotle that the main goal of human beings should be to lead a virtuous life. These two ethical objections still seemed crucial two millennia later, when seventeenth-century Protestants decided that Epicurus's atomism did make sense, but then found themselves in a moral maze about its implications. Atomism may seem obvious now, but Aristotelian continuity reigned supreme for many centuries, protected by being wrapped up in a philosophical package that meshed with Christian beliefs.

7 Technology

Who built Thebes of the seven gates?
In the books you will find the names of kings.
Did the kings haul up the lumps of rock?...
Where, the evening that the wall of China was finished
Did the masons go?

—Bertolt Brecht, 'Questions from a
worker who reads' (1935)

'Eureka!' yelled Archimedes, as he leapt out of his bath and dashed through the streets (still dripping?) to announce that he'd solved the problem of measuring the amount of gold in the king's crown. An unlikely story, but one that's become a classic example of inspired scientific genius. Archimedes is also renowned for his technical inventions, some of which sound suspiciously successful, such as the giant mirror with which he supposedly set fire to Roman ships, or the giant screw which he might (or might not) have invented for lifting water from one level to another.

So is Archimedes being mythologized as a hero of science or of technology? And which is more important—does the theory in the laboratory come first, or the invention in the factory? One approach to the relationships between science and technology is to look at words. When the first systematic dictionary of English was being drawn up in the eighteenth century, its compiler, Samuel Johnson, declared that his main motive was 'to embalm his language, and secure it from corruption and decay'.[6] Rather like modern European purists who resent Americanization, Johnson tried but failed to ossify English and preserve it in upper-class formality forever. Eventually (again resembling later linguistic conservationists), Johnson was forced to recognize that change is not necessarily a bad thing. By the time he'd finished his dictionary, Johnson had acknowledged that new inventions and new activities demand new words to describe them.

In practice, imported or invented vocabulary is less confusing than old words which look the same for centuries but gradually shift their meaning. Amongst such slippery terms, 'science' is one of the most insidious. Although its roots are classical—from the Latin word *scientia*, meaning knowledge—'science' could not possibly have been used by Johnson, let alone the Romans, in anything like its modern sense. Even the more recent word 'technology' presents problems. Coined in the nineteenth century, it stems from the Greek *techne*, which implied knowledge gained through practical work. But because *techne* originated long before heavy industry existed, it referred to manual skill rather than to mechanical efficiency; and as a consequence, 'technology' used to be far closer to the arts than it is now.

Both words—'science' and 'technology'—have been imbued with various social distinctions as well as disciplinary ones. 'Science' used to mean something closer to the type of learned knowledge that scholars gain from books—even in Johnson's time, it still made sense to talk about the 'science of language' or the 'science of ethics'. This means that scientific knowledge was restricted to rich and well-educated people, mostly men. Their condescending attitudes towards labourers were perpetuated into the Victorian era, when scientists looked down on engineers who worked with their hands and gained money from their inventions. Similarly, privileged Greek philosophers gave *techne* a pejorative spin by associating manual dexterity with the need to earn a living. Sculptors, artists, and craftsmen were paid for their physical skills, and enjoyed nothing like the status they acquired far later in Renaissance Europe.

Archimedes was neither a scientist nor a technologist, since no such people existed when he was living in Sicily during the third century BCE. Instead, Archimedes more closely resembled a different modern stereotype—the armchair philosopher. The social and scholarly landscapes of ancient Greece were dramatically different from those familiar today. Broadly speaking, two sectors of Greek society influenced what would later become science. Only individuals belonging to the smaller group are celebrated—the wealthy philosophers who thought profoundly about the Universe and its occupants, but generally felt that hands-on experimental research was beneath them and irrelevant.

In contrast, the far greater number of people from lower social orders have largely been forgotten, even though they were also vital for the course of future science. Science is a practical subject as well as a theoretical one: abstract models are important, but they need to be tested experimentally and compared with observations of the real world. Although many theoretical concepts do derive from Greek philosophers, other aspects of science originate from less privileged people who used their expertise to keep themselves alive—miners who developed ore-refining techniques, farmers familiar with weather patterns, textile workers who relied on chemical reactions.

Many practical men were skilled mathematicians. What later became the science of mechanics developed from solving the problems involved in making things work—erecting bridges, constructing irrigation systems, devising energy-efficient pulley systems, designing effective military weapons. While philosophers pondered how to triangulate the Universe, builders developed the basic trigonometry needed to make a wall vertical. These mechanical experts belonged to a different social background from the leisured theoreticians, but they also had different aims. Philosophers wanted to explain the world, whereas practical mathematicians were more interested in describing it. If you're building a house, you need to measure the planks, not wonder why the tree grew.

When Archimedes lounged in his bath or his armchair, his mind was preoccupied not so much with mundane matters of lifting heavy weights or squeezing olives, but more with contriving ingenious gadgets that would demonstrate mathematical principles. His books were about his mathematical innovations, not his technical inventions. For his elite colleagues, provoking wonder was a valuable activity in its own right, one that advertised its creator's virtuosity. They learnt to impress with magic vessels constantly filling up from a hidden reservoir, temple doors opening and shutting automatically, theatres with puppets apparently sawing wood or hammering nails. Although extremely clever, these toy-like devices were not intended to have any practical applications.

Perhaps the most famous amongst them is Hero's so-called steam engine, in which steam from a large cauldron is forced through pipes into a small hollow ball to make it rotate. Hero and his colleagues probably never thought about converting this model into a working machine, but even if they had engaged in such wishful thinking, they would have found it impossible to achieve. Technological change depends on practical feasibility, political will, and commercial stimulus as much as on scientific knowledge. Although the Greeks inherited fine metal crafts from the Babylonians and the Egyptians, they relied mainly on wood and knew little about iron production. Scaling up Hero's steam-driven sphere to industrial proportions would have demanded not only multiple technical capabilities—casting large cylinders, making pistons steam-tight—but also the organizational infrastructure essential for establishing and maintaining complex manufacturing systems.

Elite Greek philosophers claimed that they were the founders of civilization. As though perched on top of a historical iceberg, they concealed their submerged foundations, which included their inheritance from the past as well as their dependence on the workers who outnumbered them. Although Ptolemy boasted that his armillary sphere had introduced accuracy into astronomy, he made no reference to the craftsmen who physically made the instruments he handled. Just as he eclipsed his theoretical predecessors, so too Ptolemy failed to mention not only the skills of Greek artisans, but also their dependence on older techniques originating in Mesopotamia and Egypt.

Shadowing each famous Greek hero is a penumbra of scarcely discernible informants and colleagues who were also vital for science's origins. Aristotle was unusual in carrying out his own dissections, but for much of his detailed research he depended on beekeepers, farmers, and horse-trainers—men who needed accurate biological information in order to survive, and who provided him with what would now be regarded as scientific data. Occasionally Aristotle mentioned them explicitly, although not by name; for instance, he explained that experienced fishermen were so well informed about the grey mullet's mating habits that they knew where to plant male fish as decoys to catch the females (and vice versa). More commonly, Aristotle made it appear as though the observations were entirely his own, even when it seems more likely that local experts had fed him the details.

Philosophical heroes do not owe their celebrity purely to their brilliance; similarly, significant achievements do not in themselves guarantee fame. Several strategies for ensuring a favourable posthumous reputation have emerged. One reliable tactic is to die dramatically. Socrates left no writing behind him but is remembered for drinking hemlock, while Hypatia of Alexandria became a feminist icon not because of her mathematical work, but because she was (allegedly) torn to bits by an angry mob—although exactly which one and why remains unclear. Archimedes secured his place in posterity by dying the death of a romantic philosopher, (supposedly) so obsessed with finishing a geometrical diagram in the sand that an enraged soldier killed him with a sword.

According to accepted mythology, Archimedes had planned out his grave in advance. This prestigious philosopher wanted to be renowned as an inspired mathematician rather than as a pragmatic inventor, so he asked for his memorial to show not a screw or a catapult, but a sphere packed into a cylinder, along with mathematical formulae for comparing their volumes. Neither scientists nor technologists had yet been invented, but the foundations for the hierarchical distinctions between them had already been laid.

Interactions

There is no single form of science—what counts as science depends on where and when you're looking. Information, skills, and objects constantly move from place to place, pass from one generation to the next, and shift as they are adapted to fit local needs and tastes. Although Renaissance scholars claimed to be reviving Greek culture, their scientific knowledge resulted from many centuries of communications and interactions between different peoples and places. Looking back from a British vantage point in the twenty-first century, three interconnected regions were particularly significant for science's future: China, the Islamic world, and mediaeval Europe. Many crucial inventions appeared first in China, which was technologically superior to Europe until the end of the eighteenth century. In contrast, Islamic interpreters played crucial roles in interpreting, modifying, and developing the Greek expertise that reached Europe in the twelfth century. Far from being merely neutral transmitters of abstract concepts, Muslim leaders encouraged science by building massive libraries, hospitals, and astronomical observatories. In Europe, scientific ideas were pursued most strongly in religious institutions, first in monasteries and subsequently in universities. Scholars transformed Islamicized versions of Greek theories into a Christianized form of Aristotelianism that profoundly influenced Renaissance investigations of mechanics, optics, and astronomy.

II

Interactions

I *Eurocentrism*

Consider, I pray, and reflect how in our time God has transformed the
Occident into the Orient. For we who were Occidentals have now
become Orientals...Words of different languages have become common property known to each nationality, and mutual faith unites those
who are ignorant of their descent.

—Fulcher of Chartres, *A History of the Expedition
to Jerusalem, 1095–1127 (c.1105–27)*

All civilizations like to map the world around themselves. Arabic Muslims perceived Baghdad as the pivot of seven climatic zones, whereas for mediaeval
Christians, Jerusalem was the globe's navel. In contrast, when the ancient Greeks
visualized the globe, they placed their familiar Mediterranean (Latin for 'middle of
the Earth') at the centre of a giant land mass divided into Asia, Libya, and Europa,
the names of three mythological step-sisters (Princess Europa was raped by Zeus
in the form of a bull). In the heyday of Athenian supremacy, Aristotle located
his fellow Greeks between Europe and Asia, awarding them the finest characteristics of both, and also finding fault with everybody else. Western Europeans
inherited not only Aristotle's philosophy, but also his self-centred arrogance.

Repeat something often enough, and people will believe it. Because Europeans
were politically and financially powerful, they placed themselves at the centre of
everything, and wrote accounts of the past that confirmed their own supposed
superiority. 'The West is the Best' view of history prevailed in Europe for centuries, even though there was plenty of evidence against it. That's the point being
made by the map in Figure 1—and what an Australian Prime Minister meant
when he declared: 'What Great Britain calls the Far East is to us the near north.'[1]

Until very recently, Eurocentrism dominated Anglo-American history of science. In wishful thinking versions of the past, science leads to Absolute Truth—
and moreover, it started in Europe. Now that the entire globe is electronically
interlinked, science is seen as the summit of human achievement and the outcome
of American/European genius. Such self-congratulation takes little account of
the possibility that other cultures may have chosen other approaches to life not

because their finest scholars were stupid, but because they had different opinions about what is important. In any case, more science does not necessarily produce better answers. After World War II, optimists declared that science would unify the world because—unlike religious faiths—its truths transcended national boundaries. Yet although the scientific enterprise may be global in its sweep, it has clearly failed to fulfil sanguine promises either of bringing peace or of deciphering nature's deepest secrets.

For centuries, few Europeans challenged the assumption that some special essence of Westernness distinguished them from the rest of the world's inhabitants. However, 'West' and 'Europe' are fabricated entities with no fixed boundaries. They came into existence slowly, and are still changing. Deceptively, they stamp uniformity onto diversity—in the past, there were greater variations than nowadays between peoples living in different parts of the European region. Even defining exactly where Europe lies is impossible, because its physical edges remain blurred as countries enter and leave. In any case, being European implies not only geographical location, but also cultural affinities.

A particularly important step towards consolidating Europe's special identity took place in the fourth century. Whereas towards the western side of the continent, Rome was losing control of the rebellious tribes that it had once dominated, the eastern end of the Mediterranean was becoming wealthier and more stable. To enhance his own position, the Roman emperor Constantine shifted his capital eastwards to the ancient city of Byzantium, which he named after himself—Constantinople, now Istanbul. As trade, agriculture, and civilization continued to flourish, the eastern Mediterranean area became closely linked with China, India, and the Arabic nations, and the Christianized realm of the world became symbolically divided into two zones—the Byzantine East and the Roman Catholic West. This split became still more firmly entrenched in 800 AD, when the Pope crowned Charlemagne, a Frankish (early French) king, as Holy Roman Emperor. Although he ruled over a disparate collection of warring states, Charlemagne was hailed as the first ruler of a united Europe. Ever since then, Westernists have placed great emphasis on his significance as Europe's founding father—and Eurocentrism prevailed for most of the nineteenth and twentieth centuries.

The notion of Europe's glorious origins was boosted during the Renaissance, when classical revivalists located the cradle of European civilization in the Athens of Plato and Aristotle. Imbuing this small and remote city-state with the quasi-mythical aura of a bygone golden age, artists, scholars, and politicians linked themselves directly to ancient Greece, and dissociated themselves from everything in between. The major casualty of this interpretation was the so-called Dark Ages, a vaguely defined era that started around the time of Constantine, and when—supposedly—nothing much happened. Tacked on to the end of this barren historical vacuum was a slightly less bleak period called the Middle Ages, which paved the way for Renaissance creativity in the fourteenth century. By ingeniously eradicating a millennium, historians made it seem that the torch of scientific knowledge had been handed on directly from ancient Greece to Renaissance Europe.

In such simplistic East–West divisions, science played a special role. Western Europeans acknowledged the Greeks' intellectual splendour, yet stressed the practical benefits of the new experimental approach they introduced during the seventeenth century. For example, they boasted about a famous trilogy of Renaissance inventions—printing, gunpowder, and the magnetic compass—which had, they maintained, transformed knowledge about the world as well as daily lives. Despite suspicions about Chinese priority (successfully minimized until the twentieth century) advocates of Western superiority claimed this trio for Europe. Their appealing vision of Renaissance creativity strengthened the myth of European supremacy. A gratifyingly neat narrative emerged. According to this Eurocentric version of human achievement, science originated in Greece, was preserved in the Islamic empire that flourished while Europe went into decline, and then entered Spain intact in the twelfth century to spread northwards.

Metaphorically, the term 'Dark Ages' was loaded with meanings, suggesting not only that the light of intellectual illumination had been dimmed (after all, to see is to know) but also that a gloomy cloud of superstition had descended to stifle rationality and originality. While Europe lolled in its Dark Ages—or so ran the conventional story—Arabic scholars acted as caretakers of Greek knowledge. Muslims were cast as neutral transmitters of European expertise, even though they were experimenters and theoreticians in their own right, who actively transformed the skills and beliefs they had gathered together from diverse cultures. Similarly, China was viewed as a remote, esoteric place, and the impact of its agricultural and industrial success on Europe remained unacknowledged.

Rewriting history is not only a question of finding more facts: it also means deciding which facts are important. If you look in the right places, there was—unsurprisingly—a great deal of activity in those forgotten centuries. Historians have often placed the birth of modern science in the sixteenth century, when Copernicus suggested that the Sun rather than the Earth should be at the centre of the Universe. However, this entailed glossing over crucial changes that happened earlier—and also elsewhere. Admittedly there were no European universities until the end of the eleventh century, but learning did thrive in rulers' courts and Christian monasteries. Just as importantly, key developments were taking place outside Europe. The Chinese economy prospered under a powerful government that encouraged agricultural and industrial innovation, while the Islamic region was also expanding and growing richer. Muslim scholars not only absorbed Greek medical and mathematical knowledge, but also modified and expanded it further through their own investigations. Non-European inventors and scholars were producing equipment and ideas that spread westwards and later became incorporated within science and technology.

Like other empires, the European bloc forged by Charlemagne gained strength by being made to appear uniform from the outside. In reality, empires comprise many smaller groups, internal minorities who speak different languages and are often looked down upon. Conversely, rulers can promote chauvinistic solidarity by castigating outsiders as inferior. Such tactics of self-promotion through differentiation have been practised by several imperial powers over the ages, including the Chinese, the Romans, and the British. Language and religion have always been crucial distinguishing factors—in many cultures, the word that came to denote 'barbarian' originally meant 'foreigner'. The Greeks and the Romans consolidated their own imperial identities by contrasting themselves with neighbouring 'barbarians', all lumped together as if there were no distinction between them. In versions of the past that trumpeted European superiority, other groups were caricatured into simplistic stereotypes. The Chinese appeared as impractical isolationists, a nation of pacifist navel-gazers with a sideline in flower pictures, while Muslims featured not as learned, devout scholars but as aggressors who had destroyed the unity of the Roman Empire.

Viewed in retrospect, empires may seem distinct, yet in reality they were ill-defined both in time and in place, sprawling unevenly over many centuries and many lands. Since distances were long and communications leisurely, power was exerted regionally as well as centrally. Individual rulers fostered particular activities depending on local interests, so that expertise and practices varied from place to place. That meant there was no single centre from which scientific knowledge was disseminated intact. Instead, different versions of skills and learning coexisted and mingled together, interacting sporadically through personal initiatives rather than as the outcome of any coordinated plan. Extensive international trading networks spread across continental Europe and Asia, so that goods, people, and knowledge gradually travelled over large distances. They may have moved slowly, but they did get there.

As people travelled from one region to another, they took with them ideas and things that changed as they moved. This principle of transformation through migration had long been familiar. In a story by the Greek poet Homer, Odysseus takes an oar from his ship and walks inland until he reaches a village where the farmers interpret the object he is carrying as a winnowing fan. In the course of their daily lives, many travellers—merchants, monks, scholars—journeyed around the Eurasian land mass, taking with them knowledge which was constantly being exchanged and adapted to fit local circumstances. For example, from a European perspective Venice appears to be the site of many innovations. However, because it traded both eastwards and westwards, the city imported and then modified techniques that had originated in China, India, or the Islamic civilization. These included not only practical devices, such as improvements in navigation, but also more effective methods of marketing and accounting. Gradually, technologies and theories filtered westwards to help stimulate an economic and intellectual revival in Europe, where monastic scholars fused together Islamicized Greek science with Christianity.

The Eurocentrism that distorted the past is now itself being relegated to the past. Because historians are illuminating Europe's supposed Dark Ages, they are also helping to set modern political agendas, confirming that diversity should be celebrated for generating richness, instead of being condemned for diluting excellence.

2 China

The reputation which the world bestows
is like the wind, that shifts now here now there,
its name changed with the quarter whence it blows.

—Dante Alighieri, *The Divine Comedy* (*c.*1310–20)

At the beginning of the eighteenth century, Europeans knew so little about China that a flamboyant French opportunist called George Psalmanazar successfully passed himself off as a Formosan. Employed by the Bishop of London to translate the Christian catechism into his supposed native language—freshly minted for the occasion—Psalmanazar also published a detailed yet fabricated guide to Formosan culture, which was snapped up enthusiastically by English gentlemen wealthy (and gullible) enough to indulge their fascination with exotica. Two hundred years later, Chinese science still seemed just as mysterious, until, that is, the arrival of an informant as unlikely sounding as Psalmanazar, an eminent embryologist who was also one of Britain's leading Morris dancers. His name was Joseph Needham, and he revolutionized not only Chinese studies, but also how historians think about the global development of science.

Needham first visited China in 1942 as an official representative of London's Royal Society. By then, he was a distinguished and politically active left-wing scientist, already obsessed by Chinese history, which he was studying with a young Chinese scientist whom he eventually married more than fifty years after they had first met. By 1950, after intermittently working and travelling in China, Needham had outlined what seemed an ambitious yet feasible project—to produce a seven-volume work called *Science and Civilisation in China.* As the research piled up, collaborators and assistants accumulated, and fifty years later there were over twenty volumes, with more forthcoming.

Instead of festering on obscure library bookshelves, Needham's research became politically contentious. His books were initially condemned for their Marxist interpretations, and Needham himself was banned from the United States after backing Chinese allegations that American forces were using biological weapons in Korea. Once it became clear that Needham was generating an immense work of fine scholarship, his critics shifted to accusing him of political naivety—an interesting judgement on this Marxist and High Anglican preacher who raised both private and public money to fund the Cambridge's independent Needham Research Institute. In contrast, in China Needham became a national hero. Reformers as well as traditionalists welcomed his project of retrieving their own scientific and technological cultural heritage, and Needham's initiative was imitated in India and other countries trying to recover from imperial subjugation.

One of Needham's coups was to rewrite the timescale of human invention. His long list of Chinese innovations currently stands at 250, ordered alphabetically from the abacus through gear wheels, toilet paper, and umbrellas to the zoetrope (a Victorian photographic device). Most famously, Needham's table includes the trilogy of inventions conventionally claimed for Renaissance Europe—gunpowder, the magnetic compass, and printing—placing them all in China and giving them all earlier dates. As he pointed out, the Silk Road not only enabled exotic goods to travel westwards, but also encouraged the migration of technological and agricultural products. Thanks to Needham, China's priority has been demonstrated for scores of innovations previously claimed as European.

Needham argued that China needed to be reassessed—instead of being a scientific backwater steeped in ancient mysticism, Chinese civilization had been technologically vibrant and way in advance of Europe's during the so-called Dark Ages. By studying China, which had seemed an esoteric speciality, Needham queried the very nature and origins of science in Europe. Writing to convert rather than to convince, Needham introduced the heretical notion that modern science is not uniquely Western but is 'ecumenical', depending on local truths which flow into it like rivers into the sea. In particular, he preached, traditional Chinese knowledge made vital contributions to the scientific enterprise of producing universal knowledge.

Traditional historians were outraged by this suggestion, and promptly provided counter-explanations. Take gunpowder. Needham and his team uncovered alchemical recipes dating from the ninth century, and showed that explosives were being reliably prepared three hundred years later. Obdurate Eurocentrics defended Western priority by claiming that in China this new discovery was used only for fireworks and mining rather than for weapons, even though Sinologists demonstrated that China had cannons before Europe. Although their interpretations differ, both sides can claim to be right, because this Chinese military invention had a far greater effect in Europe, where guns soon led to the demise of armoured knights and feudal castles. (Ironically, knights had only come into existence because of Chinese stirrups, which had revolutionized European warfare by adding equine strength to human spears.)

Historians have compiled similar stories about magnetic compasses. Although unknown to the ancient Greeks, rotating magnetic devices were used by first-century

Chinese diviners to indicate south, the favourable direction for emperors to face. Later, complicated compasses with several concentric dials were developed to help choose appropriate locations for houses and tombs. Maritime craftsmen also made compasses, but they had nothing like the same impact in China as in Venice and Spain. Whatever the technical ability of Chinese navigators, almost four hundred years went by before Christopher Columbus reached America from the other side and launched a new era of European exploration, trade, and conquest.

Printing also produced more revolutionary results in Europe than in China, even though books were being routinely made there four centuries before the Gutenberg Bible. Chinese rulers poured money into wood-block printing (which lends itself to the non-alphabetic script), and movable-type publications were also produced. However, in China, books were valued for providing storehouses of information rather than catalysts for change, and there was no tradition of building up large accessible libraries as happened in the Islamic empire.

Another way of assessing Needham's claims is to examine how China and Europe differed. Before around 1400, the scientific and technological activities of China, Europe, and the Islamic regions were in many ways more similar to each other than to those of today: they were all tackling shared questions about human relationships with the physical world. Looking back, it is possible to pick out aspects of scholarly thought and activity that seem proto-scientific, but at the time were just one part of broader approaches towards resolving the fundamental questions of existence. So although astronomers did use instruments and mathematical techniques now associated with science, in many ways they were more like astrologers than modern scientists. Similarly, the processes now associated with chemistry were developed by alchemists who were searching for spiritual rather than scientific progress, and also by craftworkers whose skills included making glass, refining metals, and so on.

As in ancient Greece, book-learning and practical expertise were separate activities practised by different groups of people. When wealthy Chinese tourists ventured westwards, they found that the European cities they visited were technologically backwards. However, their own country's technical preeminence relied not on the ingenious inventions of leisured scholars, but on craft skills handed down within families. Throughout the interconnected Eurasian continent, artisans refined traditional devices and techniques independently of the minority elite, who remained untrained in manual skills that were passed on orally. Very gradually, people in the western and eastern sides of the land mass started to regard practical expertise differently. In China, existing hierarchies prevailed, and the social barriers to knowledge transmission remained impermeable. In contrast, in Europe, trade and warfare stimulated technical change.

Unlike Europe with its autonomous universities, China's monolithic education system encouraged stability and thwarted innovation. The government imposed rigorous national examinations to guarantee that officials did not merely come from rich families, but were also intelligent and efficient. This formality was effective but quashed innovation, and the restricted syllabus remained unchanged for seven hundred years. Set texts and commentaries were intended to be memorized rather than criticized, thus imposing a narrow uniformity that effectively became state dogma. This rigidity not only quashed originality, but also meant that many scholars focused more closely on ancient ethical and philosophical debates than on contemporary problems or scientific questions.

The powerful centralized administration system, so different from the small diverse fiefdoms of Europe, quelled individual commercial and military initiatives. In western Europe, private enterprise stimulated invention. For example, merchants welcomed portable guns to protect them on their travels, and were willing to pay extra for improvements. In China, political officials sanctioned cumbersome defensive equipment to ward off invasion, but condemned personal violence as well as personal profit. Consider the fate of Wang Ho, a twelfth-century entrepreneur who started from nothing but built up an ironworks employing five hundred men. After Wang Ho and his workers used force to repel local administrators interfering with his business, he was executed. The authorities had clamped down on Wang Ho's double transgression—unauthorized fighting and economic venture.

Philosophical and religious attitudes also diverged. Unlike Christians or Muslims, Chinese cosmologists did not assume that some sort of unmoved prime mover governs the Universe through natural laws, but instead believed that the behaviour of the heavens is correlated with that of human societies below. Court advisers divided celestial phenomena into two types: those that were regular and could be incorporated into a calendar, and those that were unpredictable and should be regarded as omens. Attention focused on the emperor and his delegates. If they performed badly, then floods, meteors, and other celestial disasters would follow; but if their conduct matched the world's organic harmony, then social peace would be maintained. Complete knowledge seemed an impossible quest, since human observers could glimpse only the shadowy effects of complex yet systematic patterns. As the eleventh-century official Shen Gua explained, 'Those in the world who speak of the regularities underlying the phenomena…manage to apprehend their crude traces. But these regularities have their very subtle aspect, which those who rely on mathematical astronomy cannot know of. Still even these are nothing more than traces.'[2]

Shen Gua was far from being a scientist in any modern sense of the word, but his career does illustrate why Needham and other historians stress the importance of Chinese science. Shen Gua was a gifted administrator who rose through the examination system to become a powerful financial, military, and political adviser

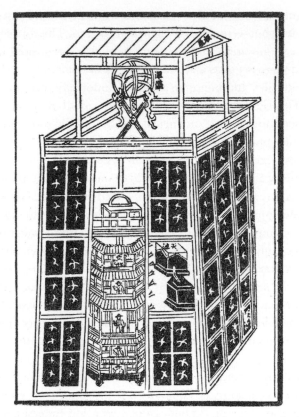

FIG. 7 Astronomical clock-tower built by Su Sung and collaborators at the Imperial Palace, Kaifeng, Homan province *c.*1090.

to the emperor, and he was Director of the Astronomy Bureau for several years. Although he later fell victim to court intrigues, Shen Gua came back into favour after making an elaborate relief map, and spent nearly the last twenty years of his life producing *Brush Talks*, which he conceived as conversations with his writing brush and ink block.

Looking back, Shen Gua might appear like a great astronomer far in advance of Europeans. He initiated a massive data collection project then undreamt of in Europe—to measure the positions of the planets three times a night for five years. Under his jurisdiction, a network of observatories was constructed, all equipped with impressive instruments. The one illustrated in Figure 7 features a large water-wheel to drive its instruments. On the roof of the two-storey tower, about 35 feet above the ground, is an ornate armillary sphere (see Figure 4), decorated with dragons and fitted with a clockwork mechanism to make it rotate. Although developed independently from the one Ptolemy described in his *Almagest*, this sphere looked similar and was used to measure the positions of stars and planets. On the first floor, a rotating globe models the movements of the heavens, and beneath this, a five-tiered pagoda houses an elaborate system of moving puppets that display and sound out the time.

Nevertheless, Shen Gua's goal was not a scientific one of deducing mathematical laws that govern the planets' behaviour. Instead, he more closely resembled an administrative astrologer who reformed the calendar in order to make better political decisions about imperial rituals. Similarly, he is celebrated for instigating many technological improvements, including an effective water drainage system constructed by 14,000 labourers. But in modern terms, Shen Gua was a fiscal expert rather than a hydraulic engineer, a bureaucrat who viewed nature as a source not of scientific evidence but of state benefit. For him, salt was not an intriguing chemical, but 'a means to wealth, profit without end emerging from the sea'.[3] Although Shen Gua's map makes him seem like an expert cartographer, he designed it not to guide travellers but to gratify the emperor by showing how much territory China controlled. And while Brush Talks does contain informative sections on astronomy, medicine, optics, and other subjects now classified as scientific, its six hundred disparate notes also bristle with court gossip, maxims and personal memories.

Despite the meticulous research carried out by Needham and countless other scholars, China's significance is still hotly debated. In the 1950s, Needham posed what became known as 'Needham's problem'. Although he never produced a completely satisfactory answer, his own discoveries enabled other scholars to continue tackling it—or to be more accurate, them, since there are two parts. First Needham asked, 'What was it, then, that happened at the Renaissance in Europe whereby mathematical natural science came into being?' And as if that were not in itself a big enough question, he continued, 'And why did this not occur in China?'[4] Rejecting any facile answer that Europeans are intrinsically superior, Needham insisted on social explanations, and—in line with his political commitments—developed a tailor-made Marxist analysis.

Although China is climatically similar to Europe, Needham argued, geographically it is very different. Whereas Europe's long, convoluted coastline invites maritime trading, China's vast land mass encourages peasant farming and internal cohesion. According to him, European science emerged in the fourteenth and fifteenth centuries when aristocratic–military feudalism evolved into capitalism. In contrast, China remained trapped in a feudal economy run by a centralized state bureaucracy focusing on production rather than defence. The Chinese empire was linked together by a vast, tightly linked administrative network which enabled officials to collect taxes and coordinate food production efficiently, but also stifled individual initiative and removed any drive to accumulate personal wealth.

At first, explained Needham, this bureaucratic structure stimulated technological development, since nationwide projects were launched to conserve water, improve transport, and promote education. However, in the absence of private incentive, this stable feudal society never moved on to the next stage of mercantile capitalism.

Needham's solution has been picked apart, because he viewed scientific knowledge as absolute and universal. Although he provoked enormous controversy by challenging Eurocentrism and emphasizing China's importance, Needham's image of flowing rivers implies that the great flood of scientific discoveries is leading inevitably towards the Ocean of Truth. Or to switch metaphors, he concluded that China reached the level of Leonardo da Vinci but never made it as far as Galileo, as though they were climbing up a mountainside towards the summit of Truth. Yet as Needham's own work confirmed, understanding science's history means thinking about social environments, not just chronicling great discoveries and theories. Knowledge about the world can appear in various forms, developed in different places for different purposes—there is no unique route towards Truth.

Nevertheless, Needham's problem refuses to go away. On the contrary, it is becoming increasingly important, because it calls into question the development of science throughout the world. Although regarded for the past fifty years as an obsession reserved for Chinese specialists, Needham's problem (which has parallels for other civilizations, such as the Islamic empire) is central to analysing the rise of global science. His findings—along with those of many others—challenge the notion of Western superiority, and force historians to recognize the importance of local practices. Science did not suddenly spring up in Renaissance Europe, but was the product of various beliefs and skills gathered together from different parts of the world over the preceding millennia. It is now clear that Needham's problem needs to be rephrased. The crucial question is not 'Why did science develop in western Europe?' but 'How did European activities lead to the form of science that now dominates the entire world?'

3 Islam

Reason is natural revelation, whereby the eternal Father of light, and Fountain of all knowledge, communicates to mankind that portion of truth which he has laid within reach of their natural faculties.

—John Locke, An Essay concerning Human Understanding (1689)

To consolidate their own identity as Europeans, Western commentators portrayed Muslims as alien beings. Stereotypes and scare stories abounded. In the eleventh century, Pope Urban II rallied his Catholic troops to attack the 'enemies of God', thus launching a crusade mentality that ignores Islam's Judaeo-Christian roots and still resonates today. Christians were scathing about Muhammad, because they believed (wrongly) that he was regarded as a pagan god rather than a human prophet who had received divine messages from God. Furthermore, they failed to recognize that for Muslims, the Qur'an supersedes both the Old and the New Testaments. Although at its peak the Islamic civilization extended along the African coast to cover Spain, many Europeans referred contemptuously to Asian Muslims as 'Saracens'—tent-dwellers—and had little idea that they were united by religious faith with the Spanish 'Moors'.

Even admirers of Islamic culture stressed difference, often concentrating on unfamiliar aspects that lent themselves to exotic renditions. For instance, the only piece of Islamic literature to have had much impact on English speakers is *The Rubaiyat of Omar Khayyam*. Set out in rhyming quatrains, the *Rubaiyat* became notorious for its fatalist outlook on life, which endorses the hedonism of a drunken philosopher soliloquizing his way through the day:

> The moving finger writes; and, having writ,
> Moves on: nor all thy piety nor wit
> Shall lure it back to cancel half a line,
> Nor all thy tears wash out a word of it.[5]

However, this version is a Victorian translation which immediately became enormously popular even though it seriously distorts the poem's original significance.

These verses are a lyrical parody of Persian philosophy, cobbled together from various sources. Far from being a pleasure-seeking libertine, Omar Khayyam was a brilliant mathematician, a Sufi sage who argued against religious hypocrisy. He has, like many other aspects of Islamic culture, been squeezed into a Western frame.

Analysed from a Western perspective, science first entered the Islamic culture in the middle of the eighth century, when the caliphs ruling Baghdad started pouring money into scholarship. Arabic, the sacred tongue of the Qur'an, effectively became an international scientific language bonding together a giant territory that stretched from the western edge of Spain along the southern Mediterranean coast and across to the borders of China. Under these conducive conditions, research flourished and theoretical knowledge reached an unprecedented level. According to the conventional story, the Islamic civilization proved unable to sustain this intellectual momentum, and science survived only by being transmitted to Western Europe around the end of the twelfth century. Put more dramatically—in the Great Race towards Scientific Truth, the Arabs fell by the wayside as Europeans picked up the Torch of Knowledge.

When viewed from within an Islamic context, this version of the past is unsatisfactory because it fails to take into account Muslim attitudes towards learning. Foreign ideas are rarely imported intact from other cultures, but instead are affected by local religious and philosophical outlooks. In Europe, Greek theories about life and the cosmos were reinterpreted from a Christian vantage point (see Figures 3 and 6). Similarly, when Muslim scholars encountered Greek ideas, they applied their own criteria of significance for fitting them into their conceptual framework. If you assume that today's science, along with its technological applications, represents the summit of human achievement, then Islamic philosophers do indeed appear to have ground to a halt after four hundred years. But for Muslims who believe that the quest for spiritual perfection is more important than dominating the material world through reason, then it is the science of Europe which took the wrong track.

The most famous book in the history of European science is Isaac Newton's *The Mathematical Principles of Natural Philosophy*. This title suggests a beginning, a mathematical foundation from which to build up knowledge of the physical world. By comparison, one of the greatest Arabic treatises is called *The Book of Healing*, and it aims to cure the reader's disease of ignorance. Instead of providing a starting point for progress, it sums up and organizes all the knowledge that a wise person needs to seek spiritual fulfilment. Paralleling contemporary studies being undertaken in Christian monasteries, *The Book of Healing* was compiled in the early eleventh century and belongs to an important tradition of classifying knowledge in massive encyclopaedias. Rather than being an arid scientific text, *The Book of Healing* is a poetic philosophical meditation that aims for comprehensiveness—it includes not only detailed explanations of Islamic Aristotelianism, but also an elaborate cosmology of angelic intelligences. After some sections were translated into Latin, *The Book of Healing* became a standard text book in Renaissance universities throughout Europe, where its author was acclaimed as a great physician—Avicenna. Although now less famous than his disciple Omar Khayyam (original author of the *Rubaiyat*), Avicenna was one of several Muslim scholars who later became authority figures in Western Europe and left a permanent mark on modern science with words such as alcohol, sugar, and alkali.

Avicenna is the Latinized form of Abu Alī al-Husain ibn Sīnā, who was renowned throughout the Islamic world for his studies not only of medicine, but also of theological philosophy. Ibn Sīnā's life and ideas were, like those of Omar Khayyam and other scholars, very different from a modern scientist's. Far from being a reclusive academic or a narrow specialist, Ibn Sīnā was an itinerant polymath who travelled between different Persian cities assuming various roles, including court physician, soldier, and political administrator—he even wrote some of his hundred-odd books on horseback using a special pannier he had invented. As well as being an expert doctor, Ibn Sīnā excelled at mathematics and music, carried out research into astronomy and optics, and produced an important treatise about music's effect on the soul. He is most famous for his *Canon of Medicine*. Over a million words long, and in Latin translation one of sixteenth-century Europe's most frequently printed books, in this classic compendium Ibn Sīnā synthesized his own observations—especially of meningitis and tuberculosis—with previous medical knowledge.

Modern science places a great premium on originality. In contrast, Ibn Sīnā's writing was valued by his contemporaries not for its novelty but for its thoroughness and systematic organization. Like Newton, Islamic scholars studied the world because they wanted to approach God—and also like Newton, whole swathes of their lives have been cut out of the history books to make them appear as proto-scientists. Ibn Sīnā preached the Islamic goal of striving for stability. For him, understanding nature was not an end in itself, since the physical, divine, and spiritual worlds are inextricably twined together. The word *islam* means both submission and peace, or being at one with God. Ibn Sīnā's aim was not to pick apart the structure of the universe, but to be led towards the unity of God.

Although science and religion are often said to be at war, Islamic holy writings stress that Muslims should acquire knowledge throughout their lives as part of their spiritual quest for perfection. Because of this sacred mission, the Islamic faith inherently fosters education. Scholars such as Ibn Sīnā divided learning into two complementary branches. Islamic revealed knowledge, which included topics such as theology, jurisprudence, and scriptural interpretation, mostly originated in the Qur'an and was handed down by direct transmission from one generation to the next. In contrast, scientific topics stemming from Greece were approached intellectually, and relied on independent thought as well as learning.

Schools were closely attached to mosques and concentrated on revealed subjects. Gradually two new types of institution emerged—observatories and hospitals—which were also associated with mosques, but included a wider range of subjects in the syllabus. They included large libraries, because Islamic teachers placed great emphasis on studying texts, especially after a new cheap material was introduced—paper, which soon supplanted papyrus and vellum. Bringing together old knowledge with new discoveries, these educational centres were spread throughout the Islamic empire and stimulated research into the natural world.

The first great centre for rational learning was the caliphs' court in Baghdad, set up in the eighth century and funded by state support as well as private patronage. Equipped with a massive library, this famous school attracted scholars from all over the Empire, who translated many of the library's Hellenistic texts into Arabic. Collecting and translating international books meant that Greek ideas were adapted and absorbed into Islamic culture from an early stage. Medical and astrological texts provided practical information about surgery, drugs, and prediction, but by the tenth century an immense range of works had been made available to Arabic experts. The caliphs subsidized this long, expensive project because it enhanced their reputation as learned patrons, and also stabilized their rule by encouraging Muslims to collaborate rather than establish ideological subgroups. Although they had no way of anticipating it, their political ambitions profoundly affected later science by ensuring the survival of Greek knowledge.

Astronomical observatories played vital roles in scientific training. The most influential was at Marāgha (in Persia, now Iran), commissioned in 1261 by the grandson of Genghis Khan and funded by a religious endowment. As well as a splendid array of accurate astronomical instruments, the Marāgha Observatory had a large library and attracted students in all the sciences. This basic plan of combining a school, observatory, and library spread throughout the Islamic world, and was later emulated by European visitors to the most accessible cities, such as Istanbul.

Teaching hospitals were another Islamic innovation that profoundly affected Europe. Like the observatories, many hospitals were built by powerful rulers or religious foundations. As well as a medical school and a library, they often incorporated a mosque and a bathhouse, because Muslims regard hygiene, health and spiritual well-being as inseparable. Rulers maintained high imperial medical standards by examining students before licensing them to practise, and the hospitals were often governed by eminent physicians who converted them into research centres. The most famous example is the hospital at Baghdad, rebuilt towards the end of the ninth century following the plan of its director, the Persian Muhammad ibn Zakariyyā al-Rāzī, author of an immense medical encyclopaedia and renowned throughout Europe as Rhazes, the Arabic Galen.

Al-Rāzī (Rhazes) has become a legendary figure, but he has different reputations. Everybody agrees that he was an exceptionally fine clinician, and he is celebrated for distinguishing small-pox from measles, using opium as an anaesthetic, and combining traditional knowledge with his own observations and remedies. He was also an excellent teacher, who followed the Islamic style of personal tuition in small groups rather than lecturing in front of a massive audience. In Europe, Rhazes's books were routinely prescribed (in Latin translation) for medical students, and historians congratulate him on revising Galen and making his own critical judgements. However, in Islamic countries, al-Rāzī was denounced as a poor philosopher and became notorious as an unorthodox Muslim who had presumed to challenge accepted authority. Practising as a physician was just one part of a scholar's religious and intellectual life, and over two-thirds of al-Rāzī's books were devoted to non-medical topics. Until recently, these aspects of al-Rāzī remained unknown to Eurocentric writers, who picked out aspects of Islamic life that contributed to their own culture.

Another example of multiple reputations is Abu al-Walīd Muhammad ibn Rushd, renowned in Europe as Averroes. For Muslims, his detailed analyses of Aristotle seemed insignificant compared with the encyclopaedic approach of Ibn Sīnā (Avicenna), whose interests covered the whole gamut of knowledge, and whose medical skills had immense practical impact. But in Europe, Averroes was regarded as a major Aristotelian commentator, and he appears alongside Plato, Aristotle, Pythagoras, and other eminent Greek philosophers in Raphael's famous Renaissance painting, 'The School of Athens'. Many Muslims objected to Ibn Rushd's philosophical views, some of which survive only in Latin and Hebrew translations because the Arabic copies were burnt—yet in Europe, Averroes was celebrated for daring to attack revealed religion.

Science is often said to have declined in the Islamic Empire after the thirteenth century, as if Muslim scholars had not been smart enough to continue occupying the intellectual territory they had gained from the Greeks. When Islamic scholars had run so successfully with the Greek torch of progress, why did they give up and fail to reach the end of the race? The regrets expressed by historians of Arabic science are similar to those of Needham in China, who tried to understand how such a scientifically advanced country failed to maintain its leadership. Like Needham, Arabic historians have generally been asking the wrong questions, because they mistakenly assume that science is a unified project striving to reach Absolute Truth.

The type of scientific research practised under Islamic rule ceased to thrive for several reasons. Political changes were extremely important. Scholarship flourished in the Islamic world when it was peaceful and prosperous, but later flagged as funds were diverted into armies and agriculture. Especially after Europe claimed the New World, trade and wealth moved steadily westwards, and the Islamic rulers lost their near-global domination. Another factor was social organization. Islamic legal and educational systems favoured local expertise handed down from one gifted scholar to another in small groups, a structure that fosters stability and preserves differences of opinion. In contrast, Europe's universities encouraged scholarly debate that challenged and overturned established knowledge, activities seen as heretical by orthodox Muslims, who valued attaining wisdom for bringing spiritual improvement.

Falsely accusing other places of backwardness has happened inside as well as outside Europe. According to the progressive Search for Truth model of science's past, at the end of the eighteenth century England fell behind France by not expressing the laws of physics in complex mathematical equations. With hindsight, it might seem that the French chose a better route towards the future. However, English men of science explicitly rejected mathematical techniques not because they were unable to understand them, but because they believed that algebraic symbols have little connection with the real world. They argued that it is a sign of false pride, condemned by God, to excel at mathematical manipulation and decipher divine mysteries that should remain forever secret. Similarly, many Islamic scholars frowned upon the pursuit of knowledge for its own sake. Having adapted all the Greek expertise that could be practically useful, they concentrated on a different type of progress—achieving happiness and spiritual perfection.

4 Scholarship

Wel knew he the oldë Esculapius
And Deyscorides, and eek Rufus,
Oldë Ypocras, Haly, and Galyen,
Serapion, Razis, and Avycen,
Averrois, Damascien, and Constantyn,
Bernard, and Gatesden, and Gilbertyn.

—Geoffrey Chaucer, Prologue,
The Canterbury Tales (*c.* 1387–1400)

By the middle of the ninth century, the caliphs' translation project in Baghdad was well under way. The aim was not merely to absorb previous expertise, but also to transform it so that it could be assimilated into Islamic culture. As one of the caliph's mathematicians explained, everything was to be done 'according to the usage of our Arabic language [and] the customs of our age'. Muslims were, he wrote, participating in a continuous endeavour of human learning, and they had two goals: 'to record in complete quotations all that the Ancients have said on the subject, secondly to complete what the Ancients have not fully expressed'.[6]

There were other 'Ancients' besides the Greeks. Because the Islamic region stretched from Andalusia to Uzbekistan, embracing Christians and Jews as well as Muslims, its inhabitants inherited a vast body of ideas blended together from different civilizations. Although the major intellectual legacy was Greek in origin, it entered the Empire through Alexandria, and so was already the product of many minds and influences. Alongside the merchants trading goods, scholars travelled round the Empire, exchanging ideas derived from many older traditions, including Persian and Indian as well as Greek—and this mixing of disparate cultures was encouraged by the annual pilgrimage to Mecca. Unsurprisingly, there was no uniform body of identical knowledge; nevertheless, the umbrella term 'Arabic sciences' is both useful and meaningful because the shared language meant that scholars could travel and exchange ideas to an extent unparalleled elsewhere.

Islamic scholars divided scientific learning into two major groups, neither of which corresponds neatly with modern scientific disciplines. One approach was to follow Pythagoras and search for mathematical order at the heart of the universe by studying four quantitative subjects—arithmetic, geometry, astronomy, and music. Although these now seem mismatched, they were also later grouped together in the European university curriculum. Other scholars were more Aristotelian and descriptive. As well as observing animals, plants, and minerals, they studied topics that now belong to physics—notably optics. From a theological perspective, they welcomed Aristotle's account of a teleological cosmos, one that has a purpose, since it corresponded to the Islamic belief that God created the universe expressly for the human race.

Algebra, algorithm, zero—three familiar mathematical words, all with Arabic origins. Muslims were attracted to Pythagorean mathematics because it meshed with their own love of harmony and their search for universal order. Traditional Islamic art and architecture clearly reveal this fascination with geometry and symmetry. In Figure 8, the roof tiles and columns of an astronomical observatory are arranged in repetitive patterns, echoed by the trees outside. This aesthetic characteristic stems from the very heart of the Islamic faith, which entails progressing up a numerical ladder from the confused world of multiplicity towards the supreme order of One, or God. When Muslims encountered the work of the Greek mathematicians in Arabic translation, they read about a quantitative spirituality paralleling their own.

Islamic mathematicians searched for cosmic as well as numerical significance in magic squares, whose rows, columns, and diagonals all add up to the same number. They also grappled with arithmetical problems that have now become classics, such as working out how many grains of wheat will pile up with one on the first square of a chess board, two on the second, four on the third, and so on, continuing to double right up to the sixty-fourth square, which would have an unmanageably immense volume of grain. That particular puzzle was devised and solved in the eleventh century by Abū Raihān al-Bīrūnī, one of the Empire's greatest intellects. At the time, he was as important as his colleague Ibn Sīnā (Avicenna), but his works were not translated into Latin, and so remained unknown in Europe. Like other Islamic scholars, al-Bīrūnī resembled a polymath more than a modern scientist. His mathematical and astronomical textbooks were used for centuries, but he was also renowned as a historian and for his comprehensive study of different religions.

Astronomers had several goals. They wanted to measure the universe more accurately and compile comprehensive star catalogues, but they also strove to achieve religious purity by demonstrating the perfection of the heavens. Many Europeans—including Newton—similarly envisaged a Pythagorean mathematical, harmonious cosmos as part of God's creation. Muslim mathematicians visualized numbers as geometrical shapes with a symbolic significance—for example, three

FIG. 8 Taqi al Din and other astronomers working in the observatory of Muradd III at Istanbul in the sixteenth century.

was connected with the triangle of harmony, four with the square of stability. From them came the idea of introducing measurement into music, an innovation that revolutionized Christian plain chant in twelfth-century Europe, when scores started to specify the time value of each note. Geometric representations and numerical relationships were bound up together with the intervals between musical notes, spiritual implications, and the proportions of the universe.

Mathematical astronomy was also important because of its practical uses, such as casting horoscopes for rulers about to take important decisions. The Islamic faith imposed special demands, and activities now seen as scientific were religiously motivated. Calendars had to satisfy the requirements of ritual fasting by ensuring that all Muslims endured the same length and conditions of privation, independently of where they lived. For different places in the Empire, Muslims needed to know the times of daily prayers as well as the direction of Mecca. Official timekeepers in observatory mosques made extensive astronomical measurements, and mathematicians such as al-Bīrūnī measured geographical coordinates to an unprecedented degree of accuracy.

As they adopted and adapted Greek astronomy, scholars relied most heavily on Ptolemy's *Almagest*, which stresses the importance of observation. To build on his work, Islamic astronomers developed sophisticated instruments, later imitated by Europeans, so that they could make more accurate measurements and compile new star catalogues. Figure 8 shows a small observatory in Istanbul. Its first major achievement was to track an exceptionally bright comet of 1577, but unfortunately the head astronomer interpreted this as a good omen. Several plagues and deaths later, the observatory was demolished as a warning against prying into holy secrets. However, because it was constructed according to traditional specifications, it provides a good illustration of Islamic astronomy.

In this geometrically constructed image, the importance of books is emphasized by the shelves at the top right. Fifteen astronomers are divided into three teams, each working together with a huge diversity of instruments, all of which became standard equipment in Renaissance Europe—including (on the far right) a mechanical clock, whose internal mechanism may have been of Chinese origin. The double circle being held up just below the books is the most famous and the most important Islamic instrument—an astrolabe, linked sets of intricately crafted rotating plates which model the heavens. The first astrolabe was reportedly invented in Greece when Ptolemy dropped an armillary sphere and his donkey trod on it—an apocryphal tale, but mathematically, a good description. This Islamic flattened version of the Greek universe became a favourite instrument of Renaissance astronomers, and was the subject of an illustrated treatise by Chaucer. Astrolabes thrived for so long because they were portable and could be used to serve many different functions, such as measuring the time, making astrological predictions, and surveying. Many beautiful brass examples were purloined by Europeans and are now preserved in Western museums.

Although Islamic astronomers followed Ptolemy's work, they also criticized it and—still more importantly—improved it. By collecting extensive and accurate data, they modified his descriptions of how the Sun and Moon appear to move, and they introduced more efficient trigonometrical methods for calculating stellar coordinates. Unsurprisingly for modern readers, Islamic scholars found it hard to reconcile their observations with Ptolemy's complicated system involving planets that speed up and slow down. One group revised Ptolemy's model by inventing similar geometric devices to those introduced by Copernicus—and modern experts are revising Eurocentric history by proving that Copernicus knew about these Islamic ideas.

There was no single discipline of Islamic astronomy. Pythagorean mathematicians and Aristotelian philosophers all turned to Ptolemy, but they approached him from different directions. Mathematicians sought to describe and quantify rather than to explain the behaviour of the world; in contrast, Aristotelian philosophers wanted to produce a more realistic, more solid version of Ptolemy's universe that would resonate better with Islamic concepts. For years, al-Bīrūnī considered the possibility of a heliocentric universe, and it's tempting to celebrate him for reaching the right answer before Copernicus. However, because he was a mathematician, al-Bīrūnī thought that it was not his role to worry about whether the Sun or the Earth lies at the centre of the Universe. As he pointed out, in terms of calculation, it makes no difference whether the Sun goes round the Earth or vice versa. Eventually, he plumped for the traditional geocentric model, leaving the cosmological problem to be pondered by philosophers.

One of the most influential Aristotelian philosophers was Abū 'Alī al-Hasan ibn al-Haitham, a tenth-century expert on optics who was known in Europe as Alhazen. After being disgraced by failing to control the Nile floods, he pretended to be insane and stayed in Egypt, quietly pursuing his own research. Ibn al-Haitham (Alhazen) tried to impose physical reality on cosmological models by introducing concentric spherical shells along the lines suggested by Aristotle. He envisaged an outer starless heaven, inside which a sphere carrying the fixed stars slowly rotates. And inside that, he claimed, each individual planet is associated with its own sets of slowly revolving spheres.

Through Latin translations of Ibn al-Haitham and other Islamic Aristotelians, variations of celestial spheres remained influential in Europe for centuries. From a scientific vantage point, such Islamic/Aristotelian/Ptolemaic fusions have many drawbacks. Most obviously, it's hard to explain how all those spheres manage to avoid banging into each other and how comets can pierce through without being deflected. But for Muslims and Christians, these models satisfactorily resolved—at least for a time—some of the questions that they considered important. Religious believers envisaged a universe that was stable, orderly, and finite, one that conformed with their holy scriptures by placing the human race at the centre of God's creation.

Ibn al-Haitham's major legacy was his long-lasting work on optics. Even during the Renaissance, experimenters still perceived Greek knowledge through Alhazenic eyes. Although he is often celebrated as the most important Muslim physicist, Ibn al-Haitham thought about the world very differently from a modern-day scientist. For one thing, his approach ignored modern disciplinary boundaries. Instead of reserving human sight for anatomists and physiologists, Ibn al-Haitham bracketed it together with atmospheric phenomena and experiments on lenses and mirrors.

Some of the standard modern theories about reflection and refraction originate with Ibn al-Haitham. Like Newton, he ground his own lenses, studied the anatomy of human eyes, and pondered over rainbows. Also like Newton, Ibn al-Haitham believed that God is the Light of the heavens and the Earth—a Qu'ranic image evoking biblical resonances. Above all, Ibn al-Haitham produced a new theory of vision. Although it now seems self-evident, Ibn al-Haitham argued that people can see because light comes from the thing they are looking at. He inherited three different Greek opinions, and synthesized them into a single unified study. Mathematicians such as Euclid drew geometrical diagrams as if light emanates outwards from our eyes—they were less interested in understanding how people manage to see than in providing a geometrical model describing what happens. In contrast, being a philosopher, Aristotle thought qualitatively and looked for causes; according to him, an object affects the medium (usually air) around it, and this alteration is then transmitted to the eye. And finally, as a physician, Galen examined the physiological structure of eyes. By combining these three approaches, Ibn al-Haitham worked out the mathematics involved when light travels in towards the eye rather than shining outwards from it.

Optics belonged to medicine as well as to science. Eye diseases were especially common in Egypt because of dust blowing in from the desert, and Ibn al-Haitham's research had important therapeutic benefits. 'Drug', 'retina', and 'cataract' all have Arabic roots, and Islamic medicine remained important in Europe right into the seventeenth century, especially through the Latinized books of al-Rāzī (Rhazes) and Ibn Sīnā (Avicenna). Islamic experts adopted many of their theories from Galen, integrating his version of Hippocratic medicine with traditional Persian and Indian practices to produce comprehensive and systematic compendia that covered every aspect of prevention, diagnosis, and treatment.

As well as improving Greek anatomy and medical philosophy, Islamic physicians also developed new pharmacological techniques, and their massive medical encyclopaedias meant that Islamic and Greek knowledge of minerals, animals, and plants reached Europe. The most important Greek source was the work of Dioscorides, beautifully illustrated manuscripts of around nine hundred drugs. By the middle of the thirteenth century, Arabic research had multiplied that number by three. Because medicines mostly came from plants, the search for more effective treatments meant that botanical knowledge was generally accurate and detailed. In contrast, descriptions of animals were often relayed by word of mouth, and drawings were often closer to mythology than reality.

Being a physician involved more than treating symptoms—a good Islamic doctor was also a virtuous man, one who fitted his patient's ailments into the patterns of the entire cosmos. In Arabic, as in many other languages, the words for breath and soul are closely related, so that restoring life to the body also implies nourishing the soul. Just as Islamic mathematicians felt an instinctive empathy with the numerical symbolism of Pythagorean cosmology, so too medical philosophers found that their theological doctrines of encouraging harmony and balance resonated well with Greek theories of humours. They picked up and developed those aspects of Greek philosophy that chimed with Islamic theology, insisting that human beings are not only part of the entire Universe, but are also miniatures of it. For Muslims, every individual is reflected in the cosmos, and the cosmos is itself a mirror image of life. This macrocosm–microcosm model of humanity now seems strange, but it was tremendously important in Renaissance Europe.

By the time of the Renaissance, physicians and natural philosophers in Europe had benefited from earlier Islamic achievements and taken them in new directions. Whereas the European economy was booming, the power of the Ottoman Empire was shrinking, and less money was available to invest in open-ended research with no clear benefits. In any case, not all Muslim philosophers agreed that deeper understanding of the natural world was necessarily an ideal worth striving for. Al-Bīrūnī shared Newton's aim of building on the Greek past. Whereas Newton wanted to see further by standing on the shoulders of giants, al-Bīrūnī urged his fellow scholars 'to confine ourselves to what the ancients have dealt with and endeavour to perfect what can be perfected'.[7]

5 Europe

And even I can remember
A day when the historians left blanks
   in their writings,
I mean for things they didn't know.

—Ezra Pound, *Draft
of XXX Cantos* (1930)

When Galileo wanted to persuade readers that his unconventional ideas were right, he created a fictional opponent called Simplicio, who was conveniently simple-minded, pedantic, and obstinate. To complete the caricature, Galileo made Simplicio a mediaeval scholar. The Middle Ages were invented in the Renaissance, and by the time that Galileo was arousing controversy in the early seventeenth century, they had been consigned to a blank page in history. Like Galileo and his contemporaries, many historians have dismissed the Middle Ages as a regrettable interlude of arcane scholasticism, an obstacle to scientific progress that started around the fifth century and eventually melted away under the blaze of Renaissance inspiration.

It all depends on where and how you look. Crucial transformations did take place during those suppressed centuries—but in fields and forges, in churches and monasteries, rather than in scholars' studies. Science is a practical subject as well as a theoretical one, and its origins lie not only in ideas, but also in things. Political, scientific, and economic changes were inseparable. After Charlemagne founded the Holy Roman Empire in 800, the European economy revived under French feudal lords, who imposed stability by ruling over vast tracts of property. They also put money into inventions that would increase their wealth and power still further. Under this new regime of competitive commercialization, technical innovations made agricultural and manufacturing methods more efficient. As profits increased, funds became available to support scholarship, so that by the end of the thirteenth century, western Europe was no longer impoverished and rural, but had become a thriving trading zone where education flourished in independent cities.

Technical inventions that now seem humble had as revolutionary an impact on society as would the steam engine several hundred years later. For instance, new horse harnesses may not sound a major breakthrough, but they dramatically reduced the human slave labour that had sustained the Greek and Roman empires. Basic mechanical developments such as gears enabled mills to tap wind and water power more effectively, while agricultural innovations—ploughs, crop rotation, animal breeding, irrigation systems—helped to ensure steadier food supplies. At the same time, metallurgical discoveries were leading to more effective weapons, while new chemical processes generated medical treatments, dyes, and household utensils. All these improved technologies stimulated scholarship by liberating people from time-consuming work and by releasing spare money.

Technological changes were introduced not because they increased knowledge, but because they were useful. Nevertheless, this enhanced practical expertise provided important foundations for the sciences of the future. Because wealthy landowners were willing to pay higher prices for better equipment, they indirectly stimulated mechanical and chemical investigations that would now be regarded as scientific. Star-gazers with little interest in astronomical theory accumulated data and techniques needed for working out the date of Easter, steering a ship, or telling the time. Similarly, skills acquired by village herbalists and monastic healers were later absorbed into pharmacy, botany, and mineralogy, while farmers were experts on meteorological, biological, and geological topics long before separate scientific disciplines were invented and named in the nineteenth century.

Both before and during Europe's economic recovery, monasteries remained the most important places where knowledge was valued for its own sake. Monks played a vital role in science's European history because they discussed secular texts as well as religious ones. Instead of rejecting classical learning as pagan, monastic scholars decided it could help them attain their goal of approaching God through deciphering the Bible, and so they debated a great variety of philosophical ideas related to their devotional studies. Although religion is often seen as being hostile to science, it was Christianity that preserved academic learning in Europe.

In particular, monks continued the Roman practice of compiling encyclopaedias. Many theological arguments revolved around how and why God had created so many life forms, and these stimulated detailed observations that later became relevant for the life sciences. Amongst the many works plundered by the monks, their most important source was Pliny, a first-century Roman military officer who became obsessed with accumulating information. His massive *Natural History*, a cut-and-paste job on a hundred other authors, contains around 20,000 facts (some rather dubious—do beavers really castrate themselves when a hunter approaches?) and is a comprehensive collection of Graeco-Roman knowledge. By adapting classical expertise and incorporating it into their own books, monastic scholars ensured that writers like Pliny remained important authorities.

As western Europe became richer and more powerful, religious centres remained vital for future science. One excellent example is Chartres Cathedral, eventually consecrated in 1260 after a series of fires, and an exceptionally fine expression of Gothic architecture's new emphasis on light and aesthetic order. The structure of this Christian building embodies Platonic ideals of an organized mathematical cosmos. For Platonists, the material world that people perceive about them is not reality itself. Whereas human designers create imperfect geometrical shapes, Plato's perfect triangles, cubes, and spheres are unchanging and eternal—and although they can never be directly seen, the existence of these ideal forms guarantees that they can be conceived mathematically. To explain these counter-intuitive ideas, Plato used the analogy of bound and blinkered prisoners, who are permitted to see objects only as flickering shadows cast onto the wall of a cave by a giant fire. If a prisoner is released, he will be blinded by the sunlight pouring through the entrance, so that the shadowy world so familiar to him will seem clearer than the real one.

Chartres Cathedral illustrates how religion, commerce, and daily life were twined together during the development of a scientific approach towards the world. Far from being simply a place to worship, the cathedral was designed to model a mediaeval vision of the Universe: God was the divine architect, and Chartres incorporated geometrical harmony to represent His creation. The human architects thought and worked geometrically, and their building's proportions mirror the harmonious God-given ratios of music and of the Universe.

The cathedral, the economy, and specialized knowledge grew together. Building Chartres was not only a religious mission but also a community project that guaranteed full employment, stimulated invention, and boosted local enterprise. Chartres is packed with technological innovations inspired by theological demands for a soaring ceiling and divine illumination. To create an immense yet stable and structured space, the cathedral's architects devised new features such as flying buttresses and also prompted a fresh interest in mathematical mechanics. Similarly, glaziers and metalworkers developed new chemical processes for producing the stained glass windows with their luminous colours and geometrical shapes. Merchants donated money both to ensure spiritual protection and to express gratitude for their profits. Paid for by local guilds, the windows not only carry scenes from the Bible, but also celebrate the town's craftsmen. For example, one sequence includes the earliest known picture of a modest yet vital invention for undertaking this ambitious construction—a wheelbarrow. In this cathedral, the mundane and the scientific blend together with the sacred.

Chartres and other mediaeval cathedrals not only dominated the activities of local villagers, but also profoundly affected the future by altering how people thought about time. Modern science and technology rely on measuring time to a minute fraction of a second, and this quantitative approach to daily experience stems from monastic rituals. Whereas Muslims and Jews based their prayers on the position of the Sun in the sky, devout Christians worshipped at regular intervals that structured their daily routine. Even before mechanical clocks were invented, cathedrals announced their services by bells, so that mediaeval life was regulated by ringing chimes summoning the faithful seven times a day.

A new concept of time emerged because religious rituals replaced the rhythms of nature. Visualizing time as something measurable that slips by at a uniform rate now seems instinctively obvious. Seven centuries ago, it was a startling concept for most people, whose lives were governed by lightness and darkness, by summer and winter, by planting and harvesting. Strange as it might seem now, traditional timekeepers such as sundials and water clocks recorded hours of varying lengths, because they measured out twelve units between sunrise and sunset. This meant that during the extended daylight period of summer, each daytime unit lasted longer than in the winter (and vice versa). Milan made horological history in 1336, when one of its church clocks struck twenty-four equal hours for the first time.

Instead of following the daily and seasonal rhythms of nature created by God, mechanical clocks artificially segment time into uniform blocks. By the end of the fourteenth century, several of Europe's cathedrals boasted prominent clock towers that aspired upwards towards God but simultaneously controlled life down below. Although not very accurate—even the better ones lost 15 minutes a day—these church clocks irrevocably altered how human beings participated in the world. Instead of responding to the patterns of natural sunlight, people started to stake out their lives in equal units determined arbitrarily and mechanically. Science relies on precision measurement and global coordination, yet this control through metered time was launched by Christian monks.

As well as ticking time away in equal blocks, clocks helped to transfer power from the Church to secular rulers—in 1370, the French king ordered all Parisian clocks to be coordinated with his own. Imposing order on time also had economic implications. 'Remember that time is money,' the thrifty electrical expert Benjamin Franklin advised traders in the eighteenth century. Now so familiar, this refrain would have appalled traditional Christians: for them, time belonged to God and could not be sold. Charging interest on loans was banned by Church authorities, who maintained that it was immoral to profit from this divine gift. But as the European economy flourished, they gradually allowed themselves to be persuaded that resistance was useless. The windows at Chartres endorsed commercialization by showing money changers counting and weighing gold coins. Whereas observant Muslims are still banned from taking out a mortgage, Christian authorities adjusted their religious principles to accommodate a capitalist credit system that pervaded Europe and underpinned their own wealth.

Mediaeval Christianity also dramatically affected future science through its scholars. The monastic school at Chartres was one of France's top educational centres for two hundred years, and it covered classical as well as Christian learning. Students followed individual teachers rather than enrolling in specific schools, and the cathedral's distinguished scholars included Bernard of Chartres, the twelfth-century source of Newton's famous quip, 'If I have seen further it is by standing on the shoulders of giants.' Bernard was thinking of two giants—not only the Bible, but also Plato (at least, a distorted, condensed version that had survived in translation)—and this mixed classical and Christian heritage typified the teaching that students received. The school gradually declined in importance not because its syllabus was narrow, but because Chartres remained a small town where education was dominated by the cathedral. In contrast, nearby Paris expanded rapidly into a major commercial city. Like other prosperous centres, Paris started to attract leading scholars, and small groups broke away from the Church to form their own independent organizations—universities.

Universities were unique institutions that made European learning different from anywhere else in the world. By 1200, Europe boasted three—first in Bologna, then Paris, and Oxford—and during the next three centuries, around seventy more were founded in cities seeking to advertise their importance. Universities became powerful institutions that could negotiate with the State as well as with the Church. Like guilds, they governed themselves, but they were also awarded exceptional privileges for their scholars, who were highly regarded as elite guardians of esoteric knowledge. This protection meant that in addition to their main role of teaching, mediaeval scholars were relatively free to discuss controversial ideas.

Even before then, some monastic scholars had already started to change their views of God. Prompted by the classical knowledge inherited through the encyclo-paedists, they were gradually moving away from the traditional view that God is the direct, immediate cause of everything that happens in the Universe. Instead, progressive theologians argued that nature is more like a harmonious machine that had been designed by God but operated independently (at least, apart from the miracles and supernatural wonders proving that He intervenes occasionally). This theological shift towards a self-regulating cosmos was important because it encouraged scholars to study the world about them as well as the Bible. To gain information about the Universe and how it works, they wanted to retrieve the ancient Greek learning that had lost significance in Latin-speaking western Europe, but had been preserved and modified in the Islamic Empire. When Latin translations became available at the end of the twelfth century, university scholars took this Greek and Arabic inheritance in new directions.

FIG. 9 Title page from Gregor Reisch, *Margarita Philosophica* (*The Pearl of Wisdom*, 1503).

There was no sudden break from the past towards a scientific style of thinking. Rather than inventing brand new courses for their students, university teachers modified the ancient Greek educational syllabus. At Chartres, a stone sculpture depicts seven figures representing the seven liberal arts, which continued to dominate university scholarship. In Figure 9, taken from a much-used encyclopaedia called *Margarita Philosophica* (*Pearl of Wisdom*), the same seven are shown clustered together in the lower half of the cycle/circle of knowledge. Each one is identified by her Latin label in the rim and by the object she holds, such as a lyre for music, dividers for geometry, and a sphere for astronomy. Aristotle appears in the lower left corner, and his work dominated university teaching: the three-headed angel represents the three divisions of Aristotelian philosophy—natural, rational, and moral.

The universities organized learning into a strict hierarchy that reflected their dual Greek and Christian origins. At the very top was theology. In the next layer down lay (rational) logic and natural philosophy, both of which had to be mastered by mediaeval students before they were allowed to enter the divinity faculty, where—sniped Church authorities—they spent years poring over texts instead of pursuing holiness. At the base lay the seven liberal arts, divided into two groups whose Latin names refer to four and three: the *quadrivium* of astronomy, geometry, arithmetic, and music; and the *trivium* of rhetoric, grammar, and logic (see Figure 9). The *quadrivium* subjects had underpinned Chartres's design as the earthly reflection of a Platonic universe. In mediaeval universities, it was transformed by the newly available Greek and Arabic translations and became especially important for later science; in contrast, the *trivium* was seen as lower, and so gave rise to the English word 'trivial'.

Scholarship was always symbolized by female muses and goddesses, even though women were banned from universities, and had little access to the seven liberal arts. These were reserved for men—and also for special men, those privileged gentlemen who studied from books, disdained manual labour, and were taught about theory rather than practice. In Latin, *liber* can mean 'free' as well as 'book', and education was intended to produce cultured citizens, not workers capable of earning a living. In the *quadrivium*, each book-bound liberal art had its mechanical equivalent suitable for the lower classes. For instance, builders used geometry to construct bridges, traders carried out arithmetical calculations, navigators used the stars for steering, and performers played musical instruments. Education was for the wealthy, and earning money through physical work was seen as demeaning, an attitude that persisted for centuries. Even in Victorian England, engineers were still seen as socially inferior to scientists, and hard labour was rated lower than intellectual speculation (unlike the United States of America, where the prolific inventor Thomas Edison was admired for declaring that 'Genius is one percent inspiration, ninety nine percent perspiration').

Paradoxically, what later became known as the sciences originated in the liberal and mechanical arts. For the derivation of 'art' in its modern sense, think of the words 'artisan' and 'artificial'; for 'science', think books. Subjects now thought of as belonging to the arts—sculpture, painting, architecture—were originally manual skills practised by the lower social classes; in contrast, mathematical scientific disciplines stem from the written, scholarly learning (Latin *scientia*) designed for the upper classes. This included not only geometrical theorems and cosmological models derived from the seven liberal arts, but also music, a subject which—like science—can be performed with instruments as well as being analysed theoretically. In addition, *scientia* embraced theology, now often said to be opposed to science. Mediaeval scholars were trying to approach God rather than to accumulate knowledge for its own sake, and they insisted that the new natural philosophy would boost rather than contradict their religious studies. As Roger Bacon, a leading spokesman for this approach, explained: 'one discipline is mistress of the others—namely, theology, for which the others are integral necessities and which cannot achieve its ends without them'.[8]

Roger Bacon was a Franciscan scholar, an experimenter, and alchemist who worked at Paris and Oxford in the second half of the thirteenth century. He is particularly well known for his work on optics, because he provides a key link figure between Alhazen (Ibn al-Haitham), the renowned Islamic authority, and Johannes Kepler, who investigated light more than three hundred years later than Bacon. Bacon's views stemmed from his religious ideas about the nature of light. In a way that now seems counter-intuitive but then made sense, he regarded light as being simultaneously transcendental and physical, divine and corporeal. For mediaeval scholars, a thing's place in the world depended on how closely it manifests God. Light was further up the spiritual hierarchy than a stone, but both of them could only be understood by perceiving their divine essence. Or to put it differently, the windows at Chartres cathedral carry pictorial scenes that explicitly deliver moral and biblical lessons, but they also act as translucent wall panels admitting sacred light to illuminate the intellect as well as the soul.

Unlike earlier monastic scholars, Bacon studied ancient texts but also carried out his own investigations by performing experiments. Instead of accepting Greek verdicts unchallenged, Bacon agreed with Alhazen—light enters the eye from other objects. In addition, he maintained that radiant light interacts with the solid world, changing as it moves and also affecting the medium it travels through. Sounding as though inspired by cathedral windows, Bacon reported that 'when a ray passes through a medium of strongly-coloured glass...there appears to us in the dark...a colour similar to the colour of that strongly-coloured body'.[9] Bacon thought from a Christian rather than a scientific perspective, and so he visualized light as an agent that linked the components of the Universe into a single divine whole. Even so, his argument that objects can act on one another was theologically controversial, because Bacon's more conservative colleagues refused to accept that the world can operate without God's continuous intervention.

Bacon influenced the later science of optics, but from his own point of view, the study of light was not so much a scientific discipline as a key to the cosmos. For him, the human route to salvation begins in the visible, tangible world of the senses, and progresses in steps upwards through the idealizations and abstractions of mathematics towards a metaphysical appreciation of divine unity. For instance, Chartres's geometrical structure stretched towards musical harmony and beauty, while the subjects of the *quadrivium* encouraged the mind to move upwards from the tangible world and approach a heavenly one. In this hierarchical route from the physical to the spiritual, geometrical optics had its roots in the mundane world, but was also an intermediary for communicating with the divine realm. Understanding how light rays interact with mirrors and prisms was not an end in itself, but a stage towards understanding God.

Galileo disparaged his predecessors, even though Europe's economic revival had stimulated the academic reforms that made his own research possible. Similarly, modern scientists dismiss the theological concepts of mediaeval scholars as irrelevant to their own work. Although Roger Bacon is often commemorated as England's first real scientist, he believed that light reveals divine aspects of God's holy creation. However alien his intellectual framework might seem, it greatly affected scientific developments that came later.

6 *Aristotle*

'I am not very well versed in Greek,' said the giant: 'Nor I neither,'
replied the philosophical mite. 'Why then do you quote that same
Aristotle in Greek?' resumed the Sirian: 'Because', answered the other,
'it is but reasonable we should quote what we do not comprehend in
a language we do not understand'.

—Voltaire, *Micromégas* (1752)

A ristotle dominated mediaeval and Renaissance scholarship. The numbers
almost speak for themselves. Around two thousand Latin manuscripts
of Aristotle's works still survive, and many, many thousands more must have
disappeared. Roughly a third were translated from Arabic rather than the original
Greek; inevitably, these were somewhat garbled since often yet another intermediary language was involved—to say nothing of scribes' transcription errors.
From the statistics of books written *about* Aristotle, one Islamic author stands out:
Averroes (Ibn Rushd), who played a vital role in persuading European scholars to
adopt Aristotelian ideas. In addition, there were the pseudo-works that had been
circulating in many different languages since soon after Aristotle's death; hundreds
of these compilations survive, repeatedly copied and mis-copied by hand, falsely
attributed to Aristotle and given beguiling titles such as the *Secret of Secrets*.

Even during the supposed Dark Ages, some Greek texts were translated into
Latin, but there was a dramatic increase in the twelfth century. By then, many travellers—traders, ambassadors, crusaders—had told Western scholars about the knowledge and skills accumulated in different parts of the Islamic Empire. Intercultural
contact had always been greatest in Spain, which had a thriving Christian community even when it belonged to the Islamic Empire. After Christian rulers regained
control in the late eleventh century, local bishops sponsored translations of the
books collected together in Spain's large Arabic libraries. Toledo became a particularly important centre, attracting translators from all over Europe who wanted to
acquire Islamic expertise as well as retrieve lost Greek works.

This was a massive yet uncoordinated international exercise, in which many
individuals were involved. Some were particularly influential. For instance, Gerard

of Cremona was an Italian scholar who migrated to Toledo in 1144 because he wanted to learn Arabic. He translated almost eighty books, and his manuscripts (later published in print) remained important for centuries—'sine' in trigonometry comes from Gerard. His most significant translation was Ptolemy's *Almagest*. The only Latin version available for three hundred years, it introduced European astronomers to a level of technical virtuosity they had never before encountered.

By the end of the twelfth century, Latin-speaking scholars in Europe could access a vast range of translated Greek texts. The early translators made usefulness their first priority, and they chose practical books on astronomy and astrology, on medicine, mathematics, and meteorology. These technical treatises presented little threat to Christian theology, and simplified versions of Ptolemy's *Almagest* were soon being taught alongside the optics of Alhazen (Ibn al-Haitham) and the medicine of Avicenna (Ibn Sīnā). Theoretical texts arrived later, and they posed greater problems. Aristotle was banned in Paris for about fifty years, because his ideas challenged Christian doctrines about the world's existence and creation.

Ideologies that had originated in Athens and Jerusalem met and clashed in western Europe. Aristotle—that is, Aristotle as presented by the Islamic Averroes (Ibn Rushd)—provoked several objections. Most importantly, Aristotle's universe is eternal, but the Christian one has a direction: the Bible opens with a detailed account of how God created the world, while the last Day of Judgement looms threateningly over the future. Divine intervention was another problem: whereas the Aristotelian cosmos is self-sufficient and law-governed, the Christian God manifests His presence through miracles and has given humanity the gift of free will. Moreover, Aristotle twined minds and bodies together, which contradicted Christian convictions that after death, souls exist independently and eternally, either in heaven or hell.

These incompatibilities were gradually resolved during the thirteenth century. At the time, the philosophical debates must have seemed to grind on endlessly, but in retrospect, three men in particular stand out. One of them was the British Franciscan Roger Bacon. By declaring that 'theology is queen of the sciences', Bacon encouraged university scholars to accept pagan ideas and subordinate them to Christian ones. While Bacon focused on optics and mathematics, his rival Albert the Great set out to analyse, explain, and supplement *all* of Aristotle's works in exhaustive detail. A German Dominican who studied theology at Paris, Albert came to be called Great because he knew about everything—not just theology, astrology, and logic, but also botany, mineralogy, and zoology (including his own observations of how partridges mate).

The scholar who did most to make Aristotle acceptable was Albert's pupil, Thomas Aquinas, an Italian aristocrat who overrode his family's protests (including being locked up at home for a year) to become a Dominican friar, renowned throughout Europe as 'Doctor Angelicus'. Although seen as a dangerous radical by his thirteenth-century contemporaries, Aquinas was later made a saint and is still revered as one of the greatest Roman Catholic theologians ever. An exceptionally hard-working man, Aquinas studied and taught at Paris and several other European universities, at the same time acting as adviser to the Pope and to one of his own relatives, the King of France. And on top of all those duties, he was constructing a synthesized version of Aristotle and Christianity (often called Thomism, from his first name) that dominated western European thought for three hundred years.

Aquinas argued that God had distinguished people from the rest of creation by giving them minds as well as five senses—only humans, he wrote, delight in beauty for its own sake. This meant that God intended truth to be obtained not only from His written words—the divine revelation of the Holy Scriptures—but also through studying the natural world. Since God would never allow truths learnt from reason and from faith to contradict each other, Christians had two books to guide them along their spiritual path: the Bible dictated by God, and Aristotle—or at least, a doctored version of Aristotle.

Mediaeval Aristotle was not Greek Aristotle. Moreover, there was not even one single version of mediaeval Aristotelianism. As Albert the Great pointed out in one of his numerous books on Aristotle, his hero was human and therefore fallible, so that Aristotle's disciples could 'expound this man in diverse ways, as suits the intention of each of them'.[10] The same is true for all the famous scientific -isms: there is no one authorized version of Cartesianism, Newtonianism, or Darwinism because followers pick out and develop only the bits they agree with and are interested in.

The cosmos provides a good example of adaptation. Aristotle himself had placed a sphere of fixed stars outside the seven planetary spheres—and beyond that nothing, not even a vacuum. This was hard for Christians to accept. What about the Bible's account of Creation, which specifies not only a heaven, but also a firmament with water above and below it? And what would happen if you stood at the very edge of the cosmos and stuck out your arm? One common way of resolving such dilemmas was to introduce three, rather than one, spheres beyond the planets, as illustrated in Figure 3. Beyond the seven planets lie the starry firmament and the transparent watery crystalline heaven, all surrounded by Aristotle's Prime Mover, a role given to the Christian God.

Down in the terrestrial realm, movement was a particularly contentious topic. Although mediaeval debates on motion sound esoteric, they are important because they affected the course of future physics. Within an Aristotelian framework, things stay still unless they are actively moved—as opposed to the Newtonian view that moving objects keep going until they are stopped. Aristotle divided motion into two kinds, natural and forced. Natural motion occurs when a body is internally impelled towards its natural place: earth and water fall downwards, fire and air float up. An object can also be forced to move in an unnatural way—a cart is pulled by an ox, an arrow is fired by an archer. But what happens after the arrow leaves the bow? Once the cause of forced motion—the bow—is no longer acting, surely natural motion should take over and make the arrow plummet to the ground? To counter that objection, Aristotle had explained that the act of shooting transforms the behaviour of air, which rushes from the front of the arrow to the back, pushing along from behind. This explanation posed some obvious difficulties. For one thing, if Aristotle were right, how could you shoot an arrow against the wind?

During the first half of the fourteenth century, mediaeval scholars set about salvaging Aristotelian motion. As Aristotelian philosophers, they focused on qualities rather than quantities, searching for causes and explanations rather than trying to derive laws and formulae. The Parisian expert was Jean Buridan, now most famous for inventing an unfortunate logical donkey who starved to death halfway between two bales of hay, unable to find a reason for choosing which one to eat first. To rescue Aristotle, Buridan claimed that it was the arrow rather than the air which was altered, and he introduced a new concept—impetus, the quality of impelling any object containing it. The act of shooting imparts impetus to an arrow, which continues upwards; similarly, a magnet gives impetus to a piece of iron, which is then internally compelled to move towards the magnet. Because Buridan's concept of impetus described effectively what happens, it dominated mechanics until overturned by Galileo and Newton.

Buridan's Oxford contemporaries approached the problem from a more mathematical angle. Based in Merton College, these scholars were called 'calculators' because they used mathematics to describe the relationships between different qualities. For instance, they suggested that doubling an arrow's velocity involves squaring the ratio between the force propelling it and the resistance obstructing it. To deduce their equations, the Merton calculators used the mental tools of logical analysis, and relied on experiments that were purely imaginary. They regarded themselves as philosophers, not as artisans who worked with instruments; in any case, clocks were still far too inaccurate for timing movement accurately. Nevertheless, because in principle their formulae could be tested, their research greatly influenced Galileo's real-life experiments three hundred years later.

For mediaeval scholars, instruments such as clocks were designed to model the Aristotelian cosmos rather than to measure it, to advertise God's magnificence by miniaturizing the cosmos as a great interconnected machine. All over Europe, craftsmen were constructing increasingly complicated and costly clocks, which incorporated not only loud bells for summoning people to work and prayers, but also intricate sets of gears for displaying the splendours of God's universe, such as the phases and eclipses of the moon, the tides and planetary motions. Large cities such as Strasbourg and Prague invested in splendid astronomical mechanisms to adorn their cathedrals, which displayed their wealth and also increased it by attracting religious pilgrims (a function now fulfilled by tourist expeditions). As technology improved, clocks did become smaller, but they were expensive status symbols, designed as elaborate ornaments rather than accurate timepieces.

These astronomical clocks exhibited the Christian–Aristotelian fusion in a physical form. By telling the time, they imposed a regular monastic schedule on the entire population. Just as significantly, these mechanical models of the cosmos made it clearly visible that life on earth is affected by the orderly operation of the heavens. People were convinced that the movements of the stars and planets affected their personalities, health, and opportunities, and the astrological vocabulary they used still survives—lunatics were swayed by the moon (*luna* in Latin), disasters descended from above (*astra* means stars), and jovial people were ruled by Jupiter.

Aristotelian Christians regarded the stars as God's intermediaries for governing the Universe. However, traditional theologians objected that if events were preordained, then the Christian concept of individual free will would be undermined. Although they accepted that stars might affect people's bodies, they insisted that the soul and the mind must be able to act independently. Thomas Aquinas cleverly resolved this conflict, arguing that wise men (and he meant men) could overcome their natural inclinations by exerting self-control. The heavenly bodies might disrupt physical health by altering the balance of the humours, but a rational Christian would never let the stars completely determine his emotional state—unlike women and labourers, he was sufficiently self-disciplined to sidestep destiny by mastering his own passions.

Astrology became compatible with Christianity by splitting into two branches. Members of one group were derided as charlatans—the personal horoscope casters who claimed to foretell particular events in an individual's life. In contrast, those who practised natural astrology made general prognostications, and were highly respected as members of an intellectual elite. Mathematical experts, they relied on the best available astronomical scholarship, such as Gerard of Cremona's Latin translation of Ptolemy's *Almagest*, and accurate tables of heavenly trajectories compiled at Toledo.

Mediaeval and Renaissance physicians routinely studied astrological medicine. Figure 10, taken from a fifteenth-century surgeons' manual, shows Zodiac Man, who embodied Aristotle's macrocosm–microcosm correspondence between human bodies and the heavens. Each part of his body is associated with a different star sign inherited from the Babylonians—Aries the Ram perches on his head, the fish of Pisces swim beneath his feet, while Taurus the Bull squats on his shoulders.

FIG. 10 Mediaeval Zodiac Man. *Guild Book of the Barber Surgeons of York* (1486).

Zodiac Man was so familiar that when Shakespeare wrote *Twelfth Night*, he knew his audience would relish the argument between two comic characters who both assigned Taurus to wrong parts of the body.[11] In addition, the planets were associated with particular humours—Mars was choleric, Saturn melancholic—and so could influence people's health by making that humour ebb and flow in their bodies like a tide. Successful treatment demanded not only restoring the humoral balance, but also intervening at the right time, when the planets were in auspicious positions.

Christianized Aristotelian astrology was an important mediaeval science. When the Black Death started sweeping through Europe in 1347, university physicians at Paris gave a rational Aristotelian explanation, pronouncing that the epidemic originated in an unusual conjunction of planets four years earlier. Warm, humid Jupiter had drawn up evil vapours, which were ignited by overheated, dry malevolent Mars and further affected by melancholic Saturn. These planetary clashes had generated hot, wet southerly winds, which corrupted the atmosphere and induced a corresponding sickness in human beings. Expressed theologically, God had manifested His displeasure through natural astrological powers. Mathematicians set to work and calculated that the next major planetary conjunction was due in 1365. In Oxford, one expert wrongly predicted that his Christian God would choose this date to destroy the infidel Saracens—presumably he failed to realize how much of his Aristotelian knowledge he owed to Islamic interpreters.

However, this devastating plague did stimulate more successful scholarship. Since it killed around a third of the European population in five years, its immediate effect was to focus scholars' minds on death and salvation rather than the present world. But afterwards, the economy began to boom. Rich families became still richer through inheriting their dead relatives' property, while the labourers who remained alive were in a strong bargaining position to insist on better pay. Trade and travel increased to meet the growing demand for luxury items—leading to new instruments, better maps, and more daring voyages of exploration. The intellectual fizz of the Renaissance had material origins.

7 Alchemy

This rod and the male and female serpents joyned in the proportion of 3, 1, 2 compose the three headed Cerberus which keeps the gates of Hell. For being fermented and digested together they resolve and grow dayly more fluid...and put on a green colour and in 40 days turn to a rotten black pouder. The green matter may be kept for ferment. Its spirit is the blood of the green Lion. The black pouder is our Pluto, the God of wealth.

—Isaac Newton, *Praxis* (*c.*1693)

Most scientists think alchemy is rubbish, yet in 1936, the nuclear physicist Ernest Rutherford gave a talk at Cambridge University describing himself as a modern-day alchemist. Instead of transmuting lead into gold, Rutherford boasted, he had changed nitrogen into oxygen by bombarding it with alpha particles. He adorned his coat-of-arms with alchemy's mythical founding father, Hermes Trismegistus (thrice-mighty Hermes), an amalgam of various ancient Egyptian sages whose name survives in the phrase 'hermetically sealed'. Rutherford may have been indulging in scientific whimsy, but he was also making a serious point. Some alchemical ambitions now seem ridiculous—such as finding the philosophers' stone or the elixir of life—yet its techniques, instruments, and attitudes lie at the heart of experimental science.

Alchemy has a long and international history. Newton and other seventeenth-century aficionados based their research on a hermetic tradition that went back to the Babylonians, was developed in Egypt and Greece as well as China and India, and arrived in Europe from the Islamic Empire in the twelfth century. As an eastern import, alchemy's specialized vocabulary stemmed from transliterated Arabic words, some of which have entered ordinary language—'alchemy' itself, as well as 'elixir' and 'mattress'. The early Latinizing translators were unfamiliar with alchemy, but they encountered it in books now famous as part of Europe's scientific heritage. When Gerard of Cremona started work on Ptolemy's *Almagest*, he discovered that it included far more than the astronomy he anticipated. Alchemical information appeared in works by many other influential writers, including Aristotle and eminent Islamic scholars such as Rhazes (al-Rāzī) and Avicenna (Ibn Sīnā). The enormously popular but pseudo-Aristotelian *Secret of Secrets* was also packed with medical, alchemical, and magical lore.

Alchemists resembled scientists in trying to alter the world, not just understand it. In contrast with mediaeval scholars, they tried to transform their environment by inventing new techniques and juggling with existing phenomena. Although alchemists wanted to help people, they also needed to make money, and they protected their discoveries by using arcane signs and symbols. At its heart, alchemy involves understanding change, which can take many forms—iron deteriorates to rust, seeds grow into trees, water freezes, the Moon assumes different shapes, alcohol evaporates but also reinvigorates, and criminals reform their characters. Strongly influenced by Aristotle, mediaeval alchemists believed in an interconnected universe of elements, qualities, and astral influences. Ardent religious believers, they constantly strove towards perfection as they searched for God. Their main goal was to find the philosophers' stone. Once identified, this would be a universal key to improvement, a sure-fire way of obtaining gold by cleansing base metals of their imperfections, of prolonging life by ridding human bodies of disease, and of achieving divine enlightenment after purifying their souls.

Figure 11 reflects many of alchemy's fundamental characteristics. Although designed by a real man—Heinrich Khunrath, a German physician who taught that God had revealed alchemical techniques in order to cure diseases—this lofty chamber with its mathematical proportions is not so much a realistic depiction as a symbolic portrait of alchemy's aims. In the centre, a table is laden with musical instruments, reminders of cosmic harmony and the Pythagorean bonds between music, astronomy, geometry, and arithmetic, the *quadrivium* subjects of the mediaeval curriculum. The spaces on either side correspond to alchemy's two main aspects. On the right, the sign on the wall indicates that this is a *laboratorium*, a place for work. The instruments ranged on the floor and shelves are for practical experiments, directed towards extracting powers and separating out pure elixirs from the gross matter of animals, plants, and minerals. As well as trying to achieve such material improvements, alchemists also searched for spiritual growth. On the left is the *oratorium*, the place for prayer, where this human sinner is trying to refine his soul and so approach nearer to God.

Alchemy may now seem ludicrous, but it expanded rapidly in mediaeval Europe because it appeared to be a coherent rational system that made sense. For many people living then, alchemy fitted in with visions of a Catholic harmonic universe governed by Aristotelian principles. After all, if bread and wine can be transubstantiated into the body and blood of Christ, and if dirty underground ores yield shiny metals, then why should it be impossible to remedy a fatal illness or convert lead into gold? Processes of transformation were built into Aristotelianism, the prevailing scholarly orthodoxy, which accounted for human behaviour and

FIG. 11 An alchemist's chamber. Heinrich Khunrath, *Amphitheatrum Sapientiæ Aeternæ* (*Amphitheatre of Eternal Wisdom,* 1598).

chemical reactions by recombining four elements—*earth, water, air, fire*—and four qualities—*hot, dry, wet, cold.* The details remained vague, leaving plenty of scope for variation without deviating from mainstream beliefs.

The most famous alchemical goal, transmuting lead into gold, had a solid theoretical basis. Like many Aristotelian commentators, alchemists believed that metals are made up of hot, dry *sulphur* and cold, wet *mercury* (idealized principles, not ordinary sulphur and mercury). Inside the womb of the Earth, these are heated up together under varying conditions, and after many years mature into different metals. To speed up this natural process, alchemists searched obsessively for chemical techniques that would short-circuit centuries of gradual transformation and produce gold directly.

Alchemists influenced the course of science in several ways. Most obviously, their research was essential for future developments in experimental chemistry and industrial technology. Over the centuries, alchemists had doggedly devised and tested apparatus for heating, distilling, and crystallizing different substances, and this tradition of innovation and refinement continued long after alchemy reached Europe. For example, hermetic practitioners invented stills for collecting liquids (including alcohol to a far higher degree of purity than ever before), perfected different types of furnace that entered chemistry as sand-baths and water-baths, and—as Figure 11 illustrates—designed a vast array of special-purpose basins and flasks, many of them still familiar to nineteenth-century chemists. Although gold remained elusive, alchemists did isolate many chemicals, including effective medical drugs and ammonium sulphate, now a staple of the artificial fertilizer industry.

Hermetic investigators also affected future science by convincing other people—university scholars, patrons, customers—that experiments could yield valuable results. Even though many alchemists were extremely learned, they mostly operated outside the universities, earning their living through offering medical treatments or developing chemical processes with practical applications. Alchemical manuscripts of secret recipes intrigued scholars because they were so different from their own theoretical tracts. Even though university physicians continued to teach and publish traditional techniques, some of them started prescribing drugs and therapies gleaned from clandestinely circulated books.

One of alchemy's most open academic advocates was Roger Bacon, who is often heralded as a pioneer of modern experimental science. In a way he was—yet his emphasis on hands-on investigation came from alchemy, which he regarded as supremely valuable because it was useful. How else could he hope to prolong life and heal sicknesses of body and soul? What better use for mathematics than working out correct medicinal proportions? Bacon invested heavily in alchemical books and equipment, and helped to make it acceptable for scholars to handle things as well as ideas. But he had to confront the opposition of his scholarly contemporaries, who generally disapproved of manipulating the natural world by changing it artificially. Instead, they adhered to the traditional 'nature knows best' point of view.

This conflict between supporters of nature and art (think artifice, artificial) continued for hundreds of years. It lay at the heart of debates about scientific method and the desirability of deploying human inventions to improve the world that had been created by God. By endorsing alchemical experimentation, Bacon placed himself firmly in the camp of those supporting art[ifice]. As he declared provocatively, 'Some there are that aske whether of these twaine bee of greatest force and efficacie, Nature or Art, whereto I make aunswere and say, that although Nature be mightie and marvailous, yet Art using Nature for an instrument, is more powerfull than naturall vertue.'[12] Rather than restricting themselves to demonstrating how nature works, alchemists wanted to effect improvements through human intervention. This twin desire to understand *and* to alter is central to scientific aspiration.

Nevertheless, there were major differences between Bacon and modern scientists. As a mediaeval scholar, Bacon was paid to think and write, not carry out experiments—he had no grant for equipment, and complained that his research ground to a halt when his own money ran out. Rather than pursuing a systematic programme of investigation, Bacon used his apparatus to confirm—not test—his theoretical ideas, which he derived from philosophical and theological preconceptions. He thought downwards from the divine to the mundane, from the abstract to the concrete, an approach very different from the scientific ideal of working upwards, of inferring general laws from particular situations. Along with his scholarly contemporaries, Bacon was searching for God rather than for any objective truth. Unlike the dry logical reasoning favoured by academic Aristotelians, alchemy put miracles back into the world and made it clear that God could still intervene. In the strictly regulated mediaeval universities, alchemy was attractive because it was anti-authoritarian, allowing a student to rely on faith rather than theological dogma.

Bacon shared with alchemists another characteristic (supposedly) alien to modern science—secrecy. According to ideological views of how science should proceed, scientists communicate knowledge freely amongst themselves. Although there are many striking counter-examples—including the atomic bomb project at Los Alamos, Charles Darwin's reluctance to release his theory of natural selection, and patents covering genetic engineering—the prevailing ethos dictates that scientific progress depends on openness, and on the readiness to be criticized and to cooperate. Resembling modern inventors protecting their rights, alchemists were loathe to let competitors access their recipes, and refused to share their knowledge beyond small circles of dedicated devotees.

To conceal their discoveries, hermeticists concocted mysterious codes of arcane signs and symbols. Some resembled chemical shorthand and were easy to decipher, such as drawing a crescent to represent silver, the metal of the Moon. Others were deliberately opaque and remain obscure: pictures of the Sun being devoured by a green dragon could be interpreted by initiates as an instruction to dissolve gold in *aqua regia* (royal water), a powerful blue-green mixture of acids. Yet like scientists applying for research funding, alchemists had to divulge enough results for convincing potential patrons of their abilities—and also like scientists, they probably sometimes adjusted their data to clinch their arguments. (This assertion that scientists fiddle their readings may seem shocking, but there are some striking examples, including the British astronomer Arthur Eddington's claim to have confirmed Albert Einstein's general relativity theory, and the American physicist Robert Millikan's measurement of the charge on an electron.)

There was a brisk trade in unofficial manuscripts that promised to reveal ancient secrets, and were snapped up not only by dedicated alchemists but also by monks and university scholars such as Bacon. Mediaeval readers were fascinated by these experimental tracts. Often of Islamic origin and falsely attributed to famous figures such as Aristotle and Albert the Great, they included herbal remedies and handy tips (amethysts ward off drunkenness, hare's intestines encourage male heirs), all allegedly tried and tested but with little theoretical justification. Repeated bans failed to quash monastic enthusiasm—one primer for priests classified sexual techniques to help them work out how many penances to dispense in the confessional booth. Yet however erotic and esoteric, these handwritten collections did emphasize the importance of providing natural explanations, even for events that seemed miraculous or magical. And unlike dogmatic university texts, they explored the world rather than insistently repeating unrealistic theories.

The alchemists' secretive behaviour resembled that of other specialized groups—including university scholars. Before printing was invented, books circulated in handwritten copies, available only to learned men who spoke Latin and could afford to pay for them. Academic friars like Bacon worked in monasteries and universities, closed hierarchical communities whose inhabitants repeatedly rehashed the ancient wisdom of Holy Scriptures and Greek philosophers. There were strong similarities between being initiated into a religious order, vetted for a university fellowship, and admitted into a clique of arcane alchemists. In Bacon's stratified world, scholars stood far above manual workers, whom he compared with goats. He warned against revealing superior knowledge to those operating outside privileged circles—why give goats lettuce when they're happy with thistles?—arguing that only the select few could be trusted not to exploit potential dangers.

Modern scientific laboratories have much in common with alchemical workshops, which provided the basic design for early chemical laboratories. Both are private places. The *laboratorium* in Figure 11 lies inside the alchemist's temple-like room, and is hidden still further behind protective curtains. Similarly, in subsequent centuries, scientific research continued to take place inside people's homes, often hidden in cool, dark cellars—even the Victorian scientist Michael Faraday carried out his experiments in the basement of the Royal Institution, concealed underground away from public view. Modern scientists communicate in arcane languages incomprehensible to the uninitiated, guarding their valuable apparatus and carefully monitoring access to exclude outsiders (women were still banned from Princeton's physics laboratory in the 1950s).

Paradoxically, universal truths are often said to be originated in specific places by particular individuals. When prophets acquire wisdom to enlighten the whole world, they meditate alone in the wilderness. Inside private studies—or under an apple tree—privileged experts formulate scientific laws that govern the entire cosmos. Similarly, solitary alchemists closeted themselves away to search for the philosophers' stone with its ubiquitous powers of purifying spirit as well as matter. They believed that any hermeticist who succeeded in ridding himself of worldly ambitions and regaining his uncorrupted nature would become a magus, a magician with a deep understanding of nature. Unlike wizards and witches who practised black magic by summoning up supernatural spirits, magi—altruistic magicians—manipulated natural powers. Often mathematically inspired, they could mediate between the celestial influences above and terrestrial existence down below by redirecting the hidden forces of the cosmos. The supreme magus was Isaac Newton, who stands Janus-faced between the Aristotelian, alchemical world of harmonic influences and the modern world of mathematical laboratory science. Scientists may abhor alchemy, but it is closer than they think.

III

Experiments

During the European Renaissance, intellectual investigation was fuelled by international exploration. Commercial trade stimulated the global exchange of skills, knowledge, and biological specimens, which circulated between different societies, changing as they travelled. Natural philosophers adapted old instruments and introduced new ones, although the experimental approach towards the world that characterizes modern science developed only gradually and intermittently. Galileo encouraged scholars to regard nature as a book written by God in the language of mathematics, but God's other book—the Bible—remained a major source of knowledge. Many innovations arose from reformulating traditional expertise rather than from inspired insights, so that ancient ideas coexisted with ones now belonging to modern science. For instance, Aristotelian principles prevailed long after Copernicus placed the Sun at the centre of the cosmos; reciprocally, magicians who relied on alchemical experiments and spiritual powers made mathematics the key to the cosmos. Perhaps the greatest magus of all was Isaac Newton, the natural philosopher and religious alchemist who wrote in the Greek language of geometry in preference to the latest mathematical techniques. Newton's Principia of 1687 now enjoys the status of scientific holy scripture, but it was rooted in the past as much as looking forwards to the future.

I. *Exploration*

We shall not cease from exploration
And the end of all our exploring
Will be to arrive where we started
And know the place for the first time

—T. S. Eliot, *Four Quartets*, 1942

'One picture is worth ten thousand words'—when this advertising slogan was invented in 1927, books were cheap, literacy was widespread, and scientists prided themselves on exchanging information freely. Four centuries earlier, when Hans Holbein painted *The Ambassadors* (Figure 12), a complex visual commentary on communication, the printed book trade was in its infancy, only the rich could read, and information about new lands, products, and processes was closely guarded. Now one of Europe's most famous pictures, *The Ambassadors* provides the only official record that its two subjects—both French diplomats—met in London. Using things rather than words, Holbein showed how travel, money, and knowledge were intimately linked.

In depicting this clandestine encounter, Holbein meditated on different ways of obtaining and transferring knowledge during the Renaissance. Traditional media—conversation, letters, hand-copied manuscripts—were being supplemented by printed books, although they could spread error as well as truth. Forty years earlier, Christopher Columbus had set sail across the Atlantic, and people based on opposite sides of the world were starting to exchange plants, animals, and raw materials as well as manufactured goods. Holbein's painting illustrates how experimental science stemmed from trade and politics rather than from any disinterested lust for knowledge.

Renaissance exploration did result in a massive increase of scientific information, but this was a side-product rather than a primary goal. International travellers were less interested in intellectual improvement than in financial profit and territorial power. They brought back unfamiliar plants and animals to use as medicines, crops, or gifts: slotting them into a global classification scheme came later. Similarly, instrument-makers wanted to earn a living, not decipher the secrets

FIG. 12   Hans Holbein, *The Ambassadors* (1533).

of nature, and equipment that might now be labelled 'scientific' was originally designed for practical uses—measuring boundaries, weighing metals, dispensing drugs, producing dyes. Navigators demanded precise details of star movements, compass readings, and wind patterns not to overthrow Ptolemy's outdated global atlases, but in order to reach their destination safely. Knowledge about the world carried great political and commercial implications, and so it was a valuable commodity to be bought, sold, and bargained over by ambassadors as well as by merchants.

In this picture replete with hidden signals, Holbein followed Aristotle's division of the universe by showing a celestial upper shelf with a terrestrial one beneath. Books and instruments were more closely related then than now, so Holbein grouped them together as complementary sources of knowledge. Above, the mathematical instruments—all identifiable and carefully recorded—were used by navigators to record star positions, measure time, and make more accurate maps. The lower shelf carries an arithmetic manual for merchants, and also a borrowed globe incorporating sensitive diplomatic information added by Holbein himself. International exploration was about profit and possession.

Yet Holbein's instruments also depict non-communication. The lute—emblem of both human and cosmological harmony—has a broken string, and the deliberately misaligned astronomical instruments carry contradictory shadows. Flanking this apparatus stand two men who have been despatched as instruments of the French state to retrieve insider gossip, but they stand silent, their non-committal countenances revealing nothing of courtly intrigues. Similarly, the precise faces of sundials and astrolabes, and the tooled leather bindings of books, provide no guarantee of internal accuracy. Printed words and intricate objects are as unreliable as human beings. Just as lenses produced distorted images, so printing made it easier to spread falsehoods, and—like dissimulating courtiers—pictures could deceive.

Printing played a major role in the growth of science, although there was no sudden transition to this new technology. Long after moveable-type printing was introduced in the 1450s, scribes continued to copy out manuscripts, and illustrations were still drawn and coloured by hand. Merchants realized that rich customers who bought sets of Aristotle's philosophy or Pliny's natural history were generally less interested in improving their minds than in enhancing their homes with artistic status symbols, and they marketed limited editions of beautiful but expensive printed books. It was not until the sixteenth century that the commercial book trade was well established, and publishers had convinced readers that printed books were affordable and identical—well, more or less.

Location was crucial for both creating and spreading knowledge. In *The Ambassadors*, Holbein carefully highlighted how his political globe originated in Nuremburg, renowned for its wealth and culture and Europe's leading centre for books, prints, and instruments during the sixteenth century. The artist Albrecht Dürer used Nuremburg as a marketing base for sending out his pictures of plants and animals. Although he self-mockingly sketched them being shovelled into ovens to be sold in bulk like cheap loaves, cheap prints had dramatic effects on scientific information. Although Dürer's famous image of an armour-plated rhinoceros may now seem endearingly laughable, it spread widely and was repeatedly reproduced, becoming rhinoceros reality for countless people who—like him—had never seen an actual specimen.

One man above all was responsible for establishing Nuremburg's scientific supremacy—the astronomer Regiomontanus (Johannes Müller). He moved there in 1471, deliberately choosing a place that not only produced fine astronomical instruments, but was also known for 'the very great ease of all sorts of communication with learned men living everywhere, since this place is regarded as the centre of Europe because of the journeys of the merchants.'[1] This prime commercial position helped convince a local businessman to invest in his new press, and Regiomontanus helped to make Nuremburg an intellectual trading post from which information could be carried all over the known world by high-quality books, instruments, and pictures.

The growth of learning depended not only on innovative authors, but also on enterprising and conscientious publishers. Regiomontanus is far less famous than another Central European astronomer, Nicolas Copernicus, the scholarly cleric who placed the Sun rather than the Earth at the centre of the universe. Yet Copernicus's celebrated book was printed in Nuremburg, and his fame depended on the publishing networks that had previously been established by Regiomontanus and his contemporaries. Without their initiatives, the revolutionary discoveries of the sixteenth century would have had far less impact.

Regiomontanus's own research was crucial for future science because his books were clear, accurate, and widespread. His ideas reached Copernicus as a student in Bologna, and also travelled to the New World with Christopher Columbus, who was searching for spice routes to eastern India. Although Columbus obstinately maintained that he had reached his original destination, his trading voyage across the Atlantic dramatically altered European visions of the world and resulted in a vast explosion of international information.

Regiomontanus initiated astronomical change by tackling Ptolemy, then well over a thousand years old but still a leading authority. He not only provided a better translation of Ptolemy's *Almagest*, but also criticized its ideas, produced new astronomical measurements, and in an influential and significant shift of emphasis, insisted that theories should match observations. To make that possible, Regiomontanus checked his proofs meticulously. Hand-written tables of older readings were riddled with mistakes, some of them due to repeated miscopying, others to poor or even invented data—the numerical equivalent of Dürer's rhinoceros. Regiomontanus made theoretical reform possible because he supplied error-free sets of astronomical measurements.

By helping to expand international trading networks, Regiomontanus ensured that books, instruments, and knowledge travelled around the globe as commercial commodities alongside silks, copper, and exotic animals. One obvious spin-off was better geographical knowledge. Just as Columbus valued Regiomontanus's astronomical data, so too merchants took advantage of improvements in instruments and printing to sail further and more safely. The Nuremburg globe in Holbein's picture—made by one of Regiomontanus's students—incorporates cartographical details purloined from Portuguese sources anxious to protect such valuable information. Armed with new measuring equipment, navigators charted the oceans and coastlines more accurately, although the continental land masses remained largely blank. As international trade expanded, merchants demanded—and got—increasingly detailed, reliable knowledge not only of the world's dimensions, but also of wind patterns, water currents, and magnetic influences.

Information, raw materials, and manufactured goods journeyed in every direction, altering the world forever. Influences ran both ways. Europeans indelibly marked the territory they settled, but their countries of origin were also permanently changed. The New World provided modern European crops such as potatoes, beans, and tomatoes; conversely, America received not only onions, cabbage, and lettuce from Europe, but also medicines, watermelons, and rice brought over by African slaves. The traders and missionaries who survived abroad were those who followed advice and adapted their behaviour, learning from local guides what to eat and wear. When travellers returned home, they took this information with them, so that European botany, agriculture, and medicine benefited from the expertise of peoples they often regarded as inferior.

Asians, Africans, and Americans took advantage of unexpected encounters with Europeans to expand their own economies by supplying food and medicine, clothes, and building materials. Recognizing that their familiar world seemed exotic to their uninvited visitors, they also provided plants and animals whose main function was to be marvelled at—the rhinoceros that Dürer claimed to have drawn from life was an Indian ruler's diplomatic gift to Portugal. Enterprising merchants soon established a thriving international trade in natural curiosities, persuading European aristocrats to display expensive wonders of nature alongside their prestigious statues and pictures. This fashion for collecting unusual animals, plants, and minerals started in the Italian courts, and then spread across Europe and into private houses. By the middle of the seventeenth century, the commercial market for curiosities was so great that the diarist John Evelyn reported visiting a Parisian souvenir shop called Noah's Ark, which sold 'all curiosities natural or artificial, Indian or European, for luxury or use, as cabinets, shells, ivory, porcelain, dried fishes, insects, birds, pictures, and a thousand exotic extravagances'.[2]

Global trade stimulated a revival of natural history that originated not in universities, but in courts, private societies, and personal collections. At the courts, princes and aristocrats acted as patrons, giving financial support to scholars who mingled with educated noblemen and discussed the latest imports. In cities, medical men and professors developed their own collections in private museums that became essential tourist stops for privileged travellers, who then reported back to learned discussion groups. Figure 13 shows Ferrante Imperato, an influential Neapolitan apothecary, chemical experimenter, and fossil expert, pointing out his spectacular exhibits to some distinguished visitors. Searching for intellectual entertainment, they have congregated here to gaze at what was often called a theatre of nature, a metaphor implying God's involvement as a superb stage manager.

This scene illustrates a new way of studying—through conversation, rather than through lectures. In universities, professors traditionally dictated classical knowledge from well-established authorities, and students were expected to absorb rather than challenge existing knowledge. But in societies and museums, scholars and aristocrats discussed questions together, reaching their own conclusions

FIG. 13 The museum of Ferrante Imperato in Naples, 1599. Ferrante Imperato, *Dell'historia naturale* (*On Natural History*, 1672 edn.).

by learning from each other as well as from natural specimens. Naturalists travelled round Europe, visiting other collectors and also incorporating imported knowledge gathered from indigenous experts all over the globe.

Despite this change in approach, classical natural history was at first expanded rather than overturned. Imperato has arranged his specimens carefully, but he has grouped them by appearance and origin rather than by any abstract classification scheme. His crocodile is stretched prominently across the ceiling not because it is scientifically significant, but because it is large, unfamiliar and expensive. Renaissance naturalists strove to describe rather than to explain, to compile before they classified, to study the particular instead of relying on grand universals. Instead of redesigning older catalogues of plants and animals, often drawn up for medical purposes, collectors slotted new specimens into existing categories which had been set up by the Greeks.

In addition to specimens, words and pictures also transmitted information about distant parts of the globe. Imperato's cabinet of curiosities includes fine books, rare and expensive sources of knowledge about nature, but their illustrations often look very different from modern scientific ones. Sometimes this is because—like Dürer and his rhinoceros—artists had never seen the animals or plants they were trying to depict. Just as significantly, illustrators often intentionally produced symbolic rather than lifelike representations. For example, after a Spanish sailor brought back a traditional Central American recipe for preparing a purgative by grinding up a local plant root, his physician drew it as a flower to indicate its importance. Even though wrongly illustrated, apothecaries all over Europe started to sell this imported drug as a safe, effective, and profitable remedy.

Realism might seem the only useful style for natural history, but old pictures of plants often appear crudely drawn. This was not because illustrators had yet to acquire the requisite skills, but because they were depicting hidden meanings rather than superficial appearances. Images were regarded as potentially deceptive. The encyclopaedist Pliny told the story of Zeuxis, who reproduced grapes so faithfully that birds attempted to eat them. He was, however, out-manoeuvred by his rival Parrahasius, whose painted curtain fooled Zeuxis into demanding that it be drawn back to reveal the picture behind.

Aristotle, Pliny, and the other classical compilers felt that artificial images were unsuitable for revealing nature's secrets, and so they described in words, not in pictures. In any case, scholars labouring with their heads had little social contact with artists, who were regarded as manual workers. Natural history illustrations were found not in academic texts, but in monastic illuminated manuscripts and in practical guides designed to help healers identify medicinal plants and prepare drugs. Collectors still relied on herbals that had originally been prepared by Greek experts— Dioscorides was particularly valued—and that were repeatedly copied by hand from one generation to the next.

Natural history albums acquired a new look in the sixteenth century, when artists started producing large woodcuts in a realistic style, and naturalists began stressing the importance of close observation. Aiming to surpass their classical ancestors, collectors compiled comprehensive, detailed catalogues of God's natural world. They started with plants, valuable for agriculture and medicine, and only later turned to animals. The most famous encyclopaedia was by Conrad Gesner, a Zurich doctor whose personal collection attracted visitors from all over Europe. Gesner updated the great classical works, incorporating not only realistic illustrations of familiar animals, but also information about New World creatures, such as guinea pigs and opossums. Inevitably, this tended to be less reliable—naturalists claimed that birds of paradise never land on the ground because commercial specimen hunters routinely chopped off the legs before handing over the gutted skins for inspection.

Gesner was a humanist scholar, not a modern biologist. For him, research meant poring through countless books to compile *all* the information he could find, so that details now classified as scientific—diet, longevity, habitat—were mixed together with fables and folklore. Take the fox. As well as telling his readers about its appearance, digestibility, and medicinal uses, Gesner provides its name in many languages together with over eighty assorted fox factoids gleaned from Aristotle onwards. Almost half the article is devoted to fox symbolism. Its reputation for cunning still survives, but Gesner includes numerous unfamiliar quotations and proverbs, such as 'a fox takes no bribes'. For modern readers, his most disconcerting section describes a fox who holds up a mask in its paws, declaring 'What a fine head this is, but it has no brain'—an allegorical injunction to value brains more highly than beauty.

Incomprehensible jokes and allusions often point to fundamental cultural beliefs that no longer exist. During the Renaissance, myths and masks formed an essential component of fox-ness. Understanding a fox—or any other creature—entailed knowing about its moral significance and psychological attributes as well as its physical role in the natural world. Such symbolism made sense, because people envisaged imperceptible occult forces binding animals, plants, and human beings together in a holistic, empathetic universe. Gesner's speaking fox indicates the importance his contemporaries attached to emblems, symbolic pictures embellished with mottos, and explanatory verses. Long after realistic pictures became standard, this emblematic approach permeated representations of the natural world: new-style images were embedded in older thought patterns.

'With the benefit of hindsight'...but hindsight can be misleading. In traditional accounts of the fifteenth and sixteenth centuries, Regiomontanus and Gesner are allocated to science, whereas Holbein and Dürer belong to art. Such harsh disciplinary boundaries had not yet been created, and these four men shared a common quest to find new ways of exploring and representing the world in pictures as well as in words. Tracing science's history means forgetting about the present and trying to understand the past. Victorian anti-Darwinists found it hard to accept their animal ancestry; modern scientists should recognize that their predecessors included not only university scholars but also herbalists, navigators, witch doctors, and instrument-makers.

2 *Magic*

I have often admired the mystical way of Pythagoras, and the secret magic of numbers.

—Sir Thomas Browne, *Religio Medici* (1643)

In 1947, the economist John Maynard Keynes shocked the academic world by announcing that 'Newton was not the first of the age of reason. He was the last of the magicians.'[3] Scientists were scandalized—they refused to believe that their greatest hero could be tainted by any association with astrology, alchemy, and other magical crafts. But historians now agree with Keynes's verdict. Newton developed rather than rejected the work of the great magi who preceded him, and magical ideas lie at the core of modern scientific knowledge.

Renaissance magicians were learned scholars who bore little resemblance to later parodies of black-caped sorcerers summoning up satanic powers. Instead, many magi were respected, well-educated men who made mathematics the key to the universe and remained influential well into Newton's lifetime. Their ideas and activities strongly affected the future course of science. In comparison with university scholars devoted to contemplating the wonders of God's creation, magi resembled modern scientists in believing that the more they understood about the world, the more they could change and control it.

Magic suffused sixteenth-century art, music, and literature. England's most eminent magus was John Dee, a university-trained mathematician hand-picked by Elizabeth I to advise her on naval affairs and political strategy, and also to calculate astrologically favourable dates for court events. Although he died poor, Dee remained a powerful icon for younger contemporaries such as William Shakespeare, who used him in *The Tempest* as a model for Prospero, the controlling magus who intervenes with nature by staging an illusory shipwreck on an island haunted by aetherial music.

Thanks to a 'most auspicious star' which is astrologically favourable, Prospero is at the peak of his powers when he stage-manages events on his sea-bound kingdom, a miniature theatre of nature that allows the audience to gaze into the magical cosmos. Prospero reforms his stranded captives by ordering Ariel, an angelic spirit, to mediate between different realms. Being an immortal intelligence, Ariel can operate in all four Aristotelian elements—earth and water as well as air and fire—whereas Prospero's human powers are restricted. In his poetic evocation of transformation, Ariel describes how a drowned man is physically upgraded into fine pearls and coral, while at the same time, his soul becomes spiritually purified and approaches God, grand Magus of the cosmos:

> Full fadom five thy father lies;
>   Of his bones are coral made;
> Those are pearls that were his eyes:
>   Nothing of him that doth fade,
> But doth suffer a sea-change,
> Into something rich and strange.
> Sea nymphs hourly ring his knell:
>
> Burthen: *Ding-dong*.[4]

Ariel's name had already appeared in the standard text on magic in Elizabethan times, Henry Agrippa's *Occult Philosophy* (first published in 1533, in Latin). Agrippan magic not only featured in great works of literature, but was also incorporated into scientific models of the Universe. A wandering magus and Germanic diplomat who studied all over Europe in the early sixteenth century, Agrippa was important not because of his own original ideas, but because he synthesized earlier European developments of the ancient Greek and Arabic inheritance.

Agrippa's most important source was Hermes Trismegistus, fictionalized amalgam of several Egyptian priests and supposed author of countless very varied Greek and Arabic manuscripts (some of them Newton's favourite reading). Although he never lived, Hermes Trismegistus was no shadowy charlatan, but a key figure in Renaissance culture, allegedly chosen to receive God's wisdom at the beginning of human history. By the end of the fifteenth century, Hermes Trismegistus had become absorbed within Renaissance religion. He was prominently portrayed in the mosaic floor of Siena Cathedral, sporting a pointed turban and a sage's beard, and flanked by Greek Sibyls prophesying Christ's arrival. At Florence Cathedral a canon called Marsilio Ficino translated his supposed writings, which had been randomly grouped together by a monk collecting Greek manuscripts for the Medicis. A devout and scholarly Catholic, Ficino interpreted the incoherent mixture of hermetic texts that he had inherited as ancient Egyptian revelations foretelling the truths of Christianity. Ficino was also studying the works of Plato and other Greek writers that had recently arrived from the Islamic Empire, and he amalgamated these disparate ancient sources to produce his own version of Renaissance Neoplatonism—a philosophical blend of Platonic, Christian, and magical ideas.

The other major influence on Agrippan magic was cabbalism, a Jewish tradition said to stem from Moses himself and imported into Florence from Spain by Pico della Mirandola, one of Ficino's Neoplatonic colleagues. Like Ficino, Pico was fascinated by hermetic beliefs, but he also placed Jewish thought at the heart of the Renaissance occult. In contrast with Ficino's natural magic, Pico's cabbalistic magic aimed to tap higher spiritual powers, enabling a magus to access God by communicating with His angels. Pico Christianized the cabbala, envisaging his own soul ascending up towards God as though climbing a cosmic-theological ladder whose rungs linked Aristotelian spheres and Hebraic archangels.

Hermeticism and cabbalism may now seem bizarre, but they had a long history. Based on solid philosophical foundations, and reinterpreted by Renaissance scholars, they contributed to the Neoplatonic ideas that had a great impact on science's future. Ficino and Pico died towards the end of the fifteenth century, but their Neoplatonism survived for a couple of centuries and became embedded within scientific thought. For instance, Copernicus is regarded as a founding figure in science because of two major innovations—placing the Sun at the centre of the Universe, and insisting on a mathematical approach to the cosmos. These were both Neoplatonic ideals. Like Hermes Trismegistus, Copernicus regarded the Sun as God made visible, and he reintroduced the geometrical cosmologies of Plato and Pythagoras that had been revived by magi.

Magicians such as Agrippa fused these hermetic and cabbalistic concepts together, incorporating Neoplatonic thought into a revised Aristotelian cosmos with zodiacal associations. They explained that God's virtue filters down from angels in the outer realm, passing through the stars and the heavens into the elemental world below. Whereas black magicians bargained with devils and controlled evil satanic forces, reputable magi tapped in to natural benign influences, searching for spiritual as well as material improvement. As Ficino put it, just as farmers till their fields in tune with the weather, so magicians are cosmic cultivators who achieve results by accommodating higher forces. By distinguishing between black and natural magic, Agrippa defused the criticisms of Catholics who denounced magi for condoning pagan rites and relying on diabolic spirits. Nevertheless, they still accused natural magicians of presumptuously taking over God's role by harnessing nature's hidden powers, by actively changing the Universe rather than passively admiring His omnipotence.

Agrippan natural magicians shared with modern scientists the goal of controlling the Universe through intervening in it. Novices started by learning how to alter the physical world through invoking innate sympathies and planetary influences. They had to grapple with some complicated learning. For instance, Leo is a solar constellation and cockerels crow at sunrise; hierarchically, the Sun is superior to both creatures, but cockerels are more powerful than lions because *air* is a higher element than *earth*. Using astral associations, natural magicians could divine the future and prepare love philtres, spells, and medicines, some of which were very effective—which is why people were willing to pay them for help. When magicians recommended students to ward off depression by countering Saturn's melancholic influence with gold of the Sun and flowers of Venus, they were dispensing sensible advice to go for a walk in the country sunshine. More advanced magicians placed great emphasis on mathematical manipulations, drawing down celestial powers from the stars and planets by mastering numerical symbolism. At the top level, a fully fledged magus performed religious ceremonies so that he could communicate with the angelic spirits of the intellectual sphere.

Magic had both theoretical and practical aspects, and so it appealed to gentlemanly scholars as well as to apothecaries, herbalists, and artisans who were already used to handling instruments and preparing potions. Like modern science, magic entailed combining intellectual dexterity with manual skills. Erudite adepts such as Agrippa wrote in Latin for educated readers, while at the other end of the social scale, craft workers used word of mouth or coded manuals to pass on their expertise. As publishing and literacy escalated in the early sixteenth century, artisans and academics exchanged secret recipes that had been developed over the centuries. Magicians' specialized knowledge was commercially valuable, and—especially in the German courts—wealthy patrons hired consultants to make their mines more profitable or to improve manufacturing techniques.

This hands-on practical expertise was excluded from conventional university curricula, but some scholars explicitly sought progress by turning away from traditional institutions to embrace magic. In the first half of the sixteenth century, one particularly influential revolutionary was Agrippa's contemporary Theophrastus von Hohenheim, who renamed himself Paracelsus (against Celsus, a Roman physician) to advertise his rejection of the classical past. This abrasive, arrogant man seemed to court opposition. Proclaiming that knowledge should be available to everyone, Paracelsus shocked university authorities by wearing a leather apron for his inaugural lecture and speaking in German. Like Agrippa, Paracelsus travelled around Europe teaching and studying, but he boasted that instead of talking to scholarly sages, he was happy to learn from tramps, old women, and barbers.

Paracelsus had a far greater influence than many of his conventional contemporaries. Because he gave public talks in small towns and villages outside the universities, his ideas spread widely and were adopted by many less-educated men and women, first in the Germanic countries and then abroad. Paracelsus reformed medicine by making it chemical, insisting that through his magical techniques, he could prepare super-medicines from ordinary matter. He also revived hermetic magic, asserting that each human being is a condensed version of the entire Universe—a microcosm of the macrocosm. A fervent Christian, Paracelsus claimed that pious therapists could decipher the correspondences linking human beings to the cosmos, and he subscribed to the doctrine of signatures, which held that nature's symbols reveal efficacious drugs for related organs, such as yellow flowers for the liver, orchids for testicles. When Shakespeare wrote *A Midsummer Night's Dream*, he knew that his audience would appreciate how a 'flower...purple with love's wound' could make Titania magically obsessed with Bottom.[5]

This insistence on searching for specific therapies to counter a particular disease was radically different from Aristotelian attempts to rebalance an individual's internal humours. Instead, it was closer to modern ideas that external agents—bacteria, viruses—attack different parts of the body. Yet despite Paracelsus's successful cures, the medical elite bitterly resented this bombast who boasted about overturning centuries of learning. Financial interests were also at stake: doctors risked losing their patients if Paracelsus undermined their prestige as experts, and one university dean clamped down on Paracelsus's recommendation of mercury for treating syphilis because it threatened profits from imported plant remedies. But although his name became a term of abuse amongst educated physicians, Paracelsus had an enormous impact on treatment and training because wealthy aristocrats—including Elizabeth I of England and Henri IV of France—hired Paracelsian therapists to supplement their official medical advisers. Royal physicians assimilated and adapted Paracelsus's ideas, so that although his theories lost credibility, his chemical remedies entered mainstream medicine.

These new continental ideas reached England, where the greatest English magus was John Dee. Dee devoured books on Paracelsian and Agrippan magic while he was a Cambridge undergraduate. Although he later rejected the university system, Dee became England's leading Elizabethan mathematician, employed to study the stars for making navigation safer, calculating Christian festivals, and forecasting propitious dates. Dee also learnt how to be a fully fledged magus, boasting about his communications with angels but complaining that he was unjustly vilified 'as a Companion of the Helhoundes, a Caller, and Conjurer of wicked and damned Spirites'.[6]

Dee regarded mathematics and magic as complementary, not contradictory. For example, he insisted that architects should make buildings cosmologically harmonious by calculating their proportions to match human dimensions—much like Leonardo's famous drawing of a man spanning a circle and a square. An early convert to Copernicus's ideas, Dee calculated the Earth's movements around the Sun, but also pledged his faith in a hierarchical, magical universe bound together by occult powers. Patronized by the Queen and respected all over Europe for his mathematical expertise, Dee was an aspiring magician who poured money into buying instruments, hiring assistants, and—like Prospero—building up an impressive library of several thousand volumes, the largest in England. Since the universities were still teaching the traditional classical curriculum, and the Royal Society had not yet been founded, Dee's home acted as the country's major experimental research centre.

Far from being trapped in old-fashioned mysticism, Dee heralded the future of science. His books had more immediate impact than the theoretical debates of sedentary scholars because he was interested in practical problems, such as lifting weights, surveying land, and designing optical instruments. As a Neoplatonist, Dee believed that mathematics was crucial to understanding the cosmos. For him, numbers and shapes carried religious as well as scientific significance, and he regarded them as abstract entities that mediated between the physical, material world and the angelic realm. Writing in Latin for his peers and in English to reach practical men, Dee explained how numbers were essential not only for tracking the stars, but also for mundane activities—planning military tactics, taking legal decisions, making pulleys, maps, and clocks.

Operating outside the universities, Dee combined theoretical research with laboratory investigation and practical applications—important features of modern science. Dee also pioneered a new scientific life-style for gentlemen, because he earned money by working in his own home. English scholars at this time were mostly single, secluded in monasteries or universities, while even continental magi avoided marriage in order to keep their souls pure. Dee broke away from all these conventions by living at home, marrying, and trying to support his family from his scientific investigations.

Dee and his wife Jane Fromond, a former lady-in-waiting to Elizabeth I, were forced to negotiate ground rules for a new kind of experimental partnership. The lines of authority were blurred. Unusually, he was working inside traditionally female domestic territory, while her tasks now included looking after his live-in assistants and entertaining his scholarly guests. Unsurprisingly, as money ran short, they often argued about the need for so many expensive books and instruments, to say nothing of the paid apprentices who increasingly dominated family life. During the next couple of centuries, this type of domestic collaboration became common amongst scientific entrepreneurs who converted their homes into schools, workshops, and research centres. It was only in the Victorian era that scientists started routinely working in large communal laboratories attached to universities or factories. Before then, science was often a home-based activity that could involve the entire family.

Modern science stems not only from university scholarship, but also from everyday trades, crafts, and magical practices. At the end of *The Tempest*, Prospero

relinquishes his special powers, but only after his charms have permanently transformed the island's inhabitants. Similarly, even though John Dee and the other
great magi were later denounced as charlatans, their influence survived. Magicians
and artisans taught natural philosophers to use their hands as well as their heads—
if they wanted to control the world, they had to leave their sequestered studies and
engage with physical reality.

3 *Astronomy*

In 1939, the German playwright Bertolt Brecht castigated Nazi policies by writing a play about Galileo's heroic behaviour under prosecution by the Catholic Inquisition. To construct his political parallel, Brecht drew on the appealing mythology of a prolonged war in which scientific revolutionaries—Nicolas Copernicus, Johannes Kepler, Galileo Galilei—battle against religious bigots to place the Sun at the centre of the Universe. Victorian scientists depicted these men as martyrs to reason who sacrificed themselves to keep the flame of truth alight, an image of confrontation between science and religion that still remains popular. Yet when viewed from their own perspective, all three were deeply religious, and more concerned about their own lives than in carving out any straight road towards the future.

Brecht might have considered fictionalizing another of these astronomical heroes—Copernicus, then more familiar to his German audiences, but also more contentious—because of the long-standing tussles between German and Polish chauvinists to claim him for their own. Rivalry reached a head in 1943, the four-hundredth anniversary of both Copernicus's death and the publication of his Sun-centred cosmology, *On the Revolutions of the Heavenly Spheres*. After the Nazi regime circulated stamps showing Copernicus with a border of swastikas, Polish exiles in New York retaliated with their own propaganda campaign, commissioning the artist Arthur Szyk to produce a commemorative image of Copernicus as their national hero.

Packed with Polish symbolism—the national colours of red and white, the royal eagle, Kraków university's coat-of-arms—the resulting, vibrant icon (Figure 14) idolizes Copernicus by distorting his scientific significance. Although he worked as a church official, this Copernicus sports an academic's chain and fur-trimmed cap, and also clasps dividers, the conventional symbol of an astronomer. The lantern implies that he was a keen observer, whereas Copernicus was primarily a theoretician who

FIG. 14 A Polish Copernicus. Coloured miniature by Arthur Szyk (1942).

took his ideas from ancient books rather than the stars. Misleadingly, the Latin and Polish epithets claim that Copernicus triumphed overnight. In reality, there were few converts even fifty years later, and the so-called 'Copernican Revolution' was a long process involving many participants. The planetary diagram is headed 'Copernicus died, but science was born', even though—like Newton and many other celebrated innovators—Copernicus had revived ancient wisdom to create new knowledge.

Far from being a prominent academic, Copernicus was an undistinguished church administrator protected by his wealthy uncle. After studying at Kraków University, remote but renowned for its astronomy, Copernicus travelled and lectured in Italy, where he encountered Ficino's Neoplatonic legacy as well as the accurate observations of Regiomontanus and his followers. Copernicus was primarily a desk-bound scholar with his heart in the past. Well-educated in the classics, he used traditional rhetorical techniques and wrote for—not against—his church colleagues.

Copernicus searched for order. Like Ficino and Pico, he believed in a harmonious, mathematically structured universe, and he applied his Neoplatonic soul to the practical problem of improving forecasts from the stars. Astronomers wanted to find better ways of keeping the calendar accurate and making medical prognoses, but found that Ptolemy's system (see 'Cosmos' in Chapter 1) was complicated and sometimes clashed with observations. Just as seriously for Copernicus, Ptolemy's epicycles were aesthetically unpleasing. By placing the Sun near (although not precisely at) the centre of the revolving planets, Copernicus resolved many of these difficulties. He satisfied his Neoplatonic idealism by giving the Sun the most important position and retaining perfectly circular orbits. By eliminating Ptolemy's cumbersome geometrical fudges, he could sequence the planets in the same order as the times they took to complete their orbits. And of prime importance for working astronomers, Copernicus proved that his model was as effective as Ptolemy's for providing predictions.

Scientific propaganda makes Copernicus's book seem momentous, but there was no great roar of protest at the time. Although it was dedicated to the Pope, His Holiness took little notice of this complicated book by a minor Polish functionary. A heliocentric universe seemed dangerous not because it contradicted the Bible—those objections came later—but because it transgressed common sense and also overturned Aristotelian physics, with its fundamental distinction between the chaotic, corrupt terrestrial realm and the unchanging perfection of the heavens.

At this stage, astronomers made few objections because they regarded Copernicus's model as simply that—a model for calculating planetary positions, not a description of how the Universe really is. Yet Copernicus had smuggled in a new way of defining the extent of astronomers' work, because he unobtrusively suggested that they might consider the truthfulness as well as the usefulness of their cosmological schemes. Hiding behind a rhetorical veil of false naivety, Copernicus apologized for using techniques that had been developed by mathematicians in order to answer questions about reality previously reserved for their intellectual superiors, natural philosophers. Bringing together mathematicians and natural philosophers was a fundamental shift, one that involved social changes as well as intellectual ones. Over a hundred years went by before Newton fused their two approaches together in his book on gravity, making astronomy a mathematical science that aimed both to describe and to explain the cosmos.

Traditionally, astronomers operated in two separate locations. In the universities, astronomy—like arithmetic and geometry—belonged to the mediaeval *quadrivium*; adopting a mathematical approach, scholars focused on teaching students and calculating accurate predictions—searching for truth was beyond their remit. Outside these scholarly enclaves, the cities provided centres for a great variety of astrological astronomers as well as for craft entrepreneurs who, like Regiomontanus, made instruments and printed tables. Following Copernicus's innovations, a new form of astronomy emerged within a third setting—the courts. Supported by wealthy princes, educated noblemen built expensive instruments not only to make calculations, but also to discover how the Universe works. In return, their royal patrons gained prestige. Complementing their museums crammed with expensive curiosities, aristocrats' observatories displayed the wealth they had invested in intellectual pursuits.

The most important of these new-style court astronomers was Tycho Brahe, a Danish nobleman who angered his parents by choosing to study low-status astronomy and then leaving the university system altogether. He eventually managed to get royal funding for a massive observatory on the island of Hven (now a Danish heritage site on Swedish territory). Using the King's money, Tycho brought together measurement and theory as he sought out the true structure of the cosmos. Behaving like a feudal lord in his island fiefdom, he ruled over his mathematical entourage, building instruments and setting up his own printing press to distribute his results. By the 1590s, half a century after Copernicus's death, Tycho had compiled impressive sets of accurate data and also devised his own suggestion for the structure of the Universe.

Unlike the scholarly Copernicus, Tycho repeatedly designed, tested, and modified his instruments. Figure 15 shows his giant quadrant, a brass quarter-arc around two metres in height and fixed to the wall, used to measure the precise position of a star as it passes by the small sight on the top left. Most of this picture is itself a picture—Tycho and his snoozing dog are part of a mural painted within the quadrant. Behind his outstretched arm lie emblematic illustrations of his observatory's

FIG. 15 Tycho Brahe's mural quadrant.
Tycho Brahe, *Astronomiæ instauratiæ mechanica* (*Machines of the New Astronomy*, 1587).

three floors, each framed by triumphal arches: the rooftop for making night-time observations, the library with its immense celestial globe, and the basement devoted to carrying out experiments (including alchemical tests on the best alloy to replace the tip of his nose that had been sliced off in a duel). The real observer is just visible on the right, calling out to his assistants who coordinate measurements of a moving star's time and position.

While he grappled with technical hitches, Tycho was also wrestling with a theoretical dilemma. It seemed clear to him that the Bible supports the common sense, Aristotelian assumption that the Earth is stationary. How could he maintain the pleasing harmony of Copernicus's system while putting the Earth back at the centre of the cosmos? Eventually Tycho found a lopsided answer. According to him, the Sun and the Moon revolve about the Earth, and the other planets circle around the orbiting Sun. Weird as that solution may sound, it explained many observations just as satisfactorily as the Copernican and the Ptolemaic systems. Operationally, it was hard to choose between them, and at the end of the sixteenth century, the three models coexisted, each with its vociferous band of supporters.

Like other beneficiaries, Tycho discovered that a king's backing is valuable but risky. When his royal patrons lost interest, Tycho was forced to leave Hven, but he found a new employer—the emperor in Prague. After Tycho died in 1601 (allegedly from a burst bladder at an imperial feast), he was succeeded by his assistant Johannes Kepler, an impoverished astrologer and former university teacher who believed—like Copernicus and the Neoplatonists—in a geometrical cosmos with a central Sun. Inheriting Tycho's accurate data, and taking advantage of the greater intellectual freedom available at the court than within the academic system, Kepler brought astronomy and reality closer together by showing how the Danish observations corresponded to elliptical rather than circular planetary orbits. He was inspired to reach this apparently scientific conclusion from a vision that now seems alien—a musical cosmos structured to mirror God's perfect geometrical shapes and bound together by hidden magnetic powers.

In Kepler's harmonious scheme, God spaced out the planetary spheres so that symmetrical Platonic shapes could be nested between them. Figure 16 shows the drawing he made, so large that the sheet was folded to fit inside its book. The outermost sphere of Saturn is separated from its neighbour—Jupiter—by a cube; moving inwards, a pyramid lies between Jupiter and Mars; similarly, other shapes define the orbits of Earth, Venus, and Mercury around the Sun. To the fury of Catholics, Kepler identified the central Sun with God the Father, the external stationary sphere with God the Son, and the space in between with God the Holy Ghost. He also made his philosophical model aesthetically appealing—its dimensions corresponded to measured distances, and the further a planet from the Sun, the longer its orbiting time.

Launching his own new approach to the Universe, Kepler decided that this divine harmony also had a physical influence—the Sun itself, which must be affecting the motion of the planets. He started by tackling the astrological God of War, Mars. This planet's orbit clearly deviated from circular perfection, a discrepancy made even more glaring by the accuracy of Tycho's data. For help, Kepler turned to a contemporary English expert, the physician William Gilbert. Objecting to the way Aristotelians viewed the Earth as inferior to the rest of the Universe, in 1600 Gilbert claimed—citing Hermes Trismegistus—that the entire Universe is an animate being with a magnetic soul. Kepler used Gilbert's ideas to envisage the Sun as a giant magnet that attracts and repels the planets to control their paths

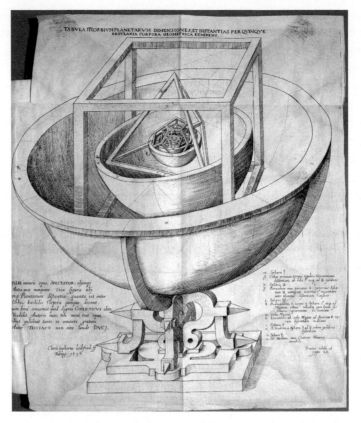

FIG. 16 Kepler's nested scheme of planetary spheres and perfect solids. Johannes Kepler, *Mysterium cosmographicum* (*The Secret of the Universe*, 1596).

through the skies. This cosmology was so influential that when Newton started investigating comets seventy years later, he thought they moved magnetically.

After many tortuous calculations and blind alleys, Kepler demonstrated that the orbit of Mars is an ellipse, with the Sun lying not at its centre, but asymmetrically at one focus. However, what might now seem like a great scientific leap forwards was ignored for several decades. Not satisfied with solving the Mars problem, Kepler tried to unify the entire Solar System by proving that Pythagoras had been right—the numerical ratios of the cosmos are musically harmonious. Attributing celestial tones to each planet (low for Saturn, high for Venus), he declared that 'The heavenly motions are nothing but a continual music of several voices, which can be comprehended by the intellect, not by the ear.'[7] Divine aesthetics and astrological influences mattered for Kepler—and however bizarre his approach might seem now, it was these musical calculations that enabled him to complete the trio of elliptical planetary laws later incorporated by Newton into modern astronomical physics.

Astronomers judged theories by their ability to predict, and Kepler spent years compiling a new set of planetary calculations, tactfully named the *Rudolphine Tables* to gratify his Prague patron, Emperor Rudolf. It was only in 1631 that Kepler's elliptical model was observationally vindicated, when Mercury passed in front of the Sun exactly as predicted. By then, Kepler had already died, but another enthusiast was defending the Copernican cause—Galileo, seven years older and a far more effective publicist than Kepler, even though he never relinquished his faith in circular orbits. Galileo's physical evidence was repeatedly contested, but he persuaded many astronomers that Copernicus had been right.

Like Tycho and Kepler, Galileo moved from the university environment to the courts, abandoning his poorly paid teaching with relief after successfully soliciting support from the wealthy Medici princes in Florence. Galileo advertised the importance of instruments for finding out the true structure of the Universe, but instead of using bulky Tychonic apparatus that measured angles, Galileo relied on a recently invented optical instrument—the telescope. After hearing reports of this Dutch device, Galileo designed his own more effective version, rapidly impressing the Venetian navy with how far he could see, and discovering myriads of stars that had previously been invisible. But contrary to scientific mythology, Galileo's telescopic images did not immediately convince Copernican critics. Being able to spot ships was very different from making cosmological claims. The blurred views were ambiguous, and Aristotelians objected that a humble terrestrial tube was inappropriate for divining cosmic perfection. Rather than winning automatic acclaim for his results, Galileo gained power tactically, adroitly upgrading himself from a mathematician into a philosophical courtier.

As he manoeuvred himself into the position of court astronomer to Florence's Medici family, Galileo adopted several different strategies. Most obviously, he used his telescope to attack—but not to disprove—the traditional Earth-centred model of the universe. He reasoned by analogy and probability, never managing to produce incontestable evidence that silenced his opponents. To undermine the objection that a giant body like the Earth could hardly be hurtling through space, Galileo claimed that his indistinct pictures of the Moon revealed a rocky surface, nothing like the smooth celestial sphere promised by Aristotle. Instead, he maintained that it resembled Bohemia—and if the Bohemian Moon could move, then why not the Earth? To demonstrate that the Earth–Moon duo is not unique—a drawback of the Copernican system—Galileo found satellites orbiting around Jupiter. His strongest physical argument was to show that Venus sometimes appears to be a circular disc like the full Moon, an impossibility under Ptolemy's model. However, even this phenomenon did not convince Galileo's adversaries, because it is compatible with Tycho's geocentric scheme.

Galileo was an astute campaigner. To ingratiate himself with his aristocratic patrons, he ingeniously named Jupiter's satellites the Medicean stars because—he claimed—they foretold the successful rise of the family dynasty. To spread his ideas more widely, Galileo made flamboyant speeches at dinner parties and wrote persuasive polemical books. Whereas Copernicus had dithered before modestly addressing a complicated mathematical treatise to the Pope, Galileo drew large audiences with his pithy propaganda pieces, leaving out the formulae and boasting with a conjuror's panache of 'unfolding great and very wonderful sights…which, unknown by anyone until this day, the author was recently first to detect.'[8] Even after the Pope warned him to keep quiet, Galileo tried to attract still more supporters by publishing his provocative *Dialogue Concerning the Two World Systems* (1632). This book was revolutionary in style as well as in content—Galileo wrote in Italian instead of Latin, and presented his arguments through a conversation between three thinly disguised fictional characters, caricaturing the mediaeval Aristotelian representative as a simpleton.

Voicing the Pope's own objections through the gullible Simplicio was not a sensible move, and the Pope clamped down on the philosopher who had flagrantly disobeyed his orders, summoning him to Rome for a trial. This was a far less controversial step than it would be nowadays. The principle of free speech had not yet become a political issue, and throughout Europe rebels were routinely silenced in the interests of maintaining stability. Notoriously, Galileo was found guilty, although this was to some extent a show case. His punishment was the mildest possible for an elderly man. Placed under lax house arrest, Galileo continued his research in a comfortable Florentine villa, palming off on his daughter the weekly duty of reciting penitential psalms.

Rather than being a head-on confrontation between science and religion, or even between Galileo and the Pope, this was a complex conflict involving rival factions within and outside the Church. Making the Earth move does contradict several passages in the Bible, but not all Christians thought that this mattered—after all, Galileo was himself a devout Catholic, and he had supporters right through the Church hierarchy. Opinion was similarly divided amongst the supposed 'scientific opposition'. Many astronomers continued to defend either Tycho's or Ptolemy's cosmos, reiterating Aristotle's simple yet persuasive proof that the Earth is stationary: an arrow fired straight upwards lands where it started. Faced with this disagreement amongst scholars, perhaps the Church behaved sensibly by going along with majority opinion and retaining Biblical certainty? Personal ambitions and rivalries were at stake, and if Galileo had manoeuvred more diplomatically, he might have contrived to defend his Sun-centred Universe without being officially condemned. It was only in the nineteenth century that scientists converted Galileo into a martyr during their own struggles for power. Brecht served his own rhetorical ends by perpetuating their simplified view of a hero battling against Catholic oppression, but it was scientific propagandists who launched the notion that science and religion must inevitably be at war.

4 *Bodies*

What thous seest in me is a body exhausted by the labours of the mind. I have found in Dame nature not indeed an unkind, but a very coy Mistress: Watchful nights, anxious days, slender meals, and endless labours, must be the lot of all who pursue her, through her labyrinths and meanders.

—Alexander Pope, *Memoirs of Martin Scriblerus* (1741)

While Nicolas Copernicus searched for God in the stars, Andreas Vesalius was making the human body God's temple on Earth. The Polish astronomer and the Flemish anatomist, two northerners who studied in Italy, regarded human beings as microcosms of the Universe, sympathetically linked together as complementary parts of God's harmonious whole. Their great books—one on cosmology, the other on anatomy—both appeared in 1543, and both authors are now celebrated as scientific revolutionaries. Nevertheless, they looked back towards the past. Like their humanist contemporaries who were reviving classical art and literature, Copernicus and Vesalius wanted to restore ancient knowledge.

Vesalius urged physicians to follow the example set by Galen over a thousand years earlier. Instead of depending on abstract scholarship, he recommended them to study for themselves the best text available—the human body. The son of an apothecary, Vesalius insisted that elite physicians should acquire the skills of working surgeons. Like many other reformers—Roger Bacon, John Dee, Tycho Brahe—he helped to make science possible by encouraging gentlemanly scholars who worked with their heads to recognize the expertise of craftsmen who worked with their hands. When Vesalius sat for his portrait, he posed with a giant dissected human forearm to emphasize his message that doctors should rely on their hands, whose inner beauty he had himself laid bare with his anatomist's scalpel.

As the new Galen, Vesalius had one great advantage over the original: he could dissect human corpses. By following Galen's own advice to look for himself, Vesalius discovered major discrepancies between traditional Galenic anatomy—much of which was based on animals rather than people—and the human corpses that he examined with meticulous care. By using Galen's own methods, Vesalius revised important errors that had been passed down through the centuries by men who pledged their faith in books rather than trusting the evidence revealed by their own scalpels.

Traditionally, medical students trained by listening and watching, not by hands-on experience. They stood below the high official chair of a professor who read out Latin texts, while a surgeon went through the routine of dissecting a corpse, and a demonstrator pointed out important features. Vesalius's first job at Padua after graduating there in medicine was as a low status dissection demonstrator, but he soon overturned this formalized ritual with its three participants.

Vesalius was a flamboyant performer in the theatre of anatomy. As portrayed in Figure 17, the frontispiece of his huge volume of anatomical drawings—*On the Structure of the Human Body*—Vesalius made himself the single central actor. As his right hand pulls back a woman's flesh to display her inner abdomen, and his left hand points up towards God, Vesalius encourages the students to cluster round closely and learn for themselves not only how to identify organs, but also how to operate on them. This deliberately shocking exposure of a woman's body reinforces Vesalius's commitment to uncovering truth, while the skeleton is both a teaching aid and a memento mori reminder of life's brevity. According to Vesalius, the cadaver was a convicted criminal who had unsuccessfully tried to postpone execution by claiming she was pregnant, and the Latin motto at the bottom alludes to Caesar's mythical birth. Vesalius was proud of his origins. At the top, the text reminds readers that he is from Brussels, and the three weasels on his coat-of-arms refer to his pre-Latin name, Andreas Van Wesele. In this image, Vesalius also boasts about his intellectual predecessors by allying himself to the classical past. The two larger-than-life figures at the front are Aristotle, looking down at the dog waiting its turn to be dissected, and Galen, wearing a physician's prescription case on his belt.

Another of Vesalius's innovations was to combine drawings with words. He delved right inside bones and organs, labelling them with tiny letters so that he could refer to them in the written text. Dedicating the same care to his engravings as Dürer, Vesalius enabled distant students to feel like immediate witnesses because he presented his book in the same order as his actual dissections. It also included many detailed illustrations of his equipment, as well as different parts of the body at various stages of exposure. His most famous images show giant skeletons striding across beautiful landscapes or lamenting the prospect of their own death, but Vesalius also portrayed with unprecedented accuracy the divine structure to be found in nerves and veins, muscles and arteries. Using the language of a Renaissance architect, Vesalius described how God had systematically designed the body's foundation and walls; writing as a practical anatomist, he provided instructions for reassembling skeletons that had fallen apart while being boiled to remove the flesh.

FIG. 17 Frontispiece of Andreas Vesalius, *De Humani Corporis Fabrica* (*On the Structure of the Human Body*, 1543).

In his campaigns to reform medicine, Vesalius behaved like Martin Luther, whose rebellion against the Catholic Church formed the backdrop of Vesalius's northern European childhood. Just as Luther revolted against the Pope's authority and went back to the original Bible, so Vesalius rejected conventional teaching and insisted on reading the true text of the body. Whether in altar-pieces or anatomical diagrams, both reformers insisted on displaying the wonders of God's world in pictures. While one restored primitive Christianity, the other resurrected classical anatomy. Following Vesalius's innovations, Protestants produced anatomical posters and preached that God can be approached by understanding His design of the human body—the same moral that Vesalius delivered in his frontispiece of an idealized dissection, where he preached hands-on anatomy to search for God's glory within the gory, smelly confines of a non-refrigerated human corpse.

Vesalius transformed book-bound medicine by insisting that physicians use their hands to study bodies, and by producing realistic pictures. Many physicians objected to the way he challenged established tradition, maintaining that the long-standing authority of Galen was worth more than this recent visual evidence. These sixteenth-century arguments were about where truth is to be found—in books or in bodies. In contrast, modern critics accuse Vesalius of not always being able to see the truth even where it belongs, in the body. Because he thought like a Galenic physiologist, Vesalius did get some details wrong. For example, he maintained that there must be tiny holes in the wall dividing the two sides of the heart, even though he could not see them. Like other scientific heroes, Vesalius did not single-handedly bring about a theoretical revolution—but, just as importantly, he changed people's attitudes and encouraged a more body-oriented style of medicine.

Vesalius's successors ensured that the status of anatomy improved all over Europe, although Padua remained the leading medical school. Local politicians ran the university like a business, enticing wealthy foreign students by hiring the best teachers. By the end of the sixteenth century, Padua owned a well-equipped anatomical theatre, lit by candles and with raked tiers of seating arranged round the central dissecting table to give everyone a clear view. Like Vesalius fifty years earlier, the anatomy teachers taught their students to look back towards the past. But by then, their hero was not Galen, but Aristotle, who taught that the soul is a function of the body. Paduan anatomists studied the body to learn about the human soul—for them, dissection was a spiritual as well as a physical process.

This Aristotelian approach made a huge impact on a young English medical student, William Harvey. Dissatisfied with the low standards at Cambridge, Harvey went to Padua for a couple of years, where he trained under Girolamo Fabrici (Fabricius); back in London, he rapidly rose to the top of the medical profession.

By importing Paduan methods into England, Harvey rejected aspects of Galenism that Vesalius had left unchanged. In Galenic physiology, there are two blood systems—the liver produces blood to supply food via the veins, while the heart heats up air and mixes it with blood for the arteries. Aristotle viewed the heart as the seat of the soul, and Harvey replaced Galen's double model with one single system, establishing that the heart constantly circulates blood around the body.

Worried about being attacked for overthrowing centuries of belief, Harvey experimented on a huge variety of animals for almost thirty years before publishing his *De motu cordis* (*Movement of the heart and blood in animals*) in 1628. Although in appearance an unlikely revolutionary tract—slim, written in substandard Latin, poorly printed—it reorganized the body. Taught by Fabricius, who had inherited Vesalius's insistence on close observation, Harvey followed Galen's own injunction to look for himself. Fabricius had already discovered valves in the veins, but with vision clouded by Galenic theory, he decided that they monitored the supply of food-carrying blood travelling along the veins from the liver. Harvey reinterpreted Fabricius's valves, incorporating them in his unified system as tiny one-way gates ensuring that blood would return from the veins to the heart to be recirculated through the arteries.

Harvey is celebrated as a scientific revolutionary, but he was an Aristotelian who saw an underlying unity between the circular motions of the planets in the heavens, of air and rain in the sky, and of blood inside bodies. As he put it, the heart 'deserves to be styled the starting point of life and the sun of our microcosm just as much as the sun deserves to be styled the heart of the world.'[9] Although now an iconic hero, Harvey's suggestions were only slowly accepted and he had many critics. Traditional physicians immediately condemned his newfangled ideas that made a nonsense of their expert blood-letting techniques, and Harvey's patients deserted their apparently demented doctor, who retreated into his research.

Harvey inherited another Aristotelian project from Fabricius—reproduction. Unlike many of his contemporaries, Harvey rejected spontaneous generation, arguing that living organisms could not suddenly be created as if from nowhere out of ordinary matter. As an ardent royalist, Harvey had access to the King's deer park, where he could observe (and interrupt) sexual activities to find out how new individuals are formed. He also literally studied the chicken and egg problem, carrying out careful experiments to conclude that everything starts with the egg. In Harvey's opinion, an activating power within male semen prompts germ material inside eggs to start following a preordained pattern of development. His opponents, the preformationists, maintained that the new organism is already fully formed and exists at birth in one of the parents (either the sperm or the egg)—a Russian doll scenario implying that the world's entire population had been pre-created by God. Sensible as Harvey's conclusions might seem, spontaneous generation and preformation remained topics of serious scientific debate until the end of the nineteenth century.

Following the gender conventions of his peers, Harvey viewed women as passive partners in the process of conception, and he perpetuated older beliefs by attributing many illnesses to women's wombs, which were held responsible for mental as well as physical complaints (the word 'hysteria' comes from the Greek for 'womb'). According to prevailing Aristotelian concepts, women were weaker thinkers than men, governed by *cold* and *wet* humours and forced to release their excess fluids as dirty menstrual blood. In contrast, men were *hot*, *dry*, rational beings whose actions were governed by their brains.

Attitudes changed only slowly. Disconcertingly for modern feminists, women also thought that their soft brains prevented them from being mathematicians or poets, and mothers routinely advised their daughters that, ruled by their wombs, they should accede to the superior wisdom of their menfolk. Medical opinion started to shift in the 1660s, when a skilled anatomist called Thomas Willis introduced the subversive notion that women are, just like men, dominated by their brains. Moreover, he used his dissections to show that there are no substantial differences between male and female brains. Discrimination did continue, but was no longer supported by Aristotelian rationales.

Willis admired the older Harvey enormously, and in the mid-seventeenth century, both men were living in Oxford, a hugely important city for science because it attracted many radical experimenters. They converged there for political as well as academic reasons. After Charles I was forced by parliamentary rebels to leave London in 1642, Oxford became the major base for royalist supporters until Charles II was restored to the throne in 1660. Allegiances were as vital for natural philosophers as for diplomats, and there are mysterious gaps in the lives of Willis and many other famous figures who effectively disappeared—perhaps abroad—until they judged it safe to resurface. Harvey strategically dedicated his book on circulation to Charles I, portraying the king as 'the sun of his microcosm, the heart of the state; from him all power rises and all grace stems'[10]—an influential stacked image of the Sun/king/heart lying at the centre of the Universe/nation/body that flourished for the next couple of hundred years.

Some of those who survived the political struggles and worked in Oxford went on to become science's most celebrated pioneers—such as the chemist Robert Boyle, the architect Christopher Wren, the naval reformer William Petty, and Willis's own assistant, the inventor Robert Hooke. Willis himself had started out as a desperately poor royalist physician, and he first became renowned for helping to rescue a hanged woman from the anatomy slab and restore her to life. As the university expanded its science education, Willis participated in the circle of young, enthusiastic experimenters, the generation following Harvey that brought fresh ideas into medicine by exploring through doing rather than through read-ing. They learnt from each other, exchanging information about their research into a dazzling variety of topics—including blood transfusion, alchemical trans-formation, weather prediction, wheat germination, microscope improvement, and magnetic variation.

The members of this Oxford group are often said to have laid the foundations of modern science. They owed much to Harvey, the royal physician who placed the English monarchy at the heart of the human body and urged his disciples to observe with their own eyes and experiment with their own hands. Ironically, his successors reformed medicine by supporting controversial French ideas that Harvey loathed, the atomic theories of the French philosopher René Descartes, an early convert to blood circulation. Harvey is celebrated as a radical reformer, but he remained an old-fashioned traditionalist who, reported gossipy John Aubrey, 'bid me goe to the fountain head and read Aristotle…and did call the neoteriques [upstart philosophers] shitt-breeches.'[11]

5 *Machines*

A machine evolves by becoming more efficient, that is, more foolproof; hence the objective of mechanical progress is a foolproof world— which may or may not mean a world inhabited by fools.

—George Orwell, *The Road to Wigan Pier* (1937)

When Copernicus and Vesalius rebelled against their predecessors, they went back to the Greeks and Romans. A century later, René Descartes aimed to make an even cleaner sweep. His policy was to doubt everything, to bulldoze the existing fortress of knowledge and systematically build up an entirely new system from solid, certain foundations. Instead of drawing organic analogies between a healthy body and a perfect cosmos, Descartes used mechanical terminology of billiard balls, whirlpools, and screws. Whereas Harvey viewed the human heart as the sun of the human microcosm, for Descartes—who spent a winter dissecting ox organs acquired from his local butcher—hearts are pumps that drive living machines, and our Solar System is just one amongst a vast number of clockwork universes.

Descartes agreed with Galileo, his older contemporary, that God's Book of Nature is written in the language of mathematics, and he introduced the geometric approach still named after him—Cartesian coordinates. Both men wanted to incorporate practical mathematics within philosophical theories about the world, to unite the expertise of craftsmen and of scholars. Descartes is renowned as a solitary thinker, the scholar who retreated into an overheated room to dream about the certainty of his own existence and coin philosophy's most famous dictum—*I think, therefore I am*. But like a modern scientist, he also engaged with the real world, carrying out experiments and analysing everyday phenomena such as light and the weather.

Just as twentieth-century scientists compared brains first with telephone switchboards and then with computers, so Descartes referred to his own contemporary technology—clocks. Clocks had been introduced at the end of the thirteenth century to ring out religious rituals, but they increasingly regulated daily life and played a vital role in the burgeoning capitalist economy (see 'Scholarship' in

Chapter 2). By Descartes' period, events were routinely measured artificially rather than by the Sun, and technological improvements had led to increased accuracy. Time was becoming money as merchants invested in cheap stock—spices, cloths, grain—that they could store up, manipulating the market to sell at a profit later. Similarly, experimenters tried to cheat time by preserving anatomical specimens or collecting curiosities in museums to be saved, like commercial commodities, for future demand.

Clockwork imagery dominated philosophical thought during the seventeenth century. In one of science's founding myths, when Galileo's attention wandered during a church service, he used his pulse to time a swinging altar lamp—a discovery of regularity that he incorporated not only within a law of physics, but also in his design for a pendulum clock. Craftsmen produced ornate clocks with intricate internal mechanisms, designed for wealthy customers to show off as expensive ornaments. These clocks modelled the cosmos, often having four faces to display astronomical information, such as the movements of the planets. This mechanical metaphor worked both ways—the cosmos was itself said to be like a clock, its concealed mechanism operating smoothly to yield regular circular motion.

Although different natural philosophers held different ideas about how the clockwork Universe might operate, they all agreed that, just like a machine, this mechanical cosmos must have a designer—God. In Descartes's version, God first created matter and then made it move before retiring to let the Universe run automatically of its own accord. Devout Christians rejected this scheme that reduced God's role in His own Universe, and they were still more appalled when Descartes claimed that even living creatures could be explained mechanically. He justified himself with an analogical argument, pointing out that although clocks have been made by human hands, they can move by themselves.

Descartes aimed to reform philosophy, but he ventured into publication only tentatively and sporadically. Aware of Galileo's condemnation, Descartes was determined to avoid controversy and retain the solitude and tranquillity he valued so highly. Originally a French Jesuit, after several years of wandering around northern Europe and enlisting in various armies, Descartes embarked on research in his early 20s. With enough inherited money to support his frugal lifestyle, Descartes spent much of his time in Holland, carrying out experiments, reading, writing, and publishing some of the conclusions he had reached. Eventually, in 1644, after over thirty years of deep thought, experimental investigation, and several unfinished projects, Descartes published four of the six parts he had planned for his *Principia Philosophiæ* (*Principles of Philosophy*).

Intended to provide a comprehensive account of his ideas, Descartes's *Principia* presented the first mechanical version of the cosmos. What, Descartes asked himself, is matter? Trusting nothing, not even the evidence of his own senses, Descartes

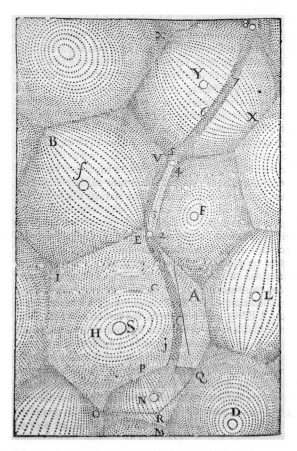

F ɪ ɢ. 18 The path of a comet through Descartes's celestial whirlpools.
René Descartes, *Principia Philosophiæ* (*Principles of Philosophy*, 1644).

concluded that the only incontrovertible thing to say about matter is that it takes up room (more formally, it has extension): matter is place is matter, so that everywhere is full of matter. Rejecting the old universe of empty space, planetary sympathies, and attractive powers, Descartes envisaged a packed cosmos in which matter can only move if it is pushed or pulled. He divided matter into three kinds. The gross third type is what we can see and feel, whereas what we perceive as empty space is filled with invisible smooth second-matter particles—and the gaps between those are plugged by superfine first matter.

Strange though the Cartesian cosmos seemed, it did explain the structure of the Solar System with its revolving planets, and its basic principles became philosophical orthodoxy, especially in France. As Figure 18 illustrates, matter swirls around in great loops or vortices, so that the cosmos is filled by celestial whirlpools formed after God set the entire system in motion. At the heart of each one lies a sun of superfine first matter, condensed there mechanically and surrounded by spinning second matter carrying great chunks of gross third matter—planets. Frenzied activity within the central suns sends out ripples through the second matter to produce light and heat. This diagram also shows how a small lump of second matter—a comet—can avoid being swept up in a vortex, and instead meander from one universe to the next.

When he redesigned the cosmos, Descartes diminished the significance of human beings by opening up the possibility of life elsewhere. The concept of multiple universes was strange and new, but by the middle of the eighteenth century many natural philosophers believed not only that they existed, but also that they provided homes for intelligent beings. Although non-committal about the appearance of these other-world inhabitants, their supporters generally assumed that they were either equal to humans or else placed higher in a spiritual chain running up towards God. Enthusiastic astronomers built gigantic telescopes, but they could detect no definitive signs of life even on the Moon, our nearest neighbour. In the absence of incontrovertible evidence, extraterrestrial life encouraged impassioned conviction on both sides.

These 'plurality of worlds' disputes were waged mainly on theological grounds. Far from being restricted to pedantic scholars, they were of great public interest, discussed in pulpits, dinner parties, and popular books. The central dilemma remained the same for over a hundred years. On the one hand, an infinite number of worlds would confirm God's magnificence, and perhaps provide homes for sinners after death. Extraterrestrial life would also make the cosmos pleasingly uniform, and provide more creatures to worship God. On the other hand, making our Earth just one amongst many raised thorny questions about Christ's uniqueness. It was hard to reconcile multiple inhabited worlds with the fundamental tenet of Christianity, that God singled out the human race for special attention.

As well as redrawing the cosmos, Descartes also changed what happens on Earth. To overthrow occult theories of mysterious invisible powers, he insisted on a mechanical approach. In Descartes' universe, things only move when they are directly pushed or pulled. Resembling billiard balls on a table, particles obey simple laws of mechanics. Although he contrived explanations for many different phenomena—heat, light, weather—he deliberately tackled magnetism, an exceptionally tricky case because it involves movement with no obvious cause. The prevailing explanation had been provided in 1600 by William Gilbert, who suggested that the Earth is a giant magnet that behaves like an animate being with a soul. To refute Gilbert's magnetic cosmology with its unexplained hidden forces, Descartes provided an elaborate mechanical scheme that sounds bizarre but became the showcase of Cartesian cosmology and survived for well over a hundred years.

Magnetism, Descartes claimed, is due to the movement of tiny first-matter particles that squeeze through narrow channels inside magnetic minerals. Like tiny screws, each particle is threaded in one of two directions, so that it can only enter a passage with the corresponding internal grooves. Streams of these corpuscles constantly pour from the Sun towards the Earth, travelling through it along their appropriate pathways—and as they loop round to repeat the circuit, they cluster through any magnets they come across, finding them easier to penetrate than the poreless air. When they stream out of two adjacent magnets, they push them apart; alternatively, attraction seems to be taking place when particles push magnets together from behind. Contrived as that explanation might seem, it sort of worked—and it was a long while before anyone had a better suggestion.

Descartes shocked Aristotelians by eliminating the Universe's magnetic soul. But the move that provoked most hostility was his separation of the human soul from its body, his conclusion that there was some mental Descartes—mind, soul, consciousness that existed independently of his body. Descartes presented organic beings as living machines in which functions such as digestion, respiration, and sexual arousal arise 'simply from the disposition of the organs as wholly naturally as the movements of a clock or other automaton follow from the disposition of its counterweights and wheels.'[12] Even more controversially, he insisted that the nervous system also operates mechanically, so that memory and deliberate actions were also covered by his clockwork model. Descartes declared that animals are machines with no soul, a claim that horrified his critics.

Descartes stopped short of suggesting that humans are mindless mechanisms. Carrying out a thought experiment on an imaginary automaton, Descartes argued that its behaviour could never perfectly imitate that of a person, who must be capable of coping with new situations impossible to cater for in advance. In his view of life, people are different from machines and also from animals because they can talk, reason, and make moral decisions. For Descartes, language was a unique distinguishing feature of human beings—and it remains so today for anti-evolutionists and philosophers unwilling to accept the growing evidence of animal communication.

By retaining the human soul, Descartes accommodated the Christian belief that life continues after death. But he had some formidable objections to overcome. Most urgently, he needed to explain how an immaterial mind can interact with a body made of matter—how can a thought be written down, or conversely, how can looking out of the window prompt a belief that it is raining? This problem was, Descartes feebly admitted, 'very difficult', and he never came up with a satisfactory answer.[13] He eventually decided that the pineal gland, tucked away inside the brain, is the place where the soul processes physical data, almost as though it were a miniature person detachedly watching the body's sensations play out on a screen in front of it. In Descartes' dual system, there is a strange gap between what experience told him—that he had a physical body—and what he claimed to have worked out logically—that he was in essence a spiritual mind. Although he never satisfactorily linked these two together, his machine view of living, feeling bodies remained important well into the nineteenth century.

Paradoxically, mechanical models triumphed because they offered a way to keep God in the Universe. Although many natural philosophers welcomed the way that Descartes had banished occult forces and Aristotelian elements, his opponents denounced him for making it possible to argue that the cosmos is entirely made up of matter, which would remove divine spirituality from the world. Descartes himself never adopted such an extreme position, but some of his successors did. To prevent atheistic heresies from spreading, Christian philosophers revised Descartes' original ideas to produce universes that operated mechanically but still guaranteed the existence of God.

One of the most influential was Robert Boyle, a wealthy aristocrat who belonged to the seventeenth-century experimental group at Oxford and later moved to London. Although he is now most famous as the chemist who gave his name to Boyle's law—which describes how gases behave—Boyle was also a theologian who believed that the whole point of doing natural philosophy was to demonstrate God's splendour and wisdom. Because he liked straightforward explanations, Boyle opted for a universe made of tiny corpuscles. When a collection of corpuscles moves, he explained, you get heat; when you suck the corpuscles out of a flask, you get a vacuum; when sharp corpuscles start interfering, you get the burning action of an acid.

Most importantly, argued Boyle, a mechanical cosmos that runs independently demonstrates God's ingenuity, so that the Great Designer can be studied in His Book of Nature as well as by poring over His Bible. As Boyle proclaimed, 'It more sets off the wisdom of God in the fabric of the universe, that he can make so vast a machine perform all those many things, which he designed it should, by the mere contrivance of brute matter managed by certain laws of local motion and upheld by his ordinary and general concourse.'[14] This is an early example of natural theology. It depends on the argument from design. When you see a watch, you know that a craftsman made it; similarly, the mechanical Universe, with its invisible clockwork mechanisms, must have been created by God.

Natural theologians continued using design arguments to prove the existence of God right through the nineteenth century. For example, when Charles Darwin presented his controversial theory of evolution, natural theologians objected that organs as intricate as the human eye could not possibly be the consequence of chance events, but must have been designed by God. Like their successors, the supporters of Intelligent Design, they found it hard to explain why God permitted short sight and cataracts.

6 Instruments

It is more important to have beauty in one's equations than to have them fit experiment.

—Paul Dirac, *Scientific American* (1963)

Aeroplanes have diminished the Atlantic Ocean to a pond, but during the sixteenth century it resembled a new and enlarged Mediterranean, a much-travelled sea that linked the lands around its edges as ships carried goods and settlers between them. Another important commodity was also being transported in every direction—knowledge. Francis Bacon, an eminent lawyer and politician, became Europe's main champion of progress through exploring and experimenting. Reactionaries remained convinced that classical Greece had been the unsurpassable pinnacle of culture, and they opted for the stable cosmos described in the Bible. Modernizers who believed that they could change the world to create a better future adopted Bacon as their patron saint, and embarked on intellectual voyages to explore the Universe.

Bacon's major manifesto for scientific research was the *Novum Organum* (*The New Organon*), designed to overthrow Aristotle's *Organon* and replace old-fashioned logic with Baconian experimental research. In its frontispiece, shown in Figure 19, two trading vessels, two ships of learning, sail through the mythological pillars of Hercules that straddle the Straits of Gibraltar, gateway between the Atlantic and the Mediterranean. 'Many will travel and knowledge will be increased,' reads the biblical quotation draped between the pillars' feet. Bacon urged reformers to leave behind the safety of Mediterranean classical scholarship, proclaiming 'it would be a disgrace to mankind if…the boundaries of the intellectual globe were confined to the discoveries and narrow limits of the ancients.'[15] Just as merchants profited by transporting goods, so, too, Europe would thrive by gathering information about nature to be packaged, printed, and marketed internationally. Experiments would, Bacon promised, transform discoveries into knowledge and create a utopian New World.

Bacon is often credited with founding modern science, but although well informed about his colleagues' activities, he was better at prescribing what should

FIG. 19 Frontispiece of Francis Bacon's Latin book *Novum Organum* (*The New Organon*, 1620).

be done than in carrying out his own investigations. The anatomist William Harvey remarked acidly that he had the eyes of a viper and wrote philosophy like a Lord Chancellor—perhaps in revenge for Bacon's scathing comments on Harvey's own work. But whatever Bacon's contemporaries felt, his pronouncements deeply influenced scientific research all over Europe. A one-time courtier and Lord Chancellor, Bacon coined the ideal slogan for converting doubters: 'Knowledge is power.' Two centuries later, this Baconian maxim was still being used to solicit government backing for scientific exploration.

Bacon set out an experimental agenda, insisting that the laws of nature could only be uncovered through collecting and organizing massive amounts of data. In contrast with Descartes, who wanted to investigate outwards from the certainty of his own mind, Bacon favoured an inductive, bottom-up approach—inferring explanations from observations untainted by theoretical preconceptions. This was to be a collaborative endeavour, one based on cooperation, communication, and state funding. Bacon envisaged a utopian island community dedicated to investigating ways of harnessing nature's powers for the benefit of society. Although he remained vague about details, Bacon recognized that good observations stem from good instruments, and he envisaged information being collected by teams of researchers organized into separate projects, such as refrigeration, metallurgy, and agriculture. In his hierarchical scheme, humble data gatherers would accumulate facts that their leaders—elite natural philosophers—would digest into scientific knowledge.

Baconian philosophers took over existing instruments, making them more accurate but not at first changing their basic design. Even by the early nineteenth century, there was still no single category of 'scientific instrument'. Instead, instrument-makers divided their wares into three groups, mathematical, optical, and philosophical. The oldest were the measuring instruments essential for everyday activities—weighing food, surveying land, navigating by the stars, assessing precious metals, telling the time, preparing herbal remedies. Made by craftsmen, these mathematical tools had been developed by artisans who needed practical information. Opticians had traditionally concentrated on reading-glasses and nautical telescopes, but in response to demand from experimenters, during the seventeenth century they expanded their range to include microscopes and astronomical telescopes. With their high magnification and better quality glass, these optical instruments revealed details of the natural world that had never before been seen. The last instruments to be devised were philosophical ones—barometers, thermometers, electrical machines, air-pumps—invented for and by experimental philosophers.

These three types of instrument are displayed as interior design accessories in Figure 20, an elegant London drawing room. To the right by the window stand a transit quadrant—used for tracking the path of the Sun, and originally devised by navigators—and an astronomical telescope fashioned for use on land rather than at sea. On the table at the back sits an air-pump, a glass globe that can be emptied of air by a pumping machine—a new and controversial invention by Bacon's seventeenth-century disciples.

The early philosophical experimenters turned to artisans for guidance. One of Bacon's most famous contemporaries was Elizabeth I's physician, William Gilbert, now celebrated as an early scientist, and then acclaimed for improving British navigation by inventing better compasses. When he started to investigate magnetism, Gilbert relied not on his scholarly colleagues, but on the maritime community. Elizabethan navigators wrote in plain English rather than learned Latin, yet their books were packed with technical instructions, Euclidean geometry, and discussions of the Earth's magnetic patterns. Far from being his own inventions, some of Gilbert's ideas and instruments were fancy versions of those he had found in a twenty-year-old book by a practical compass-maker.

Working mechanics provided the instruments and the know-how for mechanical philosophers. The prime example is Robert Hooke, an ingenious experimenter

FIG. 20 The family of John Bacon, by Arthur Devis (c.1742–3).

who worked in Oxford alongside Christopher Wren and Robert Boyle before moving to London, where his jobs including rebuilding the city after the Great Fire of 1666. Since the cosmos operates mechanically, Hooke believed, machines are essential for uncovering its internal operations. He adapted existing devices used by artisans to produce a dazzling variety of new ones intended for natural philosophers—watches, depth sounders, hygrometers, microscopes, air-pumps, scales, lamps, quadrants. For him, instruments were doubly important, because they measured the natural world, and also represented the only way that people could possibly understand it.

Hooke's accurate instruments proved important for science, but he justified them theologically. Like Bacon and many of his contemporaries, Hooke regarded human beings as fallible creatures cursed since the Fall in the Garden of Eden with imperfect senses and prejudiced minds. To perceive the world as it truly is, Hooke maintained, we need artificial aides to bypass the brain and prevent mental distortions. His new microscope would make it easy—all a flawed philosopher had to do was draw with 'a *sincere Hand*, and a *faithful Eye*, to examine, and to record, the things themselves as they appear'.[16] In *Micrographia*, Hooke showed what stunning results could be obtained.

Micrographia is an astounding collection of intricate drawings that exposed previously unimagined details of plants and insects—notably lice, those scarcely visible yet perpetual companions of seventeenth-century gentlemen. When Samuel Pepys bought his copy, he stayed awake all night, captivated by the giant fold-out pictures and Hooke's eloquent descriptions. Writing in English, not Latin, Hooke persuaded his readers to learn about God by studying the Book of Nature. He started unexpectedly, by revealing the jagged edges of razors and full-stops, human-made devices that he contrasted with the 'strength and beauty' of divinely created fleas, which are 'all over adorn'd with a curiously polish'd suit of *sable* armour, neatly jointed, and beset with multitudes of sharp pinns, shap'd almost like Porcupine's Quills, or bright conical Steel-bodkins'.[17]

Whereas Hooke's mathematical instruments measured the visible world, his optical instruments changed its normal appearance. In contrast, philosophical instruments such as air-pumps altered the world itself. The first working models were built in the late 1650s by Hooke and Boyle (although Boyle grabbed the glory), and a century later they had come to symbolize the power of scientific research. The iconic painting by Joseph Wright of Derby—mindlessly reproduced on book covers and greetings cards—portrays a darkened room, lit only by a glowing flask containing a human skull, and dominated by a magus-like natural philosopher. As his hand rests on the globe's stop-cock, his appalled yet fascinated spectators realize that his decision will determine the fate of the rare white cockatoo inside.

Yet the air-pump was not an immediate success. For one thing, leaks proved technically hard to prevent, making it easy for critics to accuse Hooke and Boyle of not having produced a vacuum at all. Fundamental theoretical issues were at stake. According to Cartesian orthodoxy, particles must always be touching other particles, and so it is in principle impossible to remove all the subtle matter from a globe. Moreover, vacuum experiments involved new and apparently perverse ways of arguing. How can you learn about nature by studying an unnatural state? Inside the evacuated globes that looked so ordinary, animals died, candles spluttered, and ringing bells became inaudible. Sceptical mechanical philosophers demanded direct evidence of the gears and wheels moving the hands of the clockwork Universe, not inferences relying on absence.

Boyle insisted that although the air-pump provided an artificial situation, one contrived by human experimenters, it still yielded valid information about God's natural world. Traditionally, natural philosophers had relied on reason, using theories to explain what happens. Boyle, Hooke, and their fellow Baconians wanted to reverse the logical direction, to start with observed facts and move towards explanations of how the Universe works. In their experimental ethos, instruments would demolish theoretical dogmatism by establishing reliable facts. By inventing

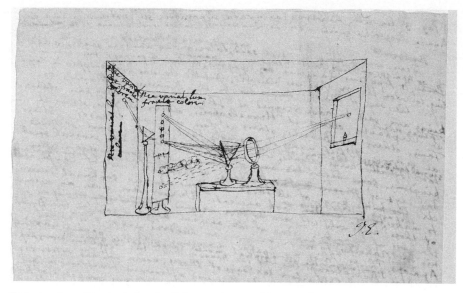

FIG. 21 Isaac Newton's sketch of his experiment with two prisms.

new research tools, they had created new knowledge—wires stretch regularly (Hooke's law), flies have multifaceted eyes, sound needs something to carry it. Developing explanations would come later.

Like books, instruments could transport information from one place to another because the same experiment would (in principle, at least) always produce the same results. Instruments displayed natural phenomena, showing or demonstrating their existence to human witnesses. But 'demonstrating' also had another meaning—verifying an existing theory. When Isaac Newton used a prism for creating a rainbow, he was not trying to reveal an unsuspected effect—aristocrats and their servants knew what chandeliers did to candlelight—but to prove or demonstrate the validity of his own explanation. After buying some cheap prisms at country fairs, Newton converted them into optical instruments for convincing his opponents that he was right.

Newton claimed that he was carrying out a crucial experiment to discriminate conclusively between his own and Descartes' rival theories. According to Descartes, objects appear coloured because they modify the light that passes through them. Newton wanted to advertise his own idea, that the full rainbow spectrum of colours already exists in the light from the Sun. In his clumsy sketch (Figure 21), sunlight streams in from the right through a small chink in his window shutters. Focused by a lens, it then passes through the prism on the table so that rays of different colours spread out to fall on a screen punched with holes. The next stage is the vital one. A coloured beam travels through the screen to a second prism, but shines on the wall unchanged, thus vindicating Newton's contention that colours originate in the light, not in the glass.

Bacon called such crucial experiments 'instances of the fingerposts' that point out the direction of truth. This decisive simplicity is often deceptive—seeing may be believing, but you shouldn't always believe what you see. Scientists claim that because experiments reveal facts, they can be repeated by anyone. However, Newton had concealed so many vital details about his prisms that critics obtained ambiguous results and were still contesting his results seventy years later. Newton wanted his experiments to be as incontrovertible as his mathematical proofs, and he tried to make instruments as persuasive as formulae and rhetorical arguments. Sometimes he felt like giving up, thrust into despair by the antagonism of his opponents and the apparent impossibility of reaching agreement. Philosophy, Newton moaned during a vituperative row with Hooke, 'is such an impertinently litigious Lady'.[18]

7 Gravity

For Newton nevertheless is more of error than of the truth,
  but I am of the WORD of GOD...
For the MAN in VACUO is a flat conceit of preposterous folly.
              —Christopher Smart, *Jubilate Agno* (1758–63)

Apples are a favourite mythological fruit. In *Don Juan*, George Byron linked Newton's inspiration in a Lincolnshire orchard with Adam's temptation in the Garden of Eden:

> When Newton saw an apple fall, he found
> In that slight startle from his contemplation...
> A mode of proving that the earth turned round
> In a most natural whirl called 'gravitation';
> And this is the sole mortal who could grapple,
> Since Adam, with a fall or with an apple.[19]

Newton originated this world-famous anecdote shortly before his death, reminiscing over a cup of tea with a younger friend, who reported that

> the notion of gravitation...was occasion'd by the fall of an apple, as he [Newton] sat in a contemplative mood. Why should that apple always descend perpendicularly to the ground, thought he to him self. Why should it not go sideways or upwards, but constantly to the earth's centre? Assuredly, the reason is, that the earth draws it...there is a power, like that we here call gravity, which extends itself thro' the universe.[20]

More than any other scientific myth, Newton's falling apple promotes the romantic notion that great geniuses make momentous discoveries suddenly and in isolation. His book on mechanics and gravity, first published in 1687, has come to symbolize the birth of mathematical science. Especially after the World War II, it was hailed as an international intellectual bible that would transcend religious differences, the product of a glorious scientific revolution ushering in the modern era. According to simplistic accounts of its impact, Newton founded modern physics by introducing gravity and simultaneously implementing two major transformations in methodology: unification and mathematization. By drawing a parallel between an apple and the Moon, he linked an everyday event on Earth with the motion of the planets through the heavens, thus eliminating the older, Aristotelian division between the terrestrial and celestial realms. As well as unifying the cosmos, Newton united mathematicians with natural philosophers, upstaging Descartes' *Principia* by making his own book a mathematical *Principia—The Mathematical Principles of Natural Philosophy*.

Although Newton was undoubtedly a brilliant man, eulogies of a lone genius fail to match events. Like all innovators, he depended on the earlier work of Kepler, Galileo, Descartes, and countless others—as he remarked snidely to the disabled Hooke, 'If I have seen further it is by standing on ye shoulders of giants.'[21] Hailing him as the creator of modern science is also misleading. Far from being a dedicated physicist in today's sense, Newton searched for God by studying alchemy and the Bible as well as the natural world. And as a further complication, natural philosophers were not immediately converted to his ideas—Newton's model of the cosmos was repeatedly criticized and modified, so that today's Newtonianism is very different from the scheme he originally proposed in the *Principia*.

The apple story was virtually unknown before Byron's time. Instead, Newton was symbolized by a comet, because he was glorified for imposing regularity on what were regarded as flaming meteors, sporadic warnings sent by God to awe a sinful world. When several comets appeared in the early 1680s—including the one now named after the astronomer Edmond Halley—Newton became obsessed by their behaviour, observing them with his own telescope, carrying out endless mathematical calculations, and engaging in vituperative correspondence with rivals such as Hooke and the Astronomer Royal. Secretive and reclusive, Newton was eventually persuaded into publication by Halley, who himself bore the printing costs of a momentous book that the Royal Society felt unable to afford.

Newton deliberately made his *Principia* unsuitable for the mathematically faint-hearted: he wanted, he remarked, 'to avoid being baited by little Smatterers in Mathematicks'.[22] Writing in Latin to reach an international audience of experts, Newton went back to a classical language—geometry. By consolidating and developing the work of Galileo and others, he set out his three laws of motion, which describe how objects such as billiard balls or bullets move and interact. Newton then applied these laws to describe the motion of the planets and minute particles, introducing the new concept of gravity as a universal attractive force stretching out through space to affect comets, apples, and atoms in exactly the same way. Just as importantly, Newton expressed gravity's effects mathematically. According to his inverse square law, the closer two objects are, and the heavier, the more strongly they attract each other.

Some mathematicians were immediately converted, but far more were bewildered. Amongst those who did understand, many were critical, and in response, Newton produced two further editions of his *Principia* (in 1713 and 1726), adding mathematical revisions and also—perhaps surprising to find in science's most famous book—explanations of God and the life-sustaining role of comets. Unlike Descartes, Newton visualized large tracts of empty space not only in the heavens but also between the tiny particles inside apparently solid matter. How, sceptics demanded to know, does gravity travel across the gaps? They accused Newton of bringing back old-fashioned occult forces that mechanical philosophers claimed to be eliminating. Still worse, gravity seemed to challenge God Himself. If gravitational power were somehow inherent in matter, then surely the clear distinction between brute matter and the spiritual world is blurred? The strongest and most enduring objections to Newton were based on religious arguments.

Now celebrated as the world's greatest scientist, Newton was a theologian and an alchemist who retained God and secret powers within his cosmology. In Newton's original philosophy, God pervaded the entire Universe, constantly involved in its activities. Rather than the inert matter favoured by Cartesians, Newton envisaged particles imbued with 'Active Principles' that kept planets revolving and blood circulating without the entire world system grinding to a halt—nature is, he had noted earlier, 'a perpetuall circulatory worker'. Newton derived these ideas not from his natural philosophy library, but from his far larger collection of works on natural magic and alchemy. For him, alchemy was not a diverting pastime, but the crucial route for understanding the Universe and achieving his own spiritual improvement.

Comets made Newton famous because he provided a mathematical explanation of how they behave. By being able to predict their return, natural philosophers gained power over astrologers who interpreted them as dire omens foretelling the end of the world. Nevertheless, Newton did—like astrologers—regard comets as God's agents, sent by Him to restore terrestrial life with special active matter in their tails. Many natural philosophers loathed this view of a God who intervened in the Universe, because it implied that He was a sloppy clockmaker whose original creation had been imperfect. 'Sir Isaac Newton, and his followers, have also a very odd opinion,' protested Newton's arch-enemy Gottfried Leibniz; 'According to their doctrine, God Almighty wants to wind up his watch from time to time: otherwise it would cease to move. He had not, it seems, sufficient foresight to make it a perpetual motion.'[23]

As well as describing the movements of the planets, Newton and his followers tried to explain the behaviour of matter on Earth. They examined a great variety of phenomena—the reflection of light, the behaviour of gases, chemical reactions, plant respiration, electrical activity, animal digestion—and set up mathematical models based on short-range attraction between minute particles. But they soon ran into theoretical problems, such as explaining how a gas of mutually attractive corpuscles can expand. From about 1740, natural philosophers started turning their attention to Newton's alternative explanation of gravity. Inspired by his alchemical investigations, Newton had suggested that special tiny repellent particles pervade the whole of space, making up an invisible, weightless medium capable of transmitting gravity or magnetism, yet rare enough to leave the planets virtually unaffected. This subtle spiritual aether eliminated the objection of action at a distance, and until the early twentieth century, versions of it were routinely summoned up to account for gravity, electricity, and other phenomena.

Newton's early supporters put the problem of gravity's cause on one side, exploring instead how the new theory might be used. Initially, most of the development work was carried out by small specialized groups, including a Scottish mathematical clique and Newton's close colleagues. Whereas Newton put little personal effort into spreading his ideas more widely, some of his followers started to lecture and publish simplified accounts of his work. By the time of his death in 1727, Newtonian Britain was at odds with much of continental Europe, which backed Descartes and Leibniz. Voltaire, an early fan, used this contrast to accuse French scholars of being behind the times: 'A Frenchman arriving in London finds things very different…He has left the world full, he finds it empty. In Paris they see the universe as composed of vortices of subtle matter, in London they see nothing of the kind.'[24] Nevertheless, France remained unconverted well into the second half of the eighteenth century.

Ironically, one of the first important books contributing to Newton's popularity was by a professor at the University of Leiden, Willem's Gravesande, who wrote in Latin and so reached students all over Europe. He commissioned Dutch craftsmen to make wooden instruments—toppling towers, cones rolling up slopes—designed to demonstrate (as opposed to testing or measuring) the principles of Newtonian mechanics. Over in London, Newton's chief experimental assistant, John Desaguliers, recognized the potential profit to be made from marketing Newtonian ideas in this way, and he set up a private school in his own home, where he invented many of his own demonstration devices, using them to teach lecturers who then set up their own Newtonian centres.

Promoting Newton's reputation resembled a marketing exercise. Attracting followers and competing with rivals, Desaguliers sold instruments, lectures, and books to earn his living. Collectively, these individual promotional initiatives helped to create a new public interest in science outside the privileged confines of the universities. Desaguliers was also an enterprising engineer, who designed fountains, mine pumps, and ventilation systems promising to rid London of its 'fuliginous vapours, arising from innumerable coal fires, and stenches from filthy

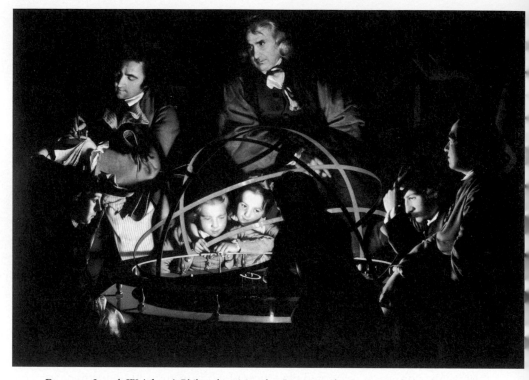

FIG. 22 Joseph Wright, *A Philosopher giving that Lecture on the Orrery, in which a lamp is put in place of the Sun* (1766).

lay-stalls and sewers'.[25] These practical inventions consolidated Newton's prestige by persuading investors and politicians that his theories could be financially rewarding—for themselves as well as for the nation.

Desaguliers and Newton's other supporters were especially proud of an instrument specifically devised to demonstrate gravity—the orrery, a display model of the planetary system. In the romanticized image of Figure 22, a Midlands family group clusters around an exceptionally fine example. The large semicircles originating from Ptolemaic armillary spheres (see Figure 4) heighten the dramatic impact, but the functional part is the flat horizontal base. A central oil lamp represents the Sun and also illuminates this scene in a private library. When the demonstrator (wearing a red banyan, philosophical leisure wear) turns the machine's handle, tiny spheres circle around the imitation sun at speeds proportional to the real planets' movement in the skies. As in other Enlightenment pictures, Newtonianism is also advertised symbolically. The patterns of light on the spectators' faces refer to the phases of the Moon and planets, and the diverse attractions between heavenly bodies are reflected in the varied human relationships—the two children are physically as well as emotionally close to each other, while the adults are loosely spaced out in a circle dominated by the lecturer.

Such visual depictions of the close ties between Newton's orderly law-governed cosmos, God's benevolent rule, and the stable hierarchy of Georgian society were supplemented by verbal expressions in poetry and philosophy. When George II was crowned in 1727, the same year that Newton died, Desaguliers published the verse equivalent of this painting, sycophantically gushing that the British monarch resembled the Sun, extending the power of attractive love over a Newtonian nation. Like Voltaire and other political reformers, Desaguliers envisaged a democratic, Newtonian society made up of free citizens drawn towards one another yet acting independently:

> That *Sol* self-pois'd in *Aether* does reside,
> And thence exerts his Virtue far and wide;
> Like Ministers attending e'ery Glance,
> Six Worlds sweep round his Throne in Mystick Dance…
> ATTRACTION now in all the Realm is seen
> To bless the Reign of GEORGE and CAROLINE.[26]

Newton had concentrated on planets and particles, but his successors carried his mathematical approach into every imaginable aspect of life on Earth. One early enthusiast was a distinguished theologian who (seriously) adapted Newton's laws to conclude that Christ's Second Coming must be before the year 3150. More influentially, natural philosophers applied Newtonian physics to describe living bodies. At first, they viewed bodies as machines operating like hydraulic pumps, but later, they developed models of nervous activity based on Newton's suggestion of an aether, describing how signals travel to and from the brain as vibrations carried through a subtle fluid inside the nerves. Naturalists tried to emulate Newton's unification of the Universe by finding a universal power that defines life. Declaring himself to be the Newton of the human mind, David Hume sought to build psychology on experimental and mathematical foundations, while Adam Smith adopted a similar approach for his economic theories.

By the end of the eighteenth century, Newtonianism dominated intellectual life with the power of religious ideology. Although individuals could—and did—chip away at loose corners of the edifice, professional survival depended on swearing allegiance. Newtonianism had become brand identification for a way of thinking, a scientific creed. For theories to be taken seriously, they had to be labelled Newtonian, even though they differed greatly from one other, let alone from what Newton himself had originally written.

Perhaps the profoundest impact of gravity was to establish the optimistic faith that simple laws govern the cosmos. Not just apples and the Moon, atoms and planets, billiard balls and galaxies, but *everything* could, in principle, be reduced to straightforward mathematical formulae—including the human psyche, the weather, crowd behaviour, chemical reactions, plant growth, and traffic flow. Newton's far-reaching influence seemed heaven-sent. 'Let us render homage to Newton,' urged a leading French medical researcher in 1801, the 'first to discover the secret of the Creator, viz. a simplicity of causes reconciled with a multiplicity of effects'.[27] Newton became the God of Enlightenment rationality, exalted as a hero by the inheritors of the French Revolution, who dreamed of future utopias fashioned along gravitational guidelines.

Institutions

Understanding how science came to form the backbone of the modern world entails describing what happened outside as well as inside laboratories and studies. Science is not just a final product, such as a theorem, chemical, or instrument, but is an integral component of society, interwoven with industry, business, warfare, government, and medicine. Old-fashioned histories of science focus on discoveries and great geniuses, misleadingly glossing over the long eighteenth century as a period when nothing much happened. In contrast, for anyone interested in appreciating how science has become so powerful, then this is the all-important period, the vital transition phase between the private experiments of a few wealthy gentlemen and the public laboratories, state funding, and industrialization of the Victorian era. Experimental entrepreneurs operated like public relations experts. By persuading their critics that investing in them was the most useful and profitable way to proceed, they promoted the societies, career structures, and funding opportunities that came to characterize international science. Institutions may lack the charisma of heroic discoverers, but they were vital for advertising scientific achievements and for attracting financial backing. Without them, the massive research centres and global projects of science would not exist.

1 Societies

Men of science argue, most violently, how the results of their own work could be applied for the benefit of mankind…Yet outside their own departments they are the most arrant diehards. They are, in fact, socialists in their laboratories and Tories in the Athenæum.

—Ritchie Calder, *The Birth of the Future* (1934)

Inspired geniuses provide appealing figureheads, but—as Karl Marx said of philosophy—the point is surely to change the world, not just interpret it. Traditionally, natural philosophers made observations to find out *why* things happen. In contrast, the new experimental investigators of the seventeenth and eighteenth centuries banded together in order to *make* things happen. By creating scientific societies, they acquired the collective power they lacked as individuals. Newton, for instance, became celebrated all over Europe by taking advantage of a ready-made promotional platform—London's Royal Society. Without the Society's support in publicizing his early inventions, experiments, and books, he would have found it hard to gain backing outside his small Cambridge circle. And for the last quarter-century of his life, Newton's position as the Society's President enabled him to dominate English research. But although Newton ruled, it was the Society that put science into society.

In *Gulliver's Travels*, Jonathan Swift won quick laughs by pouring scorn on foolish chemists who tried to make gunpowder from ice and mathematical architects who built houses from the roof downwards. But by the time the book was published in 1726, this derisive attitude had already begun to fade away. Throughout Europe more and more scientific societies were being established, all aiming to demonstrate that experiments bring results. By the nineteenth century, governments were investing heavily in scientific research and inventors were being celebrated as major contributors to the booming industrial economy. Although science was still mostly reserved for relatively well-off men, scientific societies had contributed to a dramatic change, a pervasive explosion in public science of far greater long-term significance than the individual innovations of solitary scholars.

Before the middle of the seventeenth century, intellectual activity took place in private rather than in public settings. University scholars lived in secluded communities, and even the unconventional experimental investigators at Oxford met in each others' rooms. Compared with the Victorian era, there were no public halls or lecture theatres, so scientific debates took place behind closed doors—not only in scholars' studies, but also in collectors' museums, alchemical laboratories, court chambers, artisans' workshops, aristocratic dining rooms, and magi's libraries. Very gradually, the over-riding importance of this private activity declined, and new venues that enabled people to meet together in public appeared.

Among the earliest were the English coffee houses, communal rooms that gentlemen adopted as their second homes for picking up mail, reading the papers, and discussing the latest developments away from family distractions. Other public institutions—theatres, lecture halls, gentlemen's clubs, museums, freemasonry lodges—also flourished. Along with the increasing number of daily newspapers, review journals, and books, they enabled individuals to engage in national debates by expressing their views, acquiring information, and being entertained. Although uneven, this was a Europe-wide phenomenon during the Enlightenment, an era when knowledge and power started to spread outside a narrow elite and the concept of 'public opinion'—now so familiar—started to play an influential role in decision-making. Sources of power slowly shifted. Governments started to take over from monarchs, and public organizations challenged traditional structures of intellectual domination.

The early scientific societies were created as part of this general move towards making knowledge public. Rather than being exceptional, they were just one particular type of these new institutions that allowed more people than ever before to participate in organized discussions. The first to have a major impact was London's Royal Society, initially established by members of the Oxford experimental group—Boyle, Hooke, Wren, and their colleagues—after Charles II was restored to the throne in 1660. Their early meetings took place at Gresham College, a Thames-side navigational centre renowned for its mathematics teaching. As the Fellowship consolidated, they acquired their own meeting rooms near the Strand, the centre of London's booming instrument trade.

Other European rulers soon recognized the status value of acquiring such an intellectual institution, and they encouraged the foundation of their own societies in major cities such as Paris and Berlin. In addition to these national institutions, many provincial towns set up their own smaller societies to discuss literature, science, and current affairs. By the end of the eighteenth century, there were around two hundred—varying in formality and influence—scattered right across Europe and North America. In places as far apart as St Petersburg and Philadelphia, Sweden and Sicily, enthusiasts met regularly to debate the latest scientific ideas and discoveries.

FIG. 23 Baconian ideology in the early Royal Society. Frontispiece of Thomas Sprat's *History of the Royal-Society* (1667).

Many societies modelled themselves on the early Royal Society. From the very beginning, its founders had no doubt about how 'to support our owne Enterprise'—they should 'devise all wayes to revive Lord Bacons lustre'.[1] Bacon had died some forty years earlier, but he features prominently on the right in Figure 23, the frontispiece of the Royal Society's experimental manifesto. As the Society's ideological figurehead, Bacon sits in his official Chancellor's robes, pointing to the instruments that from now on should be the source of knowledge. On the left, the Society's first President, William Brouncker, points towards King Charles II, who is being crowned with a laurel wreath by the goddess of fame. This was conventional visual flattery intended (unsuccessfully) to encourage further royal patronage. Although the shelves overflow with books by the latest scientific authors—Harvey, Copernicus, Bacon himself—the scene is dominated by instruments. The walls are adorned with modified versions of traditional mathematical devices, while in the background (near the King's right ear) lie two modern innovations, a giant optical telescope and a philosophical air-pump.

The Fellows repeatedly emphasized their Baconian ambitions. They aspired to collect observations, establish scientific laws, and use their new-found knowledge for technological inventions to benefit the nation. What happened in practice was rather different. For one thing, although they claimed to have created a democratic Society, in reality it was an elitist organization dominated by educated aristocrats and landowners who formed a new scientific priesthood. Although some instrument-makers did become Fellows, these less privileged men rarely attained positions of power—and women were essentially banned from the meeting rooms until the twentieth century.

Although most metropolitan societies followed London's lead and restricted their fellowships, they reached a large effective membership through their journals, which provided detailed reports of the latest experiments. Each copy had several readers, but even those without direct access could read summaries in the growing number of commercial periodicals (at this time, plagiarism was rife and copyright protection non-existent). By producing this written material, the societies enabled indirect members to become virtual witnesses, almost as if they had themselves been present at the original demonstrations. The societies' emphasis on spreading knowledge through print became a fundamental component of scientific activity.

Letters were also important for converting the societies' discoveries into public knowledge. Both men and women could participate in the Republic of Letters, an imaginary community that linked its intellectual citizens in an extended correspondence network. Collectors exchanged objects as well as information—interesting plants, mineral specimens, new instruments, natural curiosities. Sometimes personal letters were published to reach still wider audiences. For instance, the Methodist preacher John Wesley learnt about the medical value of electrical machines by reading the printed collection of Benjamin Franklin's letters from Philadelphia to London—and Franklin had initially become fascinated by electricity through reading a journal article about some English experiments.

As well as spreading knowledge, the societies distributed money. Traditionally, private patronage had supported natural philosophers who—like Galileo—were not independently wealthy. This moneyed influence very slowly diminished as societies gained power and started to establish other types of funding. In London, the Royal Society financed research to only a limited extent. The impecunious Hooke was employed as an Experimental Curator, but since successive kings declined to provide money, the post's salary remained small, raised by the Fellows from their annual fee. However, the French Royal Society came closer to Bacon's vision of a state-funded organization. Keen to boost his prestige, Louis XIV provided salaries for fifteen experts who met twice a week in the royal library and directed experiments towards matters of national interest. The contrasting structures of the London and Paris Societies strongly influenced the pattern of scientific development on either side of the Channel during the Enlightenment. In France, generous prizes and a secure financial base encouraged theoretical investigations and a scientifically oriented government. But in England, research was more self-interested. Wealthy aristocrats pursued their own lines of enquiry, while enterprising inventors—such as Desaguliers—focused on practical projects to generate income.

Societies gradually devised ways of levering money out of reluctant monarchs. In June 1760, the London Fellows learnt that Britain's traditional enemies, the French, had already organized several expeditions to record the following year's Transit of Venus (similar to a lunar eclipse). To justify their request of £800 from the government, the Royal Society emphasized that national honour was at stake: 'it might afford too just ground to Foreigners for reproaching this Nation [if] England should neglect to Send observers to Such places as are most proper for that purpose and Subject to the Crown of Great Britain'.[2] Although the results proved inconclusive, luckily there was another Transit eight years later, for which the Fellows demanded—and got—four thousand pounds. By the end of the century, President Joseph Banks—an aristocratic autocrat who dominated British science for forty years—was strategically loading the Society's committees with influential politicians who could help to secure state funding. By the time of his death in 1820, Banks had ensured that the Royal Society was intimately involved in Britain's imperial expansion.

Banks was only in his mid-twenties when he experienced at first hand how national interests, government policy, and scientific exploration are intertwined. In the second (1769) Transit of Venus expeditions, often presented as the founding example of scientific collaboration, many national institutions transcended political rivalries to measure the dimensions of the Solar System. However, each Society sent out its own separate team, and only later did they exchange readings. Although Britain and France were for once officially at peace, both countries wanted to control the Pacific region, which offered lucrative trade routes and strategic military bases. The British Admiralty grabbed the opportunity of combining an astronomical expedition to Tahiti with a reconnaissance mission to Australasia, and they sent the captain—James Cook—secret instructions to collect information, seize territory, and hand over all his logbooks when he returned. Banks was a paying passenger who financed his own botanical research—he knew that the government aimed to expand its political possessions rather than its scientific empire.

When historians judge scientific achievement by publications, Banks's couple of pamphlets on sheep farming eliminate him from serious consideration. But for those who think it makes more sense to measure performance by influence, Banks appears as a major innovator who made science a high-status activity permeating politics and trade. Banks introduced two new scientific role models. Through his own travels, and also by securing further funding, he consolidated the stereotype of a heroic explorer, the romantic voyager epitomized in Mary Shelley's Frankenstein: 'I voluntarily endured cold, famine, thirst, and want of sleep; I...devoted my nights to the study of mathematics, the theory of medicine, and those branches of physical science from which a naval adventure might derive the greatest practical advantage.'[3] Banks made scientific exploration glamorous— and also a sound commercial investment.

Banks also personified a scientific type whose importance continued to escalate during the nineteenth century—the scientific administrator. A wealthy landowner and confidante of George III throughout the King's intermittent attacks of insanity, Banks took science to the heart of British politics by making himself and the Society indispensable during his long reign as President. Over 20,000 letters survive, testimony to Banks's control over an international scientific empire. A skilled negotiator, Banks persuaded the East India Company to subsidize a Pacific mapping expedition, but also told the cartographers to bring back commercial information about the Indian market. Financially shrewd, he took advantage of the King's obsession with Kew Gardens to obtain royal funding for a reconnaissance mission that enabled British-owned India to undercut the Chinese tea market.

With Banks in charge, the Royal Society participated in every aspect of imperial expansion, making science inseparable from the international search for raw materials and foreign expertise. Committees to discuss colonial development were stacked with Fellows who placed research high on the agenda, so that it was often impossible to distinguish between commercial espionage, diplomatic activity, and scientific investigation. As one of the few Englishmen who had been to Australia, Banks was intimately involved in establishing the penal settlements. And as the world's most famous botanical explorer, he organized an international network of experimental gardens that transplanted crops and permanently altered the Earth's scenery by converting far-flung lands into European agricultural lookalikes supporting sheep and cows, wheat and barley.

Banks set out to improve the world along the same lines as his own country estate. Like his aristocratic colleagues, Banks believed that he was responsible for maintaining a stable, hierarchical society—he felt it his duty to improve the welfare of those beneath him by increasing his own wealth. For Banks, it was divinely ordained not only that his low-paid Lincolnshire farm workers should generate huge profits for him to spend, but also that African labourers should mine precious minerals to boost Britain's wealth. Where modern critics detect exploitation, he saw reciprocal assistance. In his view, since India was 'blessed with the advantages of Soil, Climate, Population so eminently above its Mother Country', its natural function was self-evidently to supply Britain's factories with raw materials and bind 'itself to the "Mother Country" by the strongest and most indissoluble of human ties, that of common interest & mutual advantage'.[4]

After Banks died, his Victorian successors tried to give the Royal Society a more democratic appearance by obliterating the memory of his authoritarian rule. Seeking prestigious ancestors, they linked themselves directly with Newton, Galileo, and other lone discoverers. But for many scientists, the greatest hero was Bacon, the patron saint of the scientific societies whose collective action had created public science. With his insider knowledge of politics, Bacon had coined the perfect motto to match nineteenth century ambitions: 'Knowledge is Power.'

2 Systems

Attempts to divide anything into two ought to be regarded with much
suspicion.

—C. P. Snow, *The Two Cultures* (1959)

Francis Bacon compared experimenters with ants, scurrying around to gather
up observations for the wise bees—natural philosophers—to digest. But
as books got cheaper and international travel became easier, the sheer mass of
accumulated information became unmanageable. Organization was essential.
Through imposing order, natural philosophers endeavoured to keep unruly facts
under control and convert them into scientific wisdom. The Enlightenment is
often called the Age of Classification, the period obsessed with grouping data,
objects, and knowledge into systematic categories.

Constructing these intellectual filing systems proved hard. Tristram Shandy
Senior spent three years compiling his TRISTRA-*pædia*, an organized system of
knowledge intended to educate his son, but he progressed so slowly that the
first part was out-of-date before he had finished. A less famous fictional pedant,
Dr Morosophus, wasted his life away perusing potted entries from Chambers'
encyclopædia:

> Chambers abridg'd! in sooth 'twas all he read
> From fruitful A to unproductive Z.[5]

Ephraim Chambers's *Cyclopædia*, which appeared in 1728, was England's great
Enlightenment publishing innovation, the first major attempt to marshal human
knowledge into a neat alphabetic sequence. But by the end of the eighteenth century, when Dr Morosophus was boring his colleagues, it had been superseded by
its imitators—the French *Encyclopédie* and the Scottish *Encyclopædia Britannica*.

Successive encyclopædias got bigger, but they also—at least, so claimed their
editors—got better. They chose different schemes for carving up their maps of
knowledge (a favourite Enlightenment metaphor). Chambers, a self-taught bookseller who aimed to edify citizens of the Republic of Letters, admitted that he had
somewhat arbitrarily sliced his territory into Arts and Sciences. He presented him-

self as an intellectual explorer who would guide his readers around the domains of well-mapped expertise and protect them from wandering into the wilderness of ignorance. Even so, modern travellers might soon feel bewildered. The road Chambers signposted 'Rational' leads to religion as well as to metaphysics and mathematics, while optics and astronomy are approached by the same route as falconry, alchemy, and sculpture.

Chambers might have been first, but it was his French inheritors who created the Enlightenment's definitive Bible of Reason. Like Tristram Shandy, they found their project spiralling ominously upwards, but in 1772 they eventually managed to reach the end of the Z's: they had compressed knowledge into twenty-eight volumes. Although cross references to non-existent articles did creep in, the *Encyclopédie* became the French icon of rational taxonomy, modelled as a leafy tree whose sturdy central trunk was labelled 'Reason'. The editors went back to Bacon for inspiration, drastically pruning his original scheme to squeeze theology into a tiny territory while allocating vast expanses to mathematics and natural philosophy. Over the next few decades, these intellectual boundaries were repeatedly redrawn to found the modern configuration of academic disciplines.

Squashed in between cosmology and mineralogy on the *Encyclopédie*'s plan was a relatively new science—botany. Even the word itself was only invented at the end of the seventeenth century, when naturalists first discovered that plants reproduce sexually, and collectors were being overwhelmed not only by countless new species imported from overseas, but also by recent European finds. Although many attempts were made to accommodate the many plants within Aristotelian categories, there were too many anomalies—the plant equivalents of deciding whether bats and duck-billed platypuses should be classed as birds or mammals. Unable to cope, Aristotle's original classification system was finally abandoned.

Although taxonomists proposed many new schemes, none of them satisfied everyone. Like shelving books in a library, there was no right way of organizing the natural world, no objective criteria for deciding which classification system was best. Some arguments were resolved by pulling in powerful patrons to act as referees. One French missionary tried to break down the stranglehold of the Dutch spice trade by growing nutmegs on French-owned territory. However, a rival accused him of importing a different, inferior plant with superficial resemblances. Was it or wasn't it nutmeg? The answer depended on which commercial enterprise a taxonomist belonged to. A similar problem arose in Italy, when a collector flattered his sovereign with the gift of a hermaphrodite monkey. Museum experts disagreed, insisting that it was a normal female—but with so few monkeys available for comparison, how could anyone be completely sure?

One of the earliest Enlightenment classifiers was John Ray, an ex-Cambridge scholar who relied on his friends' generosity to fund his collecting trips through

Europe and introduced some useful new words, such as 'petals' to replace 'coloured leaves'. An invalid sustained by crushed woodlice for his colic, Ray battled for thirty years to publish his massive plant compendium, and was finally forced to economize by omitting the illustrations. Struggling to reconcile conflicting opinions about category boundaries (when does a shrub become a tree?), Ray insisted on considering several characteristics simultaneously, arguing that it is impossible to penetrate beyond tangled impressions of a plant's colour, smell, and feel to discern its internal essence.

Ray suffered a fate similar to that of Chambers: although a classifying pioneer, he is less well known than his successor—Carl Linnaeus. The Swedish equivalent of Joseph Banks, Linnaeus made a couple of brief forays into the Arctic Circle, but then tried to organize the world from his own garden in Uppsala, a small university town. An expert self-publicist, Linnaeus had two major aims: to spread his plant classification system, which is still in widespread use; and to revive the national economy by producing luxuries at home. Just as Banks sat in London corresponding with botanists all over the world, so Linnaeus remained mainly in Sweden but sent out teams of apostles to bring back exotic specimens and preach his taxonomic gospel.

To his opponents' horror, Linnaeus drastically simplified the problem of classifying plants by choosing one single criterion—the number of reproductive organs. His new 'Language of Flowers' was, Linnaeus boasted, so straightforward that even women could understand it. In contrast with earlier schemes such as Ray's, which demanded skilled qualitative comparisons, Linnean taxonomy claimed to be simple and rational because it relied on counting. Linnaeus organized plants into twenty-four classes according to the number of male stamens in the flower. By taking account of the female pistils, he then subdivided each of these classes into less important orders, all numerically arranged.

Linnaeus was formulating a supposedly scientific scheme, yet his text reads like a parody of a Mills and Boon novel: 'The flowers' leaves', he gushed, 'serve as bridal beds which the Creator has so gloriously arranged, adorned with such noble bed curtains, and perfumed with so many soft scents that the bridegroom with his bride might there celebrate their nuptials with so much the greater solemnity.'[6] Although his system might seem objective, it was based on the prejudices of Enlightenment Christian moralists. The fundamental Linnean division is between male and female—the same distinction as in the highly chauvinistic society of eighteenth-century Europe.

By giving priority to male characteristics, Linnaeus imposed onto the plant kingdom the sexual discrimination that prevailed in the human world. His first level of ordering depends on the number of male stamens, whereas the subgroups are determined by the female pistils. Because this anthropomorphic way of dividing the plant kingdom appeared natural, even God-given, naturalists could then argue in reverse: since sexual hierarchies prevail in nature, male supremacy must also—so the distorted logic runs—be appropriate for people. This argument conveniently overlooks how this sexual ordering was inferred from society in the first place. Linnean classification not only mirrored social prejudice, but also reinforced it.

Linnaeus is celebrated as a taxonomist, yet he was also a religious activist, a chauvinist economist who planned to rescue Sweden by using God's laws of nature to boost the country's flagging economy. On his interpretation of the Bible, shared by many of his contemporaries, human beings had a double divine mission—to look after the world, and to exploit it for their own benefit. For many people, maximizing profits took priority over expanding knowledge, and naturalists investigated plants not just out of scientific curiosity, but also for finding ways of turning them into medicines, food, or shelter. Whereas some argued that God had scattered His riches around the Earth in order to encourage international trade, Linnaeus was convinced that God intended Sweden to prosper by growing everything it needed within its own borders.

Analysed from a Eurocentric perspective, Linnaeus ruled an international botanical empire, sending out and receiving letters, people, and specimens while remaining in his central base. But from the perspective of Asian traders selling coffee, tea, and silk, his Swedish emissaries represented gullible customers willing to pay high prices. Other aspects of imperial development during the Enlightenment can be flipped round in a similar way. In British cities, coffee houses sprung up as new social centres enabling the development of a public voice—but they were also commercial ventures initiated by enterprising African and Asian migrants, and their popularity was enhanced by the massive amounts of sugar imported from plantations worked by subjugated slaves. On one interpretation, Britain grew rich by energetically seizing colonial possessions and exploiting their unsuspected riches; on another, eastern merchants in pre-existing networks self-protectively priced British trading companies out of the market, and so pushed them into establishing their own plantations. Britain's commercial empire resembled not so much a wheel with London at its hub, but more an international network of local centres, each negotiating with those connected to it.

The world was starting to be made uniform. As opportunistic growers transplanted crops to more profitable areas, the world began to resemble a single global garden. Banks sent breadfruit from Tahiti to the Caribbean, African slaves took rice to Carolina, European growers moved coffee production from Mocha to Java. While American slaves and African chiefs wore Indian cotton, Indians were growing chilli peppers, tomatoes, and other South American crops distributed by Portuguese and Spanish invaders. In Sweden, Linnaeus persuaded the government

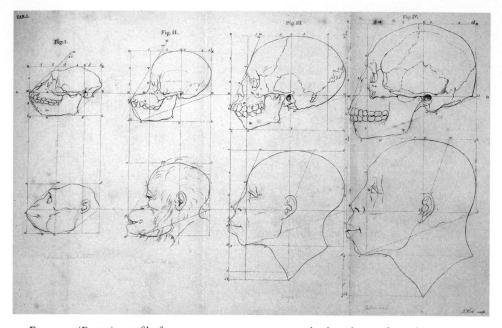

FIG. 24 'Faces in profile from apes, ourangs, negroes, and other classes of people, up to the antique.' Pieter Camper, *The Works of the late Professor Camper, on the Connexion between the Science of Anatomy and the Arts of Drawing, Painting, Statuary*...(1794).

to invest in his ambitious projects, promising northern rice paddies, cinnamon groves, and tea plantations. Linnaeus's initial success in growing Europe's first banana plant helped to win support for his futuristic visions, in which Sweden would enjoy home-grown luxuries that Britain and Holland imported from their foreign empires. Unfortunately for Sweden, Linnaeus's agricultural dreams proved less durable than his taxonomy.

Linnaeus's system prevailed not because it was inherently right, but because, together with his disciples, he persuaded naturalists that it was the most convenient. Although benefiting from powerful allies such as Banks, Linnaeus also faced forceful opponents. British gentlemen were scandalized by Linnaeus's explicitly sexual vocabulary, especially as botany was the one science deemed appropriate for women. Although French botanists could cope with the sex, they believed that it was wrong to constrain nature into artificial categories, and criticized Linnaeus for ignoring many features of a plant to focus exclusively on its flower. Their most influential spokesman was Georges Buffon, a Newtonian mathematician who was also director of the King's gardens. Buffon's 44-volume *Histoire naturelle* (*Natural History*) was the life-sciences equivalent of the *Encyclopédie*, a

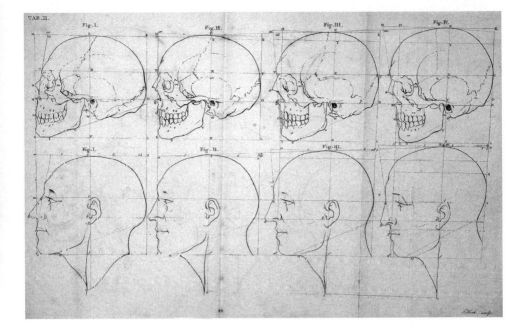

massive, lavishly illustrated compendium of information about the Earth and its inhabitants that was rapidly translated into English and admired all over Europe.

Buffon went back to Aristotle and the Great Chain of Being, envisaging a continuous hierarchy progressing from the very lowest creatures up to complex animals and humans, and then stretching on through spiritual beings up to God. Most importantly, Buffon put history into natural history. Refusing to accept literally the account given in the Bible, he expanded the Earth's past, making some form of change and evolution seem possible. Whereas Linnaeus searched for the divine order that God had imposed during His brief six-day period of Creation, Buffon contemplated a universe that altered over time. Using Newtonian arguments, he portrayed the Earth as a gradually cooling globe that supported life first in the sea and then on land. Breaking with tradition, he classified plants not by their present appearance, but by their parental origins.

Despite their differences, Buffon and Linnaeus both believed in European superiority. Although Linnaeus is heralded as the founder of modern taxonomy, his scientific convictions were rooted in his Christian faith. He regarded his botanical garden as a miniature paradise, divided into four like the original Garden of Eden, and neatly laid out as if he were displaying God's own classification scheme. When Linnaeus extended his system to humans, he grouped them into four races to correspond with the four continents, the four quarters of Paradise, and the four humours governing human health. Linnaeus's top people were the ingenious and sanguine white Europeans; the other three were the melancholy yellow Asians, the idle black Africans, and the happy-go-lucky Red Indians of America.

One obvious blow to Linnaeus's theory was the discovery of a fifth continent, Australia. Towards the end of the eighteenth century, European theories about race were transformed by encounters with other societies and by political debates about slavery. This meant that the fiercest debates were not about the number of races, but about two other related questions: Is there a definite, uncrossable boundary between humans and other primates? And are Europeans intrinsically better than other people? (And if so, in what order are black men and white women?) Abolitionists insisted that all human beings are created equal; to account for physical differences, they argued that people living in different places had gradually adapted to local climate conditions. In contrast, slave-owners tried to justify exploitation by arguing that white Europeans and black Africans were two separate species.

Naturalists set out to resolve such issues by adopting a totally new classification scheme, one based not on personal judgment, but on careful measurements. This numerical approach would, they insisted, make the study of race scientific. Despite their claims of objectivity, these quantitative taxonomists brought subjective judgment right inside discussions about race. Pieter Camper, an eminent Dutch anatomist and anti-slavery campaigner, set out to confirm that there are only superficial differences between the inhabitants of different continents; yet his diagrams support European supremacy (Figure 24). By examining skulls, Camper measured the angle at which faces slope back. After some geometrical adjustments, he ranked them in a continuous line from apes on the left, through Africans and Asians, up to living Europeans and finishing with a statue of Apollo on the right. Although apparently mathematical, an impression reinforced by the grid lines, this scale is an aesthetic one, grading humans by their relative distance from two unrealizable extremes—the grotesque primate and the perfect Greek god. Through arbitrary geometric ranking, Camper stamped scientific credibility onto Aristotle's great Chain of Being.

Camper's quantified classification scheme made racial prejudice scientifically respectable. Since then, many other human characteristics—brain size, for instance—have been measured to justify discrimination between races and sexes on the grounds of inherent physical differences. The Enlightenment is celebrated as the great Age of Classification, when science made sense of the world by organizing it into neat categories. But classifiers had differing priorities, and they could never agree on a perfect system. Like many other aspects of scientific knowledge, consensus was reached through negotiation—and the winning vote depended not only on who put forward the most compelling arguments, but also on who had the most powerful voice.

3 Careers

The Princess will build a hot greenhouse, 120 feet long, next spring
at Kew, with a view to have exotics of the hottest climate, in which
my pipes, to convey incessantly pure warm air, will probably be very
serviceable…What a scene is here opened for improvement in greenhouse vegetation!

—Stephen Hales, letter to John Ellis (1758)

English gentlemen dissociated themselves from the sordid business of earning money. 'It was not for gain,' a rich aristocrat told Parliament, that
'Newton…instructed and delighted the world; it would be unworthy of such
men to traffic with a dirty bookseller.'[7] Such high-sounding ideals were fine for
those who could afford them, but for those with neither a generous patron nor
wealthy parents, practising science entailed working out how to get paid. During
the eighteenth century, scientific entrepreneurs—lecturers, publishers, writers,
instrument-makers—devised ways of selling science for profit. Positive feedback
set in. The more effectively sellers persuaded potential purchasers that science was
useful, the more fashionable science became—and the more rapidly the number
of customers increased. First in England, and later all over Europe and America,
science expanded to become a public, commercial venture.

For the long-term future of science, the most important Enlightenment invention was not any particular instrument or theory, but the concept of a scientific
career. Nowadays, children from any background can (in principle, anyway) follow
a well-defined trajectory through school and university to acquire professional
scientific qualifications and enjoy standard benefits—a steady income, an institutional laboratory or office, subscriptions to journals and societies. No such pattern existed in the eighteenth century, when enterprising philosophers started to
experiment with their own lives and carve out possibilities for surviving through
science. Some individuals did become rich. More significantly, they helped to
create an intellectual elite that challenged the traditional aristocratic hierarchy.
Similar changes were taking place throughout Enlightenment society, as writers,
artists, and musicians struggled to establish profitable, professional positions.

Existing structures changed only slowly, and older networks of power and privilege still survived. For scientific innovators, being a Fellow of the Royal Society helped enormously. Gradually, the Society became less of a gentlemen's club and more of a serious research institution. Although it still included aristocrats and admirals, a growing number gained entry through achievement—members of the new middle classes who liked to class themselves as gentlemen, despite that demeaning need to work. Without the salaries awarded to their Parisian counterparts, many of these scientific entrepreneurs decided to market their books and inventions. Taking advantage of the prestigious initials FRS, they gained patronage and commercial contracts, using the Society to make money for themselves. Collectively, they made science important in society.

The Royal Society engineered one of England's first paid scientific posts—the directorship of the British Museum. Opening in 1759, this state-supported institution displayed not only art objects and books, but also natural curiosities, such as shells, stuffed animals, minerals, and plants. The Fellows made sure that the job of running this public establishment went to one of their own—Gowin Knight, a successful social climber. A poor clergyman's son, Knight was a physician and inventor who won a scholarship to Oxford and skilfully manoeuvred himself into the upper echelons of the Royal Society. Although criticized as a self-serving opportunist, Knight's own bids for status helped to publicize the value of innovation. Not in himself a startlingly significant individual, Knight represents many other Enlightenment entrepreneurs whose combined activities of self-promotion were vital for science's future.

Knight's life illustrates how practical innovations can matter more than ideas. His theories were turgid and convoluted—what counted were his inventions and self-marketing skills. London was the centre of the world's instrument trade, and Knight introduced high-quality steel magnets that sold for a fine profit and brought precision measurement into experimental research. Proclaiming the importance to trade and empire of improving navigation, Knight gained still more status and money by persuading the British Navy to distribute his accurate, expensive compasses. In a typical manoeuvre of mutual advantage, this naval endorsement benefited him personally, but also enabled the Royal Society to boast that science was vital for British trade. Once in power at the British Museum, Knight moulded the public face of science by organizing displays and adopting Linnean methods of classification.

Although more people were becoming interested in science, it only slowly became public rather than private. Access remained limited throughout the eighteenth century. Reflecting the prejudices of his colleagues, Knight restricted entry to the British Museum, making it harder for women and workers to learn about the latest discoveries. Membership of the Royal Society was even more tightly controlled, and depended on personal recommendations. Although some instrument-makers did manage to get elected, the Fellows turned down other applications from many other men who were scientifically knowledgeable but lacked the sycophantic skills of well-bred university graduates.

One candidate who bore a lifelong grudge over his rejection was Benjamin Martin, an influential experimentalist who did much to advertise science by inventing instruments, writing introductory books, and touring round the country giving lectures. Marketing pioneers like Martin were crucial for persuading middle-class people that science was important as well as interesting. Snooty satirists derided them as self-taught philosophers involved in dirty commerce, but despite their lack of formal education, these entrepreneurs changed the status of science in Britain by bringing it into everyday life. Figure 22 illustrates how performers captivated family audiences with orreries, air-pumps, and other devices that stimulated public demand for scientific novelty—and craftsmen responded by expanding the range of demonstration instruments they made for sale. Knowing about science became fashionable. The expensive equipment in Figure 20 is intended to display good taste, not to be used (a point sarcastically emphasized by the artist, who has shoved the globe under the table). To emphasize his cultural sophistication, this house-owner has decorated his wall with plaques of Francis Bacon and Isaac Newton (on the left; those on the right are the poets John Milton and Alexander Pope).

Public engagement affected how science developed. The Fellows of the Royal Society regarded themselves as an intellectual elite, privileged men whose scientific knowledge trickled down to the less-informed. In reality, the situation involved reciprocal interaction. Scientific customers wanted to be educated, but natural philosophers needed to convince potential purchasers that they had something to sell worth buying. This meant directing their research towards generating marketable products—not only theoretical explanations of how the Universe works, but also useful objects such as compasses that improved navigation, or theatrical orreries that educated their audiences at the same time as entertaining them. Rather than a one-way flow of information from top to bottom, producers and consumers were enmeshed in networks of mutual dependency.

Competition was fierce. To entice audiences away from conjurors and theatrical entertainers, scientific lecturers had to create spectacular performances. They soon realized that the most dramatic stage effects were generated by electricity, the scientific marketing success of the Enlightenment. As Martin enthused in one of his educational texts, electricity provided 'an Entertainment for Angels, rather than for Men'.[8] Travelling lecturers held their audiences spellbound with glowing water jets, electrified insects, and glasses of spirits set aflame by the touch of a sword. Wealthy families bought their own apparatus, enabling aristocratic ladies to titillate admirers with electric kisses. At the Hanoverian court, electrical demonstrations replaced dancing, and at Versailles, a ruthless ringmaster entertained the King by making a chain of 180 shocked soldiers leap into the air. London dinner parties were enlivened by electrified cutlery, while Americans planned a feast of turkey roasted on an electric spit.

The history of electricity is full of accidents. Overzealous experimenters suffered nosebleeds or even killed themselves, and the major discoveries were made unintentionally. Even the first electrical machine was an unanticipated by-product of Newton's research into glass and air-pumps, when his assistant—Francis Hauksbee, a draper turned scientific demonstrator—found to his surprise that a rotating evacuated globe acquired an intriguing purple glow beneath his hands. Years later, a Dutch professor was playing around with a jar of water, a gun barrel, and a version of Hauksbee's machine when he gave himself a massive shock—unwittingly, he had invented the Leyden jar, the first instrument to store static electricity. And at the end of the eighteenth century, an anatomist called Luigi Galvani happened to notice that a dead frog's leg jerked in time with a nearby electrical machine, a chance discovery that—after being taken advantage of by much hard work—led to current electricity.

Although electricity was invented inside London's Royal Society, it became important outside because commercial entrepreneurs developed entertaining tricks and useful applications. After Hauksbee published his experiments in the Royal Society's journal, a borrowed copy eventually reached Stephen Gray, a provincial dyer who decided to go to London and make electricity his vocation. Figure 25 shows one version of his most dramatic feat—suspending a small electrified boy from his bedroom ceiling to attract brass filings with his hand. News of Gray's home-based experiment spread rapidly. At first, interest was restricted to the networks of privileged natural philosophers linked with London's Royal Society, but books and journals soon made electrical excitements available to wide audiences all over Europe and north-eastern America. The picture illustrates how the Society's research projects were converted into profitable performances. On the right, one assistant turns the handle of an electrical machine, while another holds his hand on the revolving globe. With his left hand, the hanging boy electrically attracts feathers or small brass filings, while with the right, he communicates his charge to a second recruit, who is protected by an insulating stand. Sometimes girls were involved, adding a sexual tingle of attraction to electrical experiments and their paying spectators.

Another way of promoting science was to make it useful. Optimists promised all sorts of electrical benefits—prolific hens, dryer weather, larger vegetables—but two inventions were particularly important: lightning rods and shock therapy, both endorsed by the political printer Benjamin Franklin. Franklin's kite has become

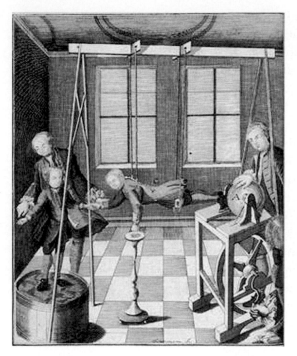

Fig. 25 The hanging boy.
William Watson, *Suite
des experiences et observations, pour
server à l'explication de la nature et
des propriétés de l'électricité* (1748).

the American equivalent of Newton's apple, a mythological story with Franklin cast as the intrepid investigator who braved a thunderstorm to hold an iron key that drew lightning down from the clouds (unlike some of his unfortunate imitators, Franklin prudently insulated his hand with a silk cloth). First in America, and later in Europe, churches, ships, and other tall constructions were—and still are—routinely protected by an iron rod to carry lightning safely into the ground.

Unlike lightning rods, using electricity to treat illnesses now seems cruel and misguided. But at the time, Franklin and other eminent investigators recommended shocks for curing all sorts of ailments, ranging from flu and toothache to insanity and paralysis. There was no established medical orthodoxy, and even the best-trained traditional physicians could do little to prevent pain or cure infections. Practitioners competed with each other for wealthy clients desperate for some sort of help—and many of them wrote affidavits testifying that electrical treatments worked. Nobody had yet officially identified the placebo effect, but electrical medicine was a profitable and respectable business at the end of the eighteenth century.

Most of the electrical physicians were men, and most of their patients were women. One reason for this gender difference is that women were said to be more susceptible to electricity's effects. More significantly, women were—along

with artisans—second-class citizens not only in political affairs, but also in intellectual activities. In the drawing room of Figure 20, the father and his elder twin
son are on the scientific side, allied with Bacon and Newton, while the mother
and daughters are in the poetic realm, along with the younger twin boy building
a fragile house of cards to indicate that chance has eliminated his inheritance. As
science became fashionable, women were cast in the role of spectators, able to
understand knowledge, but not allowed to create it. In one of Benjamin Martin's
most successful books, an Oxbridge student spends his vacations demonstrating
experiments to his sister, condescendingly providing simple explanations while
she enthuses about his brilliance. The subtext is clear: if even sisters and daughters
can understand science, then their neighbours in the intellectual class system—
poorly educated men—will also be able to follow the arguments.

Learning from the example of Martin and other popularizing authors, towards
the end of the eighteenth century some women broke with convention by deciding to write their own books and earn their own money. Abandoning Martin's
patronizing older brother, they created female authority figures, motherly
governesses who gave their young pupils moral advice, gently guiding them towards
the beauty and order of the natural world. Although women were excluded from
universities and laboratories, they played a vital role in science's development by
making information about experimental research available to a far wider range
of people than ever before. Some of their books became international bestsellers,
influencing readers who went on to become professional scientists. For example,
Michael Faraday became world-renowned for introducing electric fields; yet he
always paid tribute to Jane Marcet, the author of the chemistry textbook—
disguised as conversations between a mother and her children—that had first
convinced him to go into science.

Faraday is now remembered as the heroic founding father of the electrical industry,
but his career as a salaried scientist could never have happened without the entrepre-
neurial initiatives of the eighteenth century. In 1711, the fictional Mr Spectator had
recommended public access to science, declaring that 'I shall be ambitious to have it
said of me, that I have brought Philosophy out of Closets and Libraries, Schools and
Colleges, to dwell in Clubs and Assemblies, at Tea-Tables, and in Coffee-Houses.'[9]
Traditional hierarchies were slow to break down, but a century later his dream had
been partly realized. As a blacksmith's son, Faraday had absolutely no prospect of
going to university, but after reading Marcet's chatty book, he engineered his way
into science as the assistant of Humphry Davy, celebrated chemist and President of
London's Royal Institution, which had been set up at the very end of the eighteenth
century to encourage scientific research and education. After Davy died, Faraday
himself became President—a romantic rags-to-riches story that Mr Spectator's contemporaries in the previous century could never have contemplated.

FIG. 26 *Scientific Researches! – New Discoveries in Pneumaticks!—or—an Experimental Lecture of the Powers of Air.* Hand-coloured etching by James Gillray 1802.

Faraday was exceptional—entrenched prejudices died only slowly. Although he managed to escape his impoverished childhood and follow a scientific career, many privileged people both despised and feared upward mobility and equal opportunities. When the Royal Institution was first built in 1801, it had a discreet stone staircase enabling workers to enter separately and sit in the gallery away from their employers. This democratic stairway to higher education was soon demolished. As indicated by James Gillray's caricature of Figure 26, audiences were restricted to affluent paying customers, derided here for their assiduous note-taking at London's latest fashionable entertainment—chemistry experiments.

However solidly established science might be today, two centuries ago its status was unclear. The lecturer brandishing the bellows is Humphry Davy, now celebrated for discovering new elements and inventing the miner's safety lamp, but then frequently vilified for importing French chemistry that threatened to blow up the establishment. Gillray's scene satirizes a real event that went wrong, when a guinea pig from the audience so enjoyed the effects of laughing gas (nitrous oxide, only later used as an anaesthetic) that he refused to stop inhaling. On more successful occasions, Davy—a flamboyant performer—demonstrated that he could control the forces of nature with his chemical and electrical instruments. Working hard at self-promotion, Davy styled himself as an experimental genius and ended up as President of the prestigious Royal Society.

Nevertheless, reservations about science persevered, and there was still no fixed identity for the men who practised it—the word 'scientist' had not yet been invented. In an expression reminiscent of Bacon's desire to dominate the world by changing it, Davy boasted that experiments enabled a man 'to interrogate nature with power, not simply as a scholar, passive and seeking only to understand her operations, but rather as a master, active with his own instruments'.[10] But Davy also cautioned his audiences that ambitious scientific speculators might promise too much.

It was a woman, Mary Shelley, who best captured some of these ambiguous attitudes. After immersing herself in Davy's published *Lectures*, Shelley created Victor Frankenstein, a product of her imagination who represented the Janus faces of experimental science. Echoing Davy's warnings, Shelley captivated her readers by articulating their own ambivalent feelings towards scientific research. Nowadays, *Frankenstein* is often interpreted as a prescient warning of science's dangers, especially the atomic bomb. But Shelley was exposing the uncertain status of science in her own time. Although Knight, Martin, and countless other enterprising philosophers had sold science to the public, in the early nineteenth century many customers were still reluctant to buy.

4 Industries

> I promise to pay to Dr. Darwin, of Lichfield, one thousand pounds
> upon his delivering to me (within two years from the date hereof) an
> Instrument called an organ that is capable of pronouncing the Lord's
> Prayer, the Creed and Ten Commandments in the Vulgar Tongue, and
> his ceding to me, and me only, the property of the said invention with
> all the advantages thereto appertaining.
>
> —Matthew Boulton (3 September 1771)

By the 1830s, so many famous Britons had been buried in Westminster Abbey that space was running out. When the giant statue of James Watt was levered into place, critics protested at its incongruous style, but they were easily outnumbered by hagiographers celebrating him as a modern Archimedes. While Archimedes's 'Eureka!' moment had taken place in the bath, Watt was only a child when he watched the lid rise on a boiling kettle, an observation that (supposedly) inspired him to design steam engines for powering heavy machinery. According to the eulogists, Watt's engine had not only made Britain the world's leading industrial nation, whose cheap manufactured goods were benefiting the entire world, but also ensured her victory over France in the Napoleonic wars.

The inscription on Watt's Abbey monument—'an original genius early exercised in philosophic research'—was a compromise.[11] Fashionable Londoners found it hard to admit that the nation's wealth stemmed from northern factory owners, and they looked down on business entrepreneurs who aimed to accumulate money rather than knowledge. Instead, they preferred to think of Watt as a born genius, the self-taught son of a Scottish shipbuilder who had risen through brains and dedication. Conversely, Watt's manufacturing colleagues were concerned about the low status of engineers, and so they wanted to promote him as a serious scientific thinker. These opposed campaigners from contrasting social backgrounds eventually settled on a cross between an inspired engineer and a scholarly scientist.

Watt became the hero of industrialization, celebrated for converting steam engines into money-making machines. But one man's inventions, however important, do not in themselves explain why industrial change started far earlier in Britain than in the rest of Europe—around the middle of the eighteenth century. Part of the answer lies in the country's rich natural resources. Enterprising developers benefited from local supplies of iron, coal, and wood, the essential raw materials needed for automating manufacturing and agricultural processes. Just as significantly, Britain profited from her imperial possessions and the global circulation of people, wealth, and goods, kept running with gold mined by slave labour in Africa. To satisfy their growing overseas markets, British manufacturers had to invent more efficient ways of converting cotton and metals—cheap imports from Africa and Asia—into fine cloths and luxurious ornaments for North Americans, who paid with plantation crops produced by enslaved Africans. Britain's industrial wealth depended on oppressing not only the working classes at home, but also her colonial subjects around the world.

Britain's appearance was permanently altered in the eighteenth century. In the interests of large-scale efficiency, landowners abolished the traditional system of small personal allotments, replacing them with large open fields. To bring in raw materials and send out finished goods, factory owners commissioned cross-country canals and invested in large, paved roads. Displaced workers gravitated towards employment possibilities, so that for the first time, northern centres became larger and more important than provincial ports and cathedral cities in the south. Whereas wealth had previously depended on inheritance and agriculture, by the early nineteenth century, self-made industrialists were richer than many aristocrats.

Victorian critics expressed their horror at belching chimneys, noisy trains, and dilapidated slums, castigating prosperous employers who ignored the dirt, sickness, and poverty they inflicted on their labourers. But the eighteenth-century entrepreneurs who first introduced new manufacturing techniques were unaware that their innovations would have such deleterious effects. Although their main aim was to increase their own profits, they did also believe in progress. Machines would, they claimed, not only improve their own positions but would also bring more opportunities to their workers and to the nation. Paternalistic landowners predicted that steam automation would benefit their employees by alleviating the drudgery of manual work. It is only with hindsight that their confidence seems naively optimistic, a self-justifying excuse for exploitation.

During the early stages of industrialization, many writers and artists regarded bridges, canals, and mills as enhancing rather than despoiling the natural scenery. Their picturesque visions are epitomized by Figure 27, which shows the world's first cast-iron bridge at Coalbrookdale. The valley's coal and iron made it a natural choice for building refineries, whose produce could easily be shipped along the river Severn to the Atlantic port of Bristol. This scene is a paean to provincial

FIG. 27 William Williams, *View of Ironbridge* (1780).

progress, presenting the bridge as a wonder of the modern world and lauding iron as the versatile material of the future. The bridge's artificial structure fits harmoniously within its naturally idyllic setting, the arch's reflection exaggerated to form a circle, emblem of divine perfection. As the river meanders serenely through the gorge, its serpentine bends framed by gentle wooded slopes, the only indications of pollution are a few puffs of smoke.

In contrast with this tranquillity, the iron works at Coalbrookdale also acquired an exotic grandeur that simultaneously appalled and thrilled. In art and literature, Midlands factories were portrayed as awesome, sublime wonders—like ruined abbeys, they provided the British man-made equivalents of precipitous Alpine gorges or fiery Italian volcanoes. The home tourist industry blossomed as intrepid Londoners travelled northwards to admire the ambiguous fascination of Coalbrookdale. One southern visitor marvelled that 'the noise of the forges, mills, &c. with all their vast machinery, the flames bursting from the furnaces with the burning of the coal and the smoak of the lime kilns, are altogether sublime and would unite well with craggy and bare rocks.'[12]

Enlightenment Britain is often called the Age of Newton, a label that makes sense only if you include practical Newtonian machines alongside abstract Newtonian physics. Rationality and politeness may have prevailed at dinner parties, but this period was also marked by upheaval, dirt, and ingenuity. During the second half of the eighteenth century, while natural philosophers boasted about their orreries, electrical machines, and air-pumps, industrial researchers were advertising the far more useful and profitable products being generated from their own experiments—teapots, soap, jewellery, dyes.

While London's Royal Society was making itself indispensable for the government's programme of imperial expansion, some of its Fellows also belonged to another important select brotherhood—the Lunar Society. Gathering together from all over the Midlands, the Lunar men met at each others' homes once a month on the Monday of full moon, when their journeys home would be well lit over the unpaved roads. Founded in around 1750, this informal group had a floating membership with a central core of about a dozen close colleagues whose interests ranged over a wide range of topics, including geology, medicine, education, engines, electricity, chemistry, ballooning, botany, and silverware. No minutes of their meetings survive, but their letters reveal a fertile interchange of ideas between men who held very different interests but who were all committed to a single overriding goal—progress.

In pledging themselves to progress, the Lunar men were not exceptional, but reflected a prevailing mood of optimism that Britons were learning more and behaving better, becoming healthier as well as wealthier. Sceptics objected that too much luxury inevitably resulted in degeneracy (after all, look what happened to the Roman Empire), but they were outweighed by enthusiastic economists who argued that industrialization would benefit purchasers as well as producers. In their view, prosperity was self-reinforcing—the richer you became, the harder you worked to earn even more money. And the same was true on a national scale, so that the country as a whole would be improved thanks to the efforts of profit-seeking manufacturers.

For the Lunar Society, progress also involved improving the way society was organized. When the chemist Joseph Priestley declared that 'the English hierarchy...has equal reason to tremble even at an air pump, or an electrical machine' he threatened that technical innovations had political implications—machines would alter for ever who owned the wealth and power.[13] Fired by utopian zeal, these early industrialists promised that greater prosperity would be for the benefit of all, yet they seem to have been unconcerned about job satisfaction. The Edinburgh economist Adam Smith insisted that efficiency could be increased by dividing production up into successive stages, so that each worker was allocated a minute repetitive task rather than taking responsibility for creating a finished item. Following Smith's advice, the pottery owner Josiah Wedgwood vowed 'to make such *Machines* of the *Men* as cannot err'.[14]

The members of the Lunar Society met as equals but they are commemorated differently. Some of them have been set up as scientific founding fathers—Dr Erasmus Darwin, grandfather of Charles and also a writer on evolution;

Joseph Priestley, a self-taught clergyman who carried out chemical experiments on gases (and innocently sold his recipe for soda water to Mr Schweppes); Dr William Withering, who converted a wise woman's herbal remedy into a potent heart medicine. As 'an original genius early exercised in philosophic research', Watt straddles the scientist–engineer boundary. In contrast, their colleagues who changed the country by promoting commercial enterprise—Josiah Wedgwood, James Keir, and Matthew Boulton—are classified as manufacturers, and so have been relegated to a low place on the scientific status scale.

These divisions between science, technology, and commerce stem from the outdated snobbery of the rivals debating Watt's inscription. Although Wedgwood, Keir, and Boulton can be described as provincial industrialists, this categorization glosses over the fact that they were also elected Fellows of London's Royal Society. Wedgwood might be a commercial opportunist who advertised himself as 'Vase-Maker to the Universe', but he was also a meticulous experimenter who outpaced his rivals by systematically analysing clays, minerals, and pigments, recording the results in his secret laboratory notebooks. As the Royal Society appreciated, even though Wedgwood had originally developed his high-temperature thermometer to monitor kilns, it was valuable for a great range of scientific investigations. Keir built up a personal fortune from his soap factories, but he was an international expert on crystals whose lucrative contribution to national health and hygiene depended on detailed chemical investigations.

United by their enthusiasm and drive for improvement, the Lunar men had a major impact on British life. Boulton, a Birmingham factory owner, shared the Baconian ideals professed by the Royal Society—he explained to the Scottish writer James Boswell that 'I sell here, Sir, what all the world desires to have— Power.'[15] Steam powered Boulton's machines, which were themselves engineering shifts in social power. The Lunar men's new-found prosperity enabled them to challenge traditional hierarchies by marrying into the aristocracy, buying up land, and building luxurious houses for themselves as well as cheap accommodation for their workers. They campaigned for a democratic educational system that would make intelligence, not birth, the route to success. Darwin and some of the other members even sponsored better education for girls (although they assumed that the men would remain in charge). In his *Chemical Dictionary*, Keir aimed to make information freely available so that his readers could make up their own minds. He wanted 'the public of all nations and of all times, to decide with a full knowledge of the question'—chemistry should be for the people, not reserved for a privileged elite.[16]

Reinforcing these promises of equality, manufacturers persuaded their customers that they could buy similar products to those owned by their betters. Or to express their advertising tactics in modern jargon, they promised upward mobility through purchasing material goods. Consumer societies are based on the assumption that commodities are what make life worthwhile, and this new approach to happiness through ownership was initiated by the great marketing innovators of the eighteenth century. Wedgwood was a superb potter, but his biggest coup was to create desire, to convince customers that it made sense to replace serviceable china with his latest designs—and then to repeat the upgrade a few years later. By lowering his prices, Wedgwood constantly expanded the number of purchasers willing to work harder so they would have enough money to buy cheap imitations of aristocratic china.

Despite their democratic claims, the Lunar Society still believed that some men (them, for example) should be more privileged than others. Similarly, although they often relied on their wives' contributions to their businesses, there was no serious consideration that women should be made equal partners. Darwin celebrated automation in a long florid poem, adding extensive footnotes crammed with technical details to celebrate the innovations of his Lunar colleagues, but he omitted workers and women. This is part of Darwin's lyrical tribute to innovations in the cotton industry:

> Slow, with soft lips, the *whirling Can* acquires
> The tender skeins, and wraps in rising spires
>
> .   .   .   .   .
>
> Then fly the spoles, the rapid axles glow,
> And slowly circumvolves the labouring wheel below.[17]

In Darwin's hymn to machinery, the revolving machines resemble the harmonious planetary movements described by Newton. But it is the wheel that works. The native human labourers and the colonial slaves are glossed over, unmentioned. Darwin simply ignores the devastating effects of mechanization on skilled spinners and weavers—women as well as men—who were being replaced by a single male supervisor in charge of one machine. The factory owners who promised finer products repeatedly moaned that their workers were unreliable, uncooperative, and failed to replicate the perfection of automated equipment. When riots broke out, Watt and Wedgwood invoked the marvels of modernization, but seemed to forget their own impoverished backgrounds. They showed little sympathy for the underlying causes of unrest—hunger, long hours, unemployment.

The early industrialists were committed to progress, but by the mid-nineteenth century, reformers were campaigning for different improvements. Whereas Darwin had conveniently forgotten about the women and the workers, a new generation of writers were exposing the appalling conditions in factory slums. In 1842, when a trainee textile manager ventured into a lower-class district in Manchester, he discovered that 'the atmosphere is…laden and darkened by the smoke of a dozen tall factory chimneys. A horde of ragged women and children swarm about here, as filthy as the swine that thrive upon the garbage heaps and in the puddles.'[18] His name was Friedrich Engels, co-author with Karl Marx of *The Communist Manifesto*. Looking back to the mid-eighteenth century, Engels explained that Britain had experienced an industrial transformation whose true significance was only beginning to be understood. If he had been able to peer forwards, he might have been surprised by the revolutionary impact of his own work, which historians are now in their turn struggling to understand.

5 Revolutions

The most radical revolutionary will become a conservative on the day after the revolution.

—Hannah Arendt, *The New Yorker* (1970)

The French Revolution transformed the course of history—and it also changed how history was itself conceived. In year III of the French Revolutionary Republic (1794), a Parisian industrial spy returned from a secret reconnaissance mission into British factories to report 'that a revolution in the mechanical arts, the real precursor, the true and principal cause of political revolutions was developing in a manner frightening to the whole of Europe'.[19] In delivering this rousing message about industrial transformation, the spy gave 'revolution' its latest fashionable sense—instead of the cyclical movement of the planets around the Earth, he was referring to an abrupt, irreversible change of any kind. Since the French Revolution, many historians—political, economic, scientific—have adopted this revolutionary metaphor, constructing the past as a series of dramatic ruptures.

In analysing science's past, the Chemical Revolution often features as one of these sudden transitions. It seems doubly special because it coincided (well, more or less) with the American and French Revolutions, and its main protagonist, Antoine Lavoisier, announced at the time that he was a revolutionary. Like a political agitator, Lavoisier planned his tactics carefully, secretly recording how he intended to revolutionize science. At last, in 1789, the year the French Revolution erupted, he published a book announcing that he had overturned the old-fashioned chemical theories of Joseph Priestley and his English colleagues. Figure 28 illustrates this heroic version of Lavoisier, here gazing up at his wife, Marie Paulze, as though she were his scientific muse while he corrects the proofs of his textbook, his chemical manifesto in which he introduced new chemical names and symbols similar to those in use today. The instruments prominently displayed on the table are for producing oxygen, while those shimmering at his feet emphasize the importance of precise measurements. Meticulously painted, they symbolize Lavoisier's victory over his English rival.

Verbal equivalents of this picture are similarly dramatic, demoting Priestley into a naive blunderer who believed in an imaginary substance called phlogiston,

FIG. 28 Marie Paulze and her husband Antoine Lavoisier, by Jacques-Louis David (1788).

and elevating Lavoisier into an incisive, methodical genius who discovered oxygen and eradicated ridiculous old-fashioned concepts. Originally introduced in German mines (no coincidence that the Nazis destroyed Lavoisier's statue), phlogiston was widely used to explain burning and metal refining. Despite all the mockery, in some circumstances, the theory worked well. When ores (oxides in modern terminology) are heated with charcoal, they absorb phlogiston and turn into metals; when metals are heated, they release their phlogiston (visible as a blue shine on the surface) and turn back into ore. Problems started when chemists introduced more accurate balances, making it hard to explain why metals should gain weight when they are heated and release phlogiston—surely they should weigh less?

Lavoisier's innovation was to turn the process round, to suggest that metals absorb oxygen while ores release it. After heating some powdered mercury ore by focusing sunlight with a lens, Lavoisier collected the gas given off, tested it to eliminate other possibilities, and then invented a new name for it—oxygen. But…there are several objections to this dramatic account of his victory over Priestley. For one thing, both chemists isolated the same gas, but—like historians analysing the past—they interpreted it differently. And Priestley got there first—what Lavoisier christened 'oxygen', Priestley had already labelled 'dephlogisticated air'. Phlogiston's major source was dirty charcoal, and Priestley associated it with impurity, priding himself on having produced a refined air with marvellous life-sustaining properties (he had no qualms about timing how long it would take a mouse to suffocate in various gases).

Even Lavoisier himself thought that his revolution was about far more than identifying oxygen. He aimed to reform the whole of chemistry. A conscientious tax collector and lawyer, Lavoisier insisted on reason and order, balancing the two sides of an equation as though he were reconciling his accounts, and emphasizing the importance of precise measurement. To accompany France's new mathematical language of algebra, Lavoisier introduced a logical chemical vocabulary. Traditionally, substances had been referred to by vernacular names based on their source or their properties, but Lavoisier substituted Latinate labels that could (he claimed) be understood all over the world. Epsom salts, for instance, went international as magnesium sulphate.

British experimenters resisted Lavoisier's recommendations not because they were reactionary chauvinists, but because they favoured a different style of research. Priestley appreciated the value of unanticipated observations, and so he criticized Lavoisier for systematically planning every step in advance, thus making it impossible to learn from results along the way. In France as well as in Britain, Lavoisier's opponents accused him of leaping ahead too quickly, of proceeding deductively from a few facts to general conclusions, and relying too strongly on complicated instruments that might be introducing errors of their own. From their perspective, Lavoisier was setting himself up as a privileged expert who depended on expensive apparatus and used sophisticated words unfamiliar to people who worked with chemicals every day—apothecaries prescribing Epsom salts as a laxative, or artisans making soap and glass from ordinary soda (sodium carbonate under the new regime).

In France, Lavoisier became an icon of revolutionary chemistry not because he was indubitably right, but because he persuaded influential people that he was. Together with his wife, he waged an extensive publicity campaign, producing books and lectures, plays and pictures to defeat the opposition and promote his own ideas. After Lavoisier was guillotined by the Jacobins for his financial dealings, his followers, who had been unable (or unwilling) to save him, ensured their own futures by arguing that his new chemistry was vital for enabling France to lead the world. They celebrated Lavoisier as a revolutionary hero, even staging a mock funeral that attracted three thousand mourners. Like Galileo, Lavoisier became a mythological martyr to science, the iconic figurehead depicted in Figure 28, a dedicated chemist whose revolutionary science bore little relationship to practical concerns.

But that is not the only way of portraying Lavoisier. For example, the portfolio in the back left of this double portrait conceals his wife's drawings, which show Lavoisier not as a lone genius, but as the director of a collaborative laboratory team in which Paulze herself plays a vital role. From the Jacobins' point of view, Lavoiser was a wealthy landowner who exploited the poor—which is why they imprisoned and executed him. In contrast, his friends esteemed Lavoisier as a radical reformer so committed to improving the conditions of farmers and factory workers that he poured his own money into overhauling agricultural and manufacturing methods. And when historians look back, some represent Lavoisier as a practical innovator who improved Paris's street lighting and water supply, while others accuse him of being a dogmatic theoretician who, by modern standards, made curious mistakes—calling light and heat chemical elements, or declaring that oxygen is an essential component of all acids (a common exception is hydrochloric acid).

Heroic stories credit Lavoisier with single-handedly creating modern chemistry. More realistic versions depict him as one amongst many who gradually transformed alchemy and other skilled crafts into the scientific discipline of chemistry by inheriting and modifying their predecessors' techniques. These transitions are symbolized in Figure 29, which shows a laboratory in Kingston (near London) designed specifically for chemical research around the middle of the eighteenth century. The drawings on the wall—piped water on the left, a glasshouse in the alcove—emphasize that chemistry was important because it was useful. The left-hand side is dominated by furnaces, developed by alchemists and used for refining metals—the mining context in which phlogiston originated. Ranged on the upper shelf and the central desk with its specimen drawers are instruments stemming from alchemy, chemistry's experimental origins. Moving across the picture to the window, the researcher has set up mechanical equipment, including delicate balances to test his products' purity. They indicate how in England as well as on the Continent, precision measurement had long been essential for gold assaying, drug dispensing, and other crafts that predated scientific chemistry.

Throughout the eighteenth century, chemistry was a practical rather than a theoretical subject. Chemists gradually differentiated themselves from alchemists by rejecting arcane speculation and emphasizing the usefulness of their art (yes,

FIG. 29 A chemical laboratory in early eighteenth-century London.
Frontispiece of William Lewis, *Commercium philosophico-technicum or the philosophical commerce of the arts* (1765).

art not science—implying technical expertise, in contrast with scholarly learning). Benefiting from alchemical techniques and instruments developed over centuries, they concentrated on producing functional products—dyes, medicines, fertilizers, bleaches, cement, coal gas In England, Keir, Wedgwood, and other manufacturing entrepreneurs used their chemical research to develop new industrial processes and run profitable businesses. Across the Channel, there was more state funding, which during the Revolutionary period was directed towards military requirements. Lavoisier was in charge of Paris's gunpowder factory, responsible for producing artificially the basic ingredients that could no longer be imported because of the political situation.

Chemists introduced new theories after, not before, their search for practical applications. For instance, sulphuric acid had long been known to alchemists, but now started to be manufactured in bulk for industrial use, even though nobody could explain how the acid was being made or why it was so effective. In itself, the discovery of oxygen/dephlogisticated air was not immediately seen as momentous, because it was part of a collective search for gases from around the middle of the eighteenth century. Even the idea that ordinary air might be a mixture of other substances, not an element in its own right, originated as a by-product of the search for drugs to dissolve kidney stones. The discovery arose unexpectedly through a Priestley-like style of research, when a Scottish student called Joseph Black ignored his professors' instructions, and decided to investigate some strange discrepancies revealed by careful weighing. With no end result in mind, Black pursued the directions indicated by his experimental results to conclude that fixed air (carbon dioxide) is trapped inside some salts, but can be released by acids or heat.

By the end of the eighteenth century, chemistry was becoming an independent science. Although chemists were still using traditional techniques developed by alchemists, artisans, and apothecaries, they were starting to gain prestige and become recognized by official organizations such as the Royal Society. But their new status did not come automatically—they had to work for it. Gillray's caricature (Figure 26) mocked not only Davy himself, but also the presumptuousness of chemical experimenters. Tainted by its alchemical origins, its practical uses in industrial processes, and its links with the French Revolution, chemistry was regarded as inferior to natural philosophy. To make it respectable, and on a par with other sciences, Davy had to divest chemistry of these associations and lever himself into a position of authority.

Davy succeeded by discarding the democratic approach towards science favoured by Priestley and the Lunar chemists, instead converting himself into a Lavoisier-like figure, an expert who controlled powerful equipment. To achieve this transformation, Davy made himself indispensable to both the Royal Society and the Royal Institution. He also adopted a new instrument invented in Italy by Alessandro Volta (whose name survives in 'voltage'), an early form of electric battery that enabled Davy to break down water and to isolate new elements, such as sodium and potassium. For Davy, Volta's battery was not only a miraculous source of energy, but also 'a key which promises to lay open some of the most mysterious recesses of nature'.[20] By controlling his large, impressive apparatus to produce dramatic effects, Davy convinced his audiences that he was the ideal person to wield that key. In the scientific chemistry of the nineteenth century, spectators watched while specialists performed—only they had the authority to create and dispense scientific knowledge.

So—to summarize the Chemical Revolution: it took place...well, when did it? Was it in 1789, when Lavoisier published his new chemical creed? But many years went by before it was generally accepted—and anyway, some of it now seems wrong. The key event was...well, what was it? Lavoisier's identification of oxygen, Priestley's isolation of the same gas, Black's discovery of fixed air, or Davy's analysis of water? Such questions have more realistic but less exciting answers—no individual was uniquely responsible, there was no key moment, change took place gradually. The more you try to pin down the Chemical Revolution, the more elusive it becomes. The more information you take into account, the less significant any particular episode begins to seem. The more closely you analyse its hero, the less exceptional his behaviour appears.

As scientific revolutions go, the Chemical Revolution seems less significant than three others—the Scientific, Industrial, and Darwinian Revolutions. These now sound so familiar that they seem to be real episodes with precise beginnings and ends, but—as chemistry illustrates—scientific revolutions have such nebulous definitions that historians are now writing them out of existence. One objection is their length. The most famous, the Scientific Revolution, is generally said to have lasted from around 1550 (just after Copernicus placed the Sun at the centre of the Universe) to 1700 (a nice round date shortly following Newton's *Principia*). Similarly, although Charles Darwin has a Revolution named after him, evolutionary ideas were common even in his grandfather's time, and it was not until the 1930s that a fully fledged and rather different Darwinian theory was formulated.

Another problem is that not everything changes at once. Accounts of the Scientific Revolution (which has not featured in this book) focus on cosmology, ignore continuities in other areas such as chemistry, and imagine science (whatever that might be) operating in a cultural vacuum, unaffected by trade, politics, or social transformations. In any case, how far-reaching does a shift have to be before it counts as a Revolution? Albert Einstein claimed to revolutionize physics with his relativity theory, but many scientific disciplines (to say nothing of ordinary life) continue to operate with Newtonian mechanics. Harvey reformed physiology by showing that blood circulates, but he was also a committed Aristotelian who had little immediate impact on medical practices—traditional blood-letting continued to be a standard remedy.

Breaking the past down into revolutions does have advantages. They dramatize history, and they provide convenient signposts to major trends of the past. Most importantly, propagandists create revolutions retrospectively in order to distinguish themselves from a preceding and supposedly inferior period. Victorian economists emphasized the Industrial Revolution because they wanted to establish a definitive break between their own progressive era and the country's feudal origins. The Scientific Revolution started dominating accounts of the past only after the Second World War, when historians optimistically (and unrealistically) predicted that science would provide a universal, secular faith to unite the world.

The concept of revolutionary change has philosophical as well as historical implications. Many people regard scientific knowledge as Absolute Truth—they assume that science is cumulative and progressive, resembling a relay race or a climbing expedition in which scientists inherit the achievements of their predecessors to advance steadily onwards. In revolutionary models, on the other hand, science changes sporadically with abrupt shifts, and previous knowledge is dismissed, not incorporated as stepping-stones towards the present. An apt analogy is a branching evolutionary tree, a process with no predetermined end in which old schools of thought are jettisoned when younger researchers head off in new directions.

The main proponent of such theories was Thomas Kuhn, an American physicist and philosopher, whose 1962 book *The Structure of Scientific Revolutions* profoundly affected perceptions of science. Since Kuhn enterprisingly straddled academic disciplines, critics found his suggestions easy to attack. Philosophers liked the history but picked holes in the theories, whereas historians faulted him for simplifying facts. Kuhn's original ideas have been so drastically revised that no unreconstructed Kuhnians survive—even Kuhn himself renounced some of his early opinions. Nevertheless, his name symbolizes current views that science lurches unpredictably, a fallible human endeavour swayed like any other by local influences, personal interests, and political pressures.

Revolutions in science may or may not have happened—it all depends on how you want to think about the past. Max Planck, Germany's leading scientist in the early twentieth century, insisted that change happens slowly, not in sudden flashes: 'An important scientific innovation rarely makes its way by gradually winning over and converting its opponents: it rarely happens that Saul becomes Paul. What does happen is that its opponents gradually die out, and that the growing generation is familiarised with the ideas from the beginning.'[21] Similarly, historical truths also come and go with different generations. Revolutions are currently out of fashion for academics, even though it is hard for them to relinquish such a convenient and familiar way of structuring the past.

6 Rationality

The Church welcomes technological progress and receives it with love, for it is an indubitable fact that technological progress comes from God and, therefore, can and must lead to Him.

—Pope Pius XII, *Christmas Message* (1953)

Ebenezer Scrooge, Mr Gradgrind, Mr Micawber...the novelist Charles Dickens invented many characters who were, like their real-life contemporaries, obsessed with balance sheets, numbers, and arithmetic. Facts and figures dominated Victorian life—which is why the British government continued to pour money into the engineering dream of Charles Babbage, a Cambridge professor now celebrated as a great computer pioneer. In 1837 Babbage optimistically started designing an analytical engine, a massive machine of metal cogs that would take over the tedious work of human calculators by churning out reams of mathematical tables accurate to several decimal places, although he was to never complete a fully working model.

Babbage had started campaigning for quantification when he was a rebellious undergraduate protesting against his lecturers' old-fashioned curriculum. Cambridge was, complained Babbage and his friends, lagging sadly behind its Continental competitors, and they wanted to bring English physics up-to-date by introducing the French mathematical approach based on Leibniz's calculus. Making science mathematical might now seem like an obvious step towards modernity, but in the early nineteenth century, British men of science rejected French algebra, which dealt in abstract symbols rather than in tangible objects solidly tied to observations.

Babbage's student circle also urged their professors to stop accepting the Bible's accounts as literal truths. Instead, they favoured deism, which maintains (broadly speaking) that the Universe operates independently of God, and so can be studied rationally without relying on His written revelations. Paris's leading theoretician, Pierre-Simon Laplace, had already gone still further, eliminating God altogether. Napoleon, who enthusiastically backed scientific research, asked Laplace why God was absent from his cosmos; 'Sire,' replied Laplace (allegedly, anyway), 'I have no need of that hypothesis.'

Laplace liked to call himself 'the French Newton', but Newton himself would not have recognized Laplace's arid, force-driven Universe in which atoms whirl along predetermined paths with no divine guidance. Under Laplace's influence, French research flourished in the early nineteenth century during Napoleon's rule, later regarded by Babbage and his Victorian colleagues as a golden era for scientific achievement. Benefiting from state funding and a technologically oriented education system, a powerful group of researchers clustered around Laplace to establish a new mathematical style of physics. Modelling the Universe with equations, they systematically quantified science by making mathematics and measurement centrally important to physics and chemistry.

Rationalization originated not in Laplace's research school, but in earlier calls for social change. Even before the Revolution, while the King was still on the throne, philosophical politicians proclaimed that reason was the key to progress. They wanted to reform government by applying to France the same laws that God had devised to control nature. Just as the cosmos acted in an orderly fashion following Newtonian gravity, so too, society would advance harmoniously after similar rules had been found to describe human behaviour. Political campaigners did, of course, recognize that individual emotions and personal interests make it far harder to derive precise laws for people than for planets. To compensate for this inevitable fuzziness, mathematically minded reformers introduced probability into decision-making. Instead of relying on an individual fallible judge or an eccentric monarch, they wanted verdicts and policies to be determined collectively, and they devised formulae to calculate the risks and likelihoods involved in accepting the judgement of a majority when unanimity could not be reached. Solving such legal and administrative questions demanded new theories of probability—and these theories were later adapted to cope with scientific problems. Laplace introduced probability into physics, assessing the relative degree of plausibility he could assign to different assumptions, and estimating the errors associated with his results.

This national drive for rationality intensified in France during the 1790s. As revolutionaries divested the country of its monarchy and aristocratic institutions, they set out to reorganize daily life on democratic, rational principles. Changes were introduced by committees, regarded as ideologically preferable to individuals, although still subject to the influence of key players such as Laplace. Propaganda posters from this period show happy, well-nourished citizens measuring out cloth, wine, and wood with the new metric system, based on decimal logic. Under this short-lived regime, time was rationalized by introducing ten-day weeks divided into ten months—and clocks whose faces have ten hours, each of a hundred minutes, have survived. The committees also decimalized space, sweeping away arbitrary imperial measurements (such as gallons, pounds, and acres), and replacing them with metric units (litres, grams, hectares) based objectively on the size of the Earth. In principle, a metre was one ten-millionth of the quarter-arc from the North Pole to the equator, and this provided the essential reference for the entire metric system. The new dimensions were determined (unfortunately, rather inaccurately) by two astronomers, who set out on a hazardous seven-year expedition to measure a long section of longitude through France and Spain. On their return, a platinum metre was set up in Paris—slightly shorter than it should have been, but still a political symbol of the country's rational approach to the natural world.

Although France became more efficient through being unified, in some ways the Revolution substituted one set of rulers for another. Despite the Revolutionary rhetoric of equality, the metric system reintroduced central control by an elite group—the flip-side of unification is uniformity. Previously, different regions of France had used their own measurement methods, but when Parisian bureaucrats introduced their rational system, they eliminated local variations and placed the entire country under a single metropolitan regime. France's unique calendar and measurements not only isolated the nation from the rest of the world, but also antagonized its inhabitants. Workers objected to the longer working week imposed by the reformed calendar, Christians were horrified by the abolition of Sundays, and shoppers accused opportunistic merchants of making extra profits by fiddling the translated prices. In year XIV of the new system, Napoleon reinstated the conventional date of 1806, along with familiar units, and it was only towards the end of the nineteenth century that Europe eventually went metric.

Other rationalizing reforms also had double-edged effects. For example, the nation's health was dramatically improved by building state-subsidized hospitals that treated loyal citizens free of charge. The big, airy wards had only one patient per bed, and infections were further reduced by chemical disinfectants. By grouping patients together, doctors could time the course of an illness, record symptoms, and compare numerically the effects of various treatments. In these enlightened clinics, medical men would accumulate observations to build up expertise, acquiring a penetrative gaze that enabled them to see through surface symptoms and discern the underlying reality. On the other hand, efficient methods of diagnosis and therapy tended to reduce the amount of sympathetic one-on-one care that had formerly characterized medical treatment—patients started to become numbered cases of named diseases rather than individuals with unique imbalances in their personal humours. Professional doctors were rigorously trained and examined, but they squeezed out traditional practitioners, such as village herbalists and midwives, so that expertise became increasingly restricted to an affluent tier of male university graduates. When physicians are elevated into all-seeing experts, their pronouncements are hard to overturn.

Similarly, the state-organized educational system claimed to be democratic, but in practice remained open mainly to the privileged. Even before the Revolution, military colleges provided teaching far more mathematically oriented than that in England. Committed to technological improvement, successive governments poured funds into engineering academies, which generated highly trained men (yes, men) who brought a rational vision to many areas—architecture, communication systems, scientific research, machinery. Examinations were based on mathematical ability, which was held to provide an objective and hence democratic measure of aptitude. But high levels of skill were time-consuming and expensive to acquire, and so could effectively only be attained by students from rich families. By the early nineteenth century, the old hereditary aristocracy had been replaced by a new elite based on wealth and intelligence.

Some of these talented and mathematically trained engineering graduates were attracted to the research group run by Laplace and his close friend Claude Berthollet, a physician and chemist who conveniently lived next door in Arcueil (just outside Paris), which became the centre of Napoleonic science. Although educated before the Revolution, both men had taught in technical colleges, both were involved in Lavoisier's projects of chemical reform, and both believed that the underlying forces of nature stem from the powerful bonds between minute particles. Well-established themselves, they were able to sway selection committees and channel funds towards their own preferred acolytes—patronage remained as important under the new regime as it had been for centuries. Together, Laplace and Berthollet assembled a gifted team of disciples who rapidly extended and consolidated Laplace's mathematical approach by applying it to other phenomena. But after a few years, the reservations of outsiders turned first into challenges and then into refutations, and the Laplacian programme was abruptly abandoned.

Laplace was a forceful man in several senses. He imposed his own ideas on his followers, he moulded his models of nature to conform with his preconceived views, and he envisaged the world in terms of short-range forces. His genius was, remarked an English sceptic, like a sledgehammer that smashed open mathematical puzzles but 'gave neither finish nor beauty to the results'.[22] One of Laplace's first achievements was to make Newtonianism more perfect than Newton's own original version. Newton himself thought that, unless God intervened occasionally, the gravitational interactions between the planets would eventually make the entire system unstable. Deploying some nifty mathematics, Laplace showed that Newton had been wrong—which was why, to Napoleon's consternation, he was able to dispense with God. From then on, Laplace fashioned his results to fit his reworked brand of Newtonianism. He wanted to vindicate what he had inherited, rather than launch his own original scheme.

In the Laplacian cosmos, forces rule. Molecules attract and repel each other, and—provided you know where everything started—you can calculate where each molecule will be in the future. This is a deterministic model, in which behaviour is implacably governed by abstract forces and can be predicted mathematically. According to Laplace's version of Newtonianism, ordinary matter—metal, bone, salt—is held together by attractive forces acting over very short distances between tiny particles. In addition to these ordinary molecules, special ones make up weightless invisible fluids, such as light, heat, and electricity. Inside these aetherial substances, nearby particles mutually repel one another, although they are attracted to ordinary ones. Building on these basic concepts, Laplace aimed to formulate a sophisticated mathematical structure that would unite the whole of terrestrial physics.

Laplace worked on an impressive range of topics in physics and chemistry, and he ensured that the top jobs went to researchers who vindicated his own ideas. One outstanding example is optics. Over-ruling the objections raised by his critics, Newton had insisted that light is not a wave resembling sound, but a stream of tiny corpuscles. Laplace steered one of his most brilliant students, Étienne Malus, towards examining Iceland spar, an unusual crystal that produces double images when you look through it. As anticipated, Malus managed to confirm Laplace's Newtonian opinion by devising a mathematical, corpuscular explanation. Yet even while Malus was triumphantly confirming Arcueil's glory, experimenters in other centres beyond Laplace's direct control were rebelling against his stranglehold. From around 1815, the alternative view of light started to take over, when Augustin Fresnel used his experiments on diffraction to expose shortcomings in Malus's work and demonstrate that—contrary to Newton—light is carried by waves. As Fresnel won over converts in the closely knit Parisian scientific community, even Laplace was no longer able to persuade committees that his own candidates should prevail. And once the Laplacian view of light had been discredited, attacks mounted in other areas—heat, electromagnetism, chemistry. By 1825, French scientific power was no longer based in Arcueil.

Laplace's circle disintegrated, but his mark on the future of science was indelible. Later in the nineteenth century, the metric system he had sponsored was revived, and the world's international bureau for establishing standard measurements was established in France. Nevertheless, Laplace's opponents continued to influence the pattern of French research, and they rejected his bold, hypothetical approach to focus instead on meticulous observation. France gradually ceased to lead the world in theoretical physics. In contrast, the campaign launched by Babbage and his Cambridge colleagues proved successful, so that although British scientists abandoned Laplace's model of short-range forces, they did adopt his mathematical approach. Ironically, they also developed his work on probability theory to yield a new type of physics based on statistics and chance events—Laplace's careful assessments of experimental evidence eventually undermined his own totally predictable cosmos.

The rise and fall of Pierre-Simon Laplace depended not only on his theories, but also on the manoeuvres of his allies and enemies. Like any other human activity, scientific practices are affected by ambition, complacency, and opportunism. Seeking preeminence and fast results, Laplace dominated his own colleagues, manipulated scientific committees to promote his disciples, and took advantage of France's administrative centralization to ensure that his doctrines were perpetuated in textbooks and examination syllabuses. Outside Arcueil, beyond his direct control, Laplace's critics deployed equivalent tactics to consolidate his defeat—editing influential journals, lobbying during scientific elections, securing major teaching positions. The fate of this single individual is less significant than his long-term impact, since his rational, mathematical approach was adopted by British and German physicists during the nineteenth century and still permeates science today.

7 Disciplines

> Why is England a great nation? Is it because her sons are brave? No, for
> so are the savage denizens of Polynesia: She is great because their brav-
> ery is fortified by discipline, and discipline is the offshoot of Science.
>
> —William Grove, *On the Progress of the Physical Sciences* (1842)

'Every savage can dance,' declared Jane Austen's Mr Darcy in *Pride and Prejudice*. His antagonist's riposte now seems odd—'I doubt not that you are an adept in the science yourself, Mr Darcy.'[23] 'Science' is among the most slippery words in the English language, because although it has been in use for hundreds of years, its meanings constantly shift and are impossible to pin down. That plural (meanings) was deliberate. In the early nineteenth century, when Austen casually mentioned the science of dancing, other writers were still using 'science' for the mediaeval subjects of grammar, logic, and rhetoric. Long afterwards, 'science' could still mean any scholarly discipline, because the modern distinction between the Arts and the Sciences had not yet solidified. The Victorian art critic John Ruskin listed five subjects he thought worthwhile studying at university—the Sciences of Morals, History, Grammar, Music, and Painting—none of which feature on modern scientific syllabuses. All of them, Ruskin declared, were more intellectually demanding than chemistry, electricity, or geology.

However skilfully Mr Darcy performed his science of dancing, Austen could never have called him a scientist. That word, now so common, was not even invented until twenty years later, in 1833, when the British Association for the Advancement of Science (BAAS) was holding its third annual meeting. As the conference delegates joked about needing an umbrella term to cover their diverse interests, the poet Samuel Taylor Coleridge rejected 'philosopher', and William Whewell—one of Babbage's allies, a Cambridge mathematical astronomer—suggested 'scientist' instead.

The new word was very slow to catch on. Many Victorians insisted on keeping older expressions, such as 'man of science', or 'naturalist', or 'experimental philosopher'. Even men now seen as the nineteenth century's most eminent scientists—Darwin, Faraday, Lord Kelvin—refused to use the new term for describing themselves. Why, they demanded, should anyone bother to invent such an ugly word when perfectly adequate expressions already existed? Mistakenly, critics accused 'scientist' of being an American import, a trans-Atlantic neologism—one eminent geologist declared it was better to die 'than bestialise our tongue by such barbarisms'.[24] The debate was still raging sixty years after Whewell first introduced the idea, and it was only in the early twentieth century that 'scientist' was fully accepted.

In America, the new word was immediately adopted. But in Britain, antagonism festered for decades. Ironically, part of the problem was that experimenters like Davy had been almost too successful in establishing themselves as experts. Although they were profoundly knowledgeable about their own disciplines, they found it increasingly difficult to keep up with the latest developments in other areas. Whewell thought that expertise entailed narrowness—he was worried that as specialists burrowed deeper and deeper, they would lose sight of science's overall unity and fail to communicate with each other effectively. Reflecting nostalgically on a vanished era when individual polymaths could cover the whole gamut of natural knowledge, Whewell urged researchers to club together and retain the integrity of the scientific community. By identifying themselves as scientists, he urged, they could distinguish themselves from artists, writers, and musicians, who were also struggling to negotiate a high-status identity.

Money was a contentious issue in these debates. Supporters of the new word argued that if individuals grouped together as scientists, they would gain lobbying power for persuading the government or large commercial companies to finance their research projects, which were becoming more ambitious and expensive. On the other hand, well-connected gentlemen liked to regard themselves as members of an elite group who were pursuing knowledge for its own sake. Even those who had been born neither rich nor aristocratic affected to regard earning one's living as a rather sordid way of proceeding, and they looked down with disdain on entrepreneurs who turned their scientific activities into commercial ventures.

The nineteenth-century arguments about 'scientist' were so virulent because far more was at stake than the word itself. The new label signalled changes in class, money, and status—long-term social transformations that the privileged classes found hard to accept. In a sense, the gentlemanly men of science became victims of their own success, because it was partly through their own efforts that science became more democratic. Keen to advertise the benefits of their activities, they ensured that scientific knowledge became available to a far larger sector of the population, and slowly ceased to be the preserve of privileged gentlemen. As research grew, and education expanded, new opportunities for paid employment as laboratory assistants, museum curators, or astronomical calculators arose. Very gradually, science became a paid profession open to many, rather than an all-absorbing but expensive occupation for the leisured classes. Eventually, it became a compliment rather than a sneer to call someone a scientist.

The single term 'science' bracketed together disciplines that had very different pasts. Some subjects—astronomy, optics, mechanics—stemmed directly from mediaeval university syllabuses: although they slowly changed over the centuries, their roots can be clearly seen stretching back over time. In contrast, although chemistry was a new science, its origins lay not in abstruse scholarly studies, but in everyday practices such as alchemy, medicine, and skilled crafts. Similarly, the word 'biology' was only invented in the early nineteenth century, but the new speciality inherited a good deal of accurate knowledge from herbalists, merchants, and collectors (women as well as men)

Not all the new sciences had such ancient lineages. One freshly minted discipline was geology, whose birth was marked in 1807 by the foundation of Britain's first specialized scientific society. Geologists' desire to study the Earth's structure for its own sake, rather than for some practical gain, was relatively recent. Before then, various groups of people had accumulated their own specialized knowledge—miners who knew how to detect and identify ores, surveyors who picked out the best routes for roads, farmers who knew which crops to grow on different soils, soldiers who mapped the terrain of areas they hoped to conquer (the Ordnance Survey started in Scotland, not England, because the army wanted to repress the Jacobite rebels). It was only in the early nineteenth century that geological collecting became a fashionable craze for the middle classes, who spent many happy hours tapping rocks with their hammers to chip off mineral specimens and fossils, often freshly revealed by the cuttings made for canals and railways. But geology also became a serious science, whose challenge to the Bible's version of Creation stimulated theories of evolution.

The discipline that dominated nineteenth-century science—electromagnetism—was also a new one. Although electricity and magnetism are now inseparably tied together, they used to be completely distinct from each other. For one thing, as powers of nature, they behaved very differently—electricity flashed and hurt, whereas magnetism operated invisibly, affecting iron but leaving people unscathed. In addition, their sciences contrasted strongly. Electricity was an exciting eighteenth-century innovation, advertised by experimental philosophers who captured public attention with their spectacular performances. Magnetism, on the other hand, was one of nature's traditional mysteries, a God-given power tapped by navigators but largely ignored by natural philosophers. Although a few did conscientiously try to make sense of its vagaries, compasses and iron filings could hardly compete with the fascination of sparks and charges.

The symbolic year of change is 1820, when Hans Oersted, a physics professor in Copenhagen, contrived a dramatic demonstration to impress his students—as he passed a current through a wire, a small magnetic needle twitched in response. All over Europe, researchers set about investigating this effect, and Humphry Davy—by then President of the Royal Institution—asked his assistant, Michael Faraday, to report on progress. Within a few months, Faraday had devised a small and deceptively simple instrument that definitively linked together electricity and magnetism. Moreover, he showed them to be symmetrical powers: he could make a magnet move because of a current, but he could also make an electric wire rotate around a magnet. A new scientific discipline—electromagnetism—was forged by bringing together the electrical inventions of Enlightenment philosophers and the centuries-old magnetic expertise of mariners.

'Scientist' was an umbrella term, but not everybody was allowed to shelter beneath it. Hungry for prestige, scientists wanted the authority to declare that they were incontrovertibly right, that the knowledge they produced in their laboratories was irrefutably correct. New specializations were being invented, but not all of them were deemed worthy to be labelled science. Science was splintering into disciplines—but disciplining means controlling as well as teaching. Like police guards patrolling national borders, scientists decreed which topics should lie inside the large domain they ruled over, and which ones should be outlawed.

In retrospect, their verdicts seem straightforward, but at the time, they were not always clear-cut. Chemistry became a major scientific discipline, yet Gillray's caricature (Figure 26) illustrates how chemists were initially disparaged because of their links with alchemy, industry, and the French Revolution. Conversely, practices now widely condemned as rubbish at one time had many supporters who claimed they were legitimate sciences. In principle, determining whether these sciences were valid should have been a rational process of assessing how well they worked. But that was not necessarily the case—prejudice, prestige, and politics were often involved.

Take mesmerism, or animal magnetism, a medical therapy that flourished intermittently throughout the nineteenth century. The system was originally introduced in the 1780s by Franz Mesmer, who claimed to cure sick people by redirecting magnetic fluid through their bodies. Although his rivals denounced him as a quack, Mesmer made a quick fortune when he set up his clinic in Paris. Wealthy aristocrats, many of them women, flocked for treatment not only because mesmerism was fashionable, but also because it seemed to work. Figure 30 shows patients clustering round an oval wooden tub stuffed with iron filings, magnets, and other special ingredients. As Mesmer conducts operations from the right with his magnetic baton, the lame man on the left is tying his leg to an iron hoop to suck up the tub's magnetic fluid, while the woman collapsed in a chair has

Fig. 30 A salon in Franz Mesmer's magnetic clinic. Undated engraving by H. Thiriat.

swooned into a crisis, a controversial side effect induced by Mesmer's close proximity, intense gaze, and suggestive hand movements. Critics accused him of sexual impropriety, but he produced impressive affidavits from grateful clients testifying to his therapeutic success.

Animal magnetism had respectable antecedents. Mesmer, a fully qualified Viennese physician, gained a doctorate for his theories, which he had derived from Newtonian gravity. His techniques originated from wearing special magnets next to the skin, a medical therapy that had recently been enthusiastically recommended by an official French committee. Mesmer's nebulous magnetic fluid circulating through the atmosphere might sound bizarre, but it was conceptually no stranger than the electrical aethers being endorsed by Europe's top natural philosophers. And most importantly as far as his patients were concerned, Mesmer's soothing regimen helped to alleviate their symptoms—which is why they went on paying his high fees.

Mesmer appeared dangerous not because he was dramatically different from other physicians, but because he was similar enough to represent a real threat. Nowadays mesmerism might be classed as alternative medicine, but two hundred years ago there was no such either/or situation. Even the most highly trained physicians often had little hope to offer, and desperate patients were willing to buy whatever help they could to alleviate the symptoms of incurable conditions.

As they struggled for prestige, eminent doctors derided their less-educated competitors as charlatans, even though they themselves marketed useless nostrums at exorbitant prices. Physicians were spread out along a continuous spectrum of qualifications. At one end lay the high society doctors who had been to university, belonged to professional associations, and charged large fees; at the other were untrained men and women trying to scrape a living by caring for the poor. In between, all sorts of practitioners catered for different illnesses and budgets—surgeons, apothecaries, herbalists, midwives. Making science authoritative entailed imposing firm boundaries between establishment and quack medicine, between orthodox and pseudo-science. In the absence of abstract criteria for distinguishing between them, the decisions were often taken on social grounds.

Mesmer's competitors became very worried. Often unable to provide effective treatment themselves, they watched him taking over their wealthiest patients. Soon, splinter groups were mushrooming all over France. Because relatively uneducated men could be trained as mesmerists, magnetic medicine became linked with radical politics, and so threatened the position of traditional physicians. Castigating Mesmer as a quack was one way to get rid of him, and a royal committee was set up to justify his exclusion. After a series of investigations, they pronounced sentence—he was to be banned. Tellingly, they acknowledged that his treatments worked, but castigated him for being unable to provide any rational explanation. The committee attributed mesmeric cures to the power of imagination, a psychosomatic effect they rejected because it could not be accounted for rationally.

Along with alchemy, astrology, and many other practices, mesmerism was eventually banished from legitimate science and—despite being a traditionally educated physician—Mesmer was branded a charlatan. Even so, mesmeric societies flourished in the nineteenth century, partly because it was a democratic therapy that could be practised by ordinary people. This had revolutionary implications—magnetizers exerted power over their patients, so what would happen if control passed out of the hands of the privileged classes? Still worse, because Mesmer affected his patients' physical health by influencing their imaginations, he challenged the supremacy of reason.

This was a frightening prospect, since the ideology of scientific detachment insisted that rational men of science could use their minds to discipline their bodies. Disciplinary science was to be based on reason and order, on logic and explanation. The eighteenth century is often called the Age of Reason, and Enlightenment philosophers bequeathed this passion for rationality to the scientists who followed them. Trained as experts, and organized into specialized disciplines, the goal of nineteenth-century scientists was to unify and discipline the world by finding simple laws that described the behaviour of everything—people as well as things, minds as well as bodies.

Laws

Committed to progress, nineteenth-century scientists searched for laws to govern the human as well as the physical worlds. Establishing themselves as experts, they gradually gained prestige, wresting authority from religious leaders to create a new scientific priesthood. Yet however much scientists presented themselves as stalwart warriors of reason, theological attitudes continued to pervade debates about life and the Universe, with no sudden switch from biblical faith to scientific conviction. Scientists claimed to obtain *Absolute Truth* by recording the world as objective observers, but this view was challenged by German Romantic philosophers who stressed a unified cosmos in which human beings are integrated within the natural world. Although ultimately less influential than the champions of detached precision and mathematical laws, their approach resonates with modern environmental attitudes. Personal judgements kept creeping back into supposedly neutral science. Although instruments were designed to eliminate human error, their use inevitably entailed subjective assessments. Even the most famous innovation of the century—Charles Darwin's theory of evolution by natural selection—was no logical analysis, but depended on accumulating corroborative evidence rather than providing incontrovertible proof. Instead of spreading uniformly around the globe, science varied geographically, developing through local processes of adaptation and exchange. In principle, international scientific collaboration transcended political differences; yet standardizing time was fraught with conflict—although it did give rise to relativity, an esoteric theory rooted in practical concerns of improving telegraph systems.

I   Progress

God made man in His own image, but the Public is made by Newspapers.

—Benjamin Disraeli, *Coningsby* (1844)

One sunny autumn day in 1858, a distinguished clique of scientific, religious, and political dignitaries headed a procession through the streets of Grantham, a small English provincial town. Accompanied by a military band, the octogenarian Henry Brougham—Scottish Baron and eminent judge—climbed up onto a dais decorated in the colours of the rainbow and sat down in a battered armchair with its stuffing showing. This deliberately unrestored relic had once belonged to Isaac Newton, a local hero now elevated to national grandeur. Brougham was about to unveil a statue of Newton forged from a Russian cannon captured during the Crimean War and donated by Queen Victoria herself.

Erecting Newton's statue was such a momentous event that it hit the national press—Figure 31 was reproduced in several journals. Although sculptures of monarchs, saints, and military leaders abounded all over Europe, commemorating a scientific figurehead was something new. The carefully orchestrated ceremony at Grantham indicates how the status of science had risen during the first half of the nineteenth century, when Newton became acclaimed as an English genius, the scientific counterpart of William Shakespeare. Funded by donations from all over the country, Newton's massive monument is an early example of Britain's heritage industry, intended by scientists to rouse public enthusiasm for science and stimulate funding opportunities.

This sculpture not only imagines what Newton might have looked like, but also represents an idealized form of how Victorian physicists thought they should behave. Admired for his single-minded dedication, Newton epitomized the methodical scientist, the steadfast logical searcher for Absolute Truth. Dressed in formal university robes, this authoritarian figure is pointing at a planetary diagram, emblem of his three laws of motion and the mathematical order he had imparted to the Universe. 'Search for laws!'—Faraday's summary of a lecture he

STATUE OF SIR ISAAC NEWTON, INAUGURATED LAST WEEK AT
GRANTHAM.

FIG. 31 Isaac Newton's statue at Grantham, sculpted by William Theed (1858).
Illustrated London News, 2 Oct. 1858.

gave at the Royal Institution—was a leitmotif of nineteenth-century science. The
major goal of Victorian physicists was to explain the world in mathematical laws,
to unite its disparate branches—heat, light, mechanics, electricity—into one single
system. Similarly, scientists in other fields wanted to adopt this law-based approach
for describing how societies behave, how the Earth's landscape has changed, how

living organisms function. Just as God governs through moral laws, or rulers maintain discipline through state legislation, so Newton had imposed regularity on the cosmos by deciphering the laws of nature—a mathematical triumph that Victorian scientists aspired to emulate.

Standing in front of Newton's statue, Brougham formulated his own scientific law—his 'Law of Gradual Progress'. Like many Victorians, Brougham believed that hard work was the key to success, and he preached the virtues of steadily building up knowledge step by tiny step. Newton's sober bronze figure epitomized the rewards of dedication (and it inspired Margaret Thatcher, Grantham's other famous workaholic, on her daily walk to school). To inspire his listeners with faith in science's potential, Brougham eloquently outlined a progressive overview of human history. Newton had, he explained, inherited the achievements of his predecessors, and by steadfastly applying his genius from an early age had been able to reach further, extending the bounds of theoretical knowledge and—just as importantly—paving the way for steam engines, source of the nation's industrial supremacy. By building on Newton's legacy, declared Brougham, scientists would lead Britain forwards to a magnificent future.

Progress was a major refrain of nineteenth-century science. Campaigners forecast advance on many fronts—new laws would be formulated, unexplored parts of the globe would be surveyed and brought under control, machines would become bigger, better, and faster, the general level of education would improve...the promises multiplied. Significantly, by the 1830s, scientific theories themselves encapsulated the notion of progress, contradicting the traditional view that God created the Universe as it is now. Geologists described an Earth that had gradually cooled from its original fluid state, astronomers suggested that the Solar System had condensed out of swirling clouds, and early evolutionists dared to propose that present-day plants and animals had not always existed.

Brougham was a veteran scientific campaigner and also an astute politician. Throughout his life, he sketched out utopian schemes for making scientific knowledge available to even the poorest cottagers. When the Society for the Diffusion of Useful Knowledge (SDUK) began publishing cheap scientific books, Brougham wrote an enthusiastic introduction that sold over thirty thousand copies—a huge number at the time. But his was not a plea for equal opportunities. Rather than prompting workers to demand a university education, Brougham hoped that they would improve their performance if they understood their tasks better.

The SDUK might sound like a philanthropic organization, but its organizers acted like scientific missionaries with a hidden agenda. Even the name they chose reveals their feelings of superiority, implying that a core elite sent down to the working classes predigested information that was not necessarily intellectually demanding, but would help them carry out their work more efficiently (and so generate more profit for their employers). By convincing labourers that progress

FIG. 32 William Heath [Paul Pry], *The March of Intellect* (1829).

came through science, the privileged classes hoped to reduce the risk of political protests about low wages and poor working conditions. As one radical writer quipped, Brougham wanted all Englishmen to read Bacon, whereas what they needed was bacon on their dinner plates.

Many people not only disapproved of Brougham's plans to educate the workers, but also disagreed with his optimistic views about scientific progress. 'Lord how this world improves as we grow older' was the satirical heading of a caricature labelled *The March of Intellect*, a common catchphrase of the time (Figure 32). This panorama displays devices which appear absurd yet refer to contemporary engineering projects. George Stephenson's *Rocket* was just being launched, and early train passengers were terrified by the speed at which 'Steam Horses' (lower right) raced through the countryside. The central vignette shows passengers boarding at Greenwich for their vacuum-propelled flight to Bengal—but less than twenty years later, similar propulsion tubes were in service on several railway lines.

Steam power was making manufactured goods cheaper, but it also threatened existing hierarchies. By fantasizing about steam razors and airships, this artist poked fun at privileged critics who questioned the value of technological innovations, insisting that convenience would inevitably result in moral decadence and intellectual deterioration. After all, if the labouring masses could afford education, travel, and luxuries, then perhaps they would neglect their duties? Wealthy aristocrats feared that their power was being eroded and passing into the hands of self-made men, the industrial investors who were making rapid fortunes—hence the Latin sign 'God regards only pure hands, not full ones' above the Royal Patent Boot Cleaning Engine on the bottom right. Several scenes show workers behaving inappropriately. The owner of the boot cleaner lolls against the wall reading a French newspaper, while a dustman and his scruffy companion gorge themselves on exotic foods, scandalously ignoring an elegant lady sheltering beneath her black servant's umbrella.

Despite the satires, steam power had a dramatic impact on scientific progress. Fast trains and ships effectively shrunk the world, so that knowledge and people, specimens and instruments, could be transported more rapidly than ever before. Just as importantly, steam revolutionized publishing. Cheap books and journals meant that for the first time, wide sectors of the population could read about science. As production processes became increasingly mechanized, paper prices tumbled and printing was vastly speeded up. By the 1830s, publishers had realized that it made good marketing sense to increase profits by selling large numbers at low prices, an opportunity that had never existed previously—no coincidence that this was when the SDUK and its competitors started to flourish.

Because of cheap printing, when Brougham made his Newtonian speech eulogizing scientific progress, he knew that he was addressing the entire country. Souvenir pamphlets sold out the same day, newspapers printed summaries of his lecture, and engravings of Newton's magisterial statue enabled it to reach far beyond the confines of Grantham. These new publicity opportunities enabled scientists to promote themselves more efficiently, and so sway public opinion in favour of investing in their exploratory voyages and research projects. At the same time, as the *March of Intellect* caricature illustrates, their critics also became more visible. Instead of being restricted to a privileged minority, debates about science and its impact started to be conducted publicly.

These unprecedented media possibilities transformed science. One particularly influential organization was the British Association for the Advancement of Science (BAAS), which took advantage of cheap publishing to advertise science, aiming to increase the number of people involved in it. Founded in 1831, the BAAS encouraged researchers to speed up progress by sharing their findings at provincial meetings held once a year in different parts of the country. Backing William Whewell, who recognized the advantages of forming a scientific community, they urged experimenters to join forces as scientists, arguing that this would enable them to exert more leverage than when split into separate disciplines, and also provide the protection they needed in the absence of professional support structures.

Men who embarked on a life of science had no set path to follow, but were forced to carve out their own career routes. Lacking the job security associated with modern professional science, even some eminent men found money a perpetual problem. Thomas Huxley, for instance, is now famous for promoting Darwin's theory of evolution, but he constantly struggled financially. Many scientists worked at home. Darwin is the most famous example, but the Cambridge physicist Lord Rayleigh apparently cleared his scientific apparatus from the top of the piano every day to make way for family prayers. Even within universities, professors struggled to obtain minimal facilities—after Lord Kelvin converted a disused wine cellar into a laboratory, it was constantly permeated with coal dust from the adjacent store. It was only towards the end of the century, following many individuals' hardship and enterprise, that school leavers could aspire to become salaried professional scientists.

As well as this practical incentive to act collectively, nineteenth-century scientists were also theoretically inspired to cooperate. They believed that the sciences are interconnected, so that progress towards Ultimate Truth can only be made by drawing on a range of insights, not by depending on inherently limited advances in individual specialities. They believed that there was only one way to find the unifying mathematical laws that governed the whole of nature—searching systematically. Whatever their discipline, scientists claimed to share a common scientific method that characterized their unique approach to the world and distinguished them from non-scientists.

But making themselves special created problems. On the one hand, scientists were trying to consolidate their status by publicizing their achievements—they wanted to disseminate their ideas, improve scientific education, enlist recruits. Aware of the new power available through cheap publishing, they produced a wide range of books and magazine articles to promote their activities amongst wider and wider audiences. But at the same time, the leaders of the BAAS were privileged men who doubted whether everybody was capable of understanding the profundity of their scientific thought. Could one really expect lower class men (to say nothing of women) to follow the rigorous mental demands of the scientific method? Perversely, by insisting on their unique abilities, scientists made it impossible for everyone to share equally in the scientific endeavour.

This intellectual class system placed workers and women at the bottom of the scientific hierarchy. Even at the supposedly welcoming BAAS meetings, wives and daughters were relegated to the light-hearted evening talks. Elite Victorians prided themselves on progress, but were reluctant to acknowledge that achieving it meant depending on people excluded from the higher echelons of science. Many less privileged groups did make vital contributions, but they have been rendered almost invisible. Most obviously, countless technical assistants were concealed behind the scenes, less-educated men essential for building apparatus, organizing laboratories, and repeatedly running experiments. Similarly, eminent scientists rarely credited the editing, drawing, and collecting skills of their wives, who were often carefully picked out as potential collaborators. Mary Lyell, for instance, was an ideal scientific bride. The daughter of a rich and famous scientist (perfect for an ambitious son-in-law), she became the unacknowledged intellectual partner of her husband, the geologist Charles Lyell. Before their marriage, she agreed to learn German and so save him the bother. Subsequently, she accompanied him on a geological field trip for their honeymoon, edited and illustrated his books, organized his mineral collection, and became an expert on shell classification, even training her maid to kill and clean snails.

Although scientists set themselves up as experts, knowledge did not simply diffuse downwards from elite organizations. Instead, change often stemmed from interactions between diverse groups, and on exchanges of information rather than one-way flows. For example, some exceptionally significant fossils were dug out not by specialized London geologists, but by provincial residents who made a living through selling local finds. The most famous was Mary Anning of Lyme Regis, who was only a young girl when she discovered her first dinosaurs on the English seashore. Later, she set up a business selling fossils to rich scientists, who were baffled by these skeletons that differed so much from any living species. Many of them have ended up in museums (mostly without her name on), but although Anning's discoveries transformed geology by providing hard evidence of extinction, she never published and so failed to gain formal recognition. Instead, she became somewhat of a collectors' item herself, a provincial curiosity to be marvelled at by London visitors.

Specialists in other disciplines relied on similar networks, whose diverse participants excelled in different ways—the scientific experts did not always know best. Around Manchester, groups of weavers set up informal Botanical Societies which met in village pubs. Although not always literate, the weavers took their studies seriously, fining members who turned up drunk and carefully comparing their specimens with textbook illustrations to learn their Latin names. Scouring the local hillsides, they became extremely knowledgeable about plant distribution. Eminent botanists relied on these artisan collectors, who could gather and identify rare flowers which they would have been incapable of locating themselves.

Another way in which science relied on non-professionals was through mass publishing, which transformed the ways in which women could participate. Previously, canny authors had tried to increase sales by targeting women as potential purchasers, but during the nineteenth century, women started to write the books themselves. The most striking example is Mary Somerville, a mathematical physicist of such astounding ability that, despite the disadvantage of being unable to study at university, she carried out research original enough to be published in the *Philosophical Transactions of the Royal Society*. Even so, since she was banned from entering, her husband had to read the paper for her, although the Fellows did install her bust in the entrance hall.

Like other gifted women, although excluded from scientific laboratories and scholarly societies, Somerville had a profound influence on science through her writing. Recruited by Brougham to popularize Laplace's book on astronomy, she instead produced an expert's text, one that explained to mathematically challenged British scientists the complex calculations essential for grasping Laplace's innovations. Although her next major book was less specialized, it addressed a central theme of nineteenth-century physics—linking together apparently disparate phenomena.

Familiar with a wide range of authors, Somerville not only synthesized, but also provided a fresh interpretation that influenced later debates on light and electromagnetism. Elite scientists were impressed, general readers could cope after she introduced some diagrams, and Somerville's *On the Connexion of the Physical Sciences* (1834) became a scientific classic that did much to consolidate the public reputation of Victorian physics. By choosing unification, Somerville wrote on precisely the topic that inspired the BAAS—especially Whewell, whose new word 'scientist' made its first printed appearance in his enthusiastic review of her book.

2 Globalization

Thanks to the interstate highway system, it is now possible to travel from coast to coast without seeing anything.

—Charles Kuralt, *On the Road* (1980)

After Christopher Columbus set off for India but landed in the Bahamas, Europeans were forced to recognize the existence of another great land-mass. Keen to maintain the separation of the Old World from the New, they imagined a north–south line running down the middle of the Atlantic Ocean. Three hundred years later, a German Columbus called Alexander von Humboldt spent five years exploring Latin America, and he decided to slice the globe in a different direction—across the equator. Claiming to be more interested in climate than in history, Humboldt planned to use systematic measurements for founding a new terrestrial physics that would unite the entire globe.

In one sense, science was already global. Natural historians had long taken advantage of international trading connections and personal friendships to exchange specimens, so that plants, animals, and minerals travelled around the world in many directions. The new sciences of botany and geology depended on these global interchanges, which increased during the nineteenth century as nations expanded their empires and commercial networks. Information was also being transferred from one place to another, not just in books but also in activities—manufacturing processes, medical treatments, agricultural techniques. Merchants, emigrants, and colonial occupiers integrated their own customs with local expertise, so that knowledge was not adopted wholesale, but was transformed and assimilated before being exported to other countries. For instance, European engineers designing irrigation systems incorporated methods that had been developed in the Nile valley over centuries, while colonial doctors in the tropics tested traditional remedies to formulate powerful, portable drugs.

In addition, a new type of global science emerged. Scientists started to regard the globe as an entity to be analysed in its own right, so that the world itself became a laboratory. European explorers began to investigate natural phenomena where they occurred, in real time, rather than taking samples back home to examine them later. Humboldt, a pioneer of this field-based approach, declared himself to be a terrestrial physicist who operated very differently from naturalists. Instead of merely collecting and describing, he declared, his goal was to analyse—by building up massive data sets of precise measurements, he would derive scientific laws describing the entire globe. Linking East with West, Humboldt pictured climatic bands circling the Earth on either side of the equator, each with its own typical vegetation, landscape, and human society.

A skilled self-promoter, Humboldt took advantage of the expanding media industry to publicize his travels. Sober German scientists relegated his romanticized adventures to children's literature, but elsewhere he came to epitomize the enterprising adventurer who braves mountains, rivers, and diseases to chart the globe scientifically. As well as promoting terrestrial field sciences, Humboldt also engaged in activities often regarded as lying outside scientific territory, such as stimulating European investment and encouraging independence movements. For Central and South Americans, Humboldt became a hero not for his global physics, but for convincing Europeans that their countries mattered. Unlike most modern scientists, Humboldt was independently wealthy and had no professional brief to fulfill. As a relatively free agent, he chose to spend huge amounts of money and time on amassing accurate measurements, but he also garnered opinions from indigenous Indians and political revolutionaries. After learning how Peruvian farmers used guano, he transformed the local economy and also glorified himself by converting this traditional fertilizer into a scientific discovery that would benefit Europeans.

Armed with impressive arrays of accurate instruments, Humboldt demonstrated that accumulating meticulous measurements could reveal patterns in nature's vagaries, and so impose mathematical order on variable phenomena such as air pressure, magnetism, and plant distribution. Figure 33 shows his visual argument that there must be general laws describing how temperature varies across the Earth's surface. Humboldt's chart stretches from the east coast of America on the left over to Asia on the right, and it illustrates a new and crucially important statistical approach to nature. Instead of plotting actual temperatures on any particular day, Humboldt calculated the annual mean temperature for each place, thus amalgamating many thousands of observations into a few curved lines, called isotherms. By averaging out fluctuations, Humboldt ordained and displayed global regularity.

Humboldt was a visual innovator. Although it now seems obvious that diagrams enable scientists, advertisers, and politicians to summarize evidence and present it persuasively (if not always fairly), Figure 33 is an extremely early example. In the first half of the nineteenth century, graphs, bar charts, and so forth were only just being introduced, and they were slow to catch on. Scientists trying to

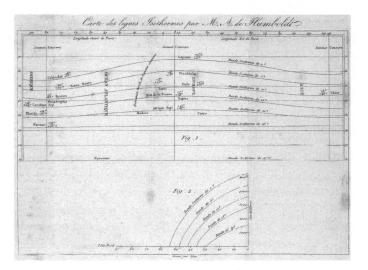

FIG. 33 Alexander von Humboldt's 'Chart of isothermal lines'. *Annales de chimie et de physique* 5 (1817).

interpret diagrammatic data had to learn a new visual language—just like reading, deciphering graphs and maps only becomes automatic with practice. Even contour lines, which directly represent actual mountain heights, seemed alien and were not routinely used until the early twentieth century. Humboldt's isotherms involved yet a further conceptual leap, because they were idealized summaries with no physical reality. By recording averages as lines, Humboldt made statistical regularities visible, short-circuiting masses of detailed numerical readings to present scientific relationships at a glance.

This impetus to think and understand through diagrams was encouraged by new printing techniques, which made it possible to reproduce images cheaply and also to incorporate them within the text rather than binding in separate sheets of paper. Gradually, ingenious visualizing techniques became important in many scientific disciplines. Faraday, for example, knew little mathematics but was an inspired three-dimensional visualizer who developed the concept of electromagnetic fields by imagining lines of force extending out through space with a quasi-real existence. Geology's great visual innovator was Darwin's friend Charles Lyell, who included an increasing number of diagrams in successive volumes of his hugely influential *Principles of Geology* (1830–3). As geologists learnt how to interpret schematic cross sections down through the Earth's crust, they gradually acquired the skill of automatically translating the vertical scale into vast expanses of time.

In his search for unifying laws, Humboldt integrated human society and the natural world. By analysing the globe environmentally, he effectively divided the American continent into two stereotypes—the northern temperate region capable of resembling Europe, and the southern tropics where nature flourishes exuberantly but high culture is impossible. With words and pictures, Humboldt portrayed equatorial America as a wild lush region where human travellers confront the full force of nature's mysteries. Piling on the drama, he evoked unstoppable torrents and invasive vegetation, rendering the local people as forest exotica, dumbly waiting to serve their civilized visitors:

> When the cornice was so narrow, that we could find no place for our feet, we descended into the torrent, crossed it by fording or on the shoulders of a slave...The Indians made incisions with their large knives in the trunks of the trees, and fixed our attention on those beautiful red and yellow golden woods, which will one day be sought for by our turners and cabinet makers.[1]

Humboldt's personal vision strongly influenced how the New and the Old Worlds viewed themselves as well as each other. These complex relationships are symbolized by Figure 34, the frontispiece of his *Atlas of America*, which portrays the multiple bonds between science, commerce, and politics. The two Europeans, the goddess of wisdom (Athena) and the god of trade (Hermes), have their arms round one another as they console the Aztec warrior they have conspired to vanquish. Emphasizing the youth of New World societies, the upturned statue (lower left) is deliberately primitive, while the scattered ruins of Mexican culture represent political turbulence and correspond to the background volcano, emblem of natural upheavals. This snowy mountain is Ecuador's Mount Chimborazo, site of Humboldt's personal glory—after nearly reaching its peak, he boasted that he had climbed higher than any other man. Its horizontal division is another of Humboldt's visual devices, indicating how he had averaged masses of data to condense Latin America's climate and agriculture into distinct environmental zones. Just as his terrestrial physics has imposed order on the young continent's powerful forces of nature, so too European civilization will tame its unruly people.

Explorers are never neutral observers. However accurately and conscientiously they record data, they select and interpret from a personal perspective. Instead of portraying a primitive continent bursting with tropical nature, Humboldt could have chosen to emphasize its well-organized agricultural cultivation. His perceptions of America were coloured by his knowledge of recent archeological expeditions to Egypt; in turn, Humboldt's own accounts of southern America pervaded his successors' attitudes towards Africa and Asia. Thanks largely to his self-promotional campaigns, Humboldt became a romantic icon who inspired Charles Darwin and many other young men to hazard their lives by travelling to remote parts of the globe. Like Humboldt, imperial explorers enticed their audiences by painting environments and peoples in lurid colours, visualizing themselves as conquerors

Voy. de Humb. et Bonpl.

FIG. 34 Frontispiece of Alexander von Humboldt's *Atlas géographique et physique du Nouveau Continent* (*Geographical and Physical Atlas of the New Continent*, 1814), engraved by Barthèlemy Roger after a drawing by François Gérard.

who had overcome nature's extremes and civilized the indigenous inhabitants. Concealing their dependence on local expertise, they took over the knowledge of their guides and presented themselves as lone scientific discoverers.

Humboldt stressed that finding global laws called for obsessive data collection, but without any systematic coordination, charting the globe proceeded erratically. Although it now seems clear that it would have made sense for countries to pool resources and exchange information, governments needed to be convinced that science was worthwhile. Funding was easier to raise for practical projects that glorified individual nations, and—as Humboldt stressed—investigating terrestrial magnetism promised to improve navigation. In the 1830s a group of British scientists, many of them associated with the BAAS, decided to collate magnetic measurements from all over the world.

The campaigners were constantly torn between the conflicting demands of collaborating with foreign informants, and of contributing to national glory by competing against their political rivals. However enthusiastically they plugged scientific progress, it was the lure of beating France and America that eventually convinced the British government to sponsor an Antarctic expedition. Financing international networks of observation posts was even more difficult. Nevertheless, all the lobbying proved effective, and by the mid-nineteenth century, places as far apart as Philadelphia, Peking, and Prague were sharing magnetic information. Gradually and sporadically, networks of laboratories came to span the globe, constantly monitoring patterns of weather, tides, and other variable phenomena. But they were spaced unevenly and staffed by observers of varying ability, so that individual initiatives remained crucial for determining whether and how measurements were taken. For investors, establishing global scientific laws had a low priority.

Global communications seemed a far more attractive investment. From the 1840s, governments and private companies poured money into electric telegraph systems, which were first used in Britain to send messages along railway lines: one of the earliest allowed a murderer to be caught at Paddington station. Subsequently, laying underwater cables enabled messages to be transmitted almost instantaneously from one country to another, initially between France and England, and later around the globe. As with so many technological innovations, there was no single eureka moment, no lone inventor who transformed international communications overnight. The most famous pioneer is the American Samuel Morse, celebrated for flashing an iconic message taken from the Bible—'What hath God wrought?'—in his dot-dash code from Washington to Baltimore. Bright ideas alone are rarely enough, and Morse benefited from knowing how to gain financial backing and exploit the patent system. Countless other early experimenters, each with their own heroic story of struggle, have been largely forgotten—the Russian inventor who linked the tsar's summer and winter palaces, the American who purloined his wife's silk underwear to insulate his electromagnets, and (unfortunately, more typically) the British electrician who fled to Australia, unable to afford the patent fees and squeezed out by his competitors.

Above all, the global telegraph system was a product of the British Empire. Britain took the definitive lead in the middle of the nineteenth century. Rival nations were unable to compete with her reservoir of electrical expertise, her massive financial investments in underwater cables, and her control over supplies of natural resources from her colonies, such as Malayan gutta-percha for insulation. Engineers insisted that for the telegraph system to operate globally, every country should use the same measurements—and because Britain dominated telegraphy, British electrical units became standard all over the world.

Science was inextricably linked into this imperial–technological–commercial complex. The new discipline of electromagnetism had inspired the inventions that made telegraphy possible; reciprocally, and just as significantly, establishing global telegraphic networks stimulated further research, and many innovations originated in colonial development sites rather than in metropolitan centres. Because telegraph scientists needed to monitor signals, they invented sensitive instruments such as resistance coils and condensers which later became standard equipment in scientific laboratories. Victorian imperialists enthused that the telegraphic network resembled a giant nervous system that connected the brain of London to remote regions like the sensitive tips of a starfish's limbs feeling out sources of food. As the Empire expanded, these electrical tentacles of communication wrapped themselves round the globe, despatching commands to ensure central control, but also depending on essential information generated overseas.

Solving the practical problems of transmitting messages over long distances encouraged scientists to develop different theories about how electricity travels. Most French and German scientists focused on the interactions between electrical particles and currents. In contrast, British physicists involved in telegraphy started to think about the role of space, the zone around the cable rather than inside it. Perplexed by strange effects, they turned to Faraday, who revived and developed his visions of electromagnetic fields extending out through an apparently empty universe. Following Faraday, field models have become central to modern theoretical physics—but Victorian electromagnetism was nurtured by practical concerns of the telegraph industry.

Humboldt thought of himself as the first terrestrial physicist, but electricity was the new global physics of the nineteenth century. As Britain became the richest and most powerful nation, she ruled her colonies with telegraphic networks, imposed her electrical units on international science, and dominated theoretical physics with field theories derived from cable telegraphy. Britain's own nerve-centre for electricity was Glasgow, where William Thomson (later Lord Kelvin) was the world's leading telegraph physicist, the economic engineer who brought government, industry, and science together when (after some failed attempts) he successfully laid down the trans-Atlantic telegraph cable in 1866. Like Humboldt, Thomson linked the New and the Old Worlds together—and Humboldt would have approved of Thomson's quantitative approach to abstract theories. 'When you can measure what you are speaking about and express it in numbers you know something about it,' Thomson insisted; 'but when you cannot measure it…you have scarcely, in your thoughts, advanced to the stage of *science*'.[2]

3　Objectivity

Mind, they say, rules the world. But what rules the mind? The body
(follow me closely here) lies at the mercy of the most omnipotent of
all potentates—the Chemist.

—Wilkie Collins, *The Woman in White* (1860)

Victorian scientists worshipped Isaac Newton—or rather, the paragon of
rationality they imagined him to be (Figure 31). Suppressing all reports of his
alchemical experiments and episodic insanity, they made him resemble 'Nietzche's
description of "the objective man", a passionless being concerned only to "reflect"
such things as he is tuned to perceive'.[3] Like a scientific instrument, Newton
supposedly recorded neutrally the world about him, and then analysed his data
with detachment. Taken to an extreme form, Newton epitomized a pervasive if
unattainable scientific stereotype—the selfless genius who measures the Universe
as though he were an external observer.

Many people questioned whether such objectivity was either possible or desirable. These doubts became especially strong in Germany during the first half of
the nineteenth century, when Romantic philosophers, writers, and artists aimed
to transcend distinctions between the physical and the human worlds, between
abstract research and inspired creativity, between science and literature. The most
outspoken enthusiasts for these idealistic views were known as the *Naturphilosophen*
(in English, this German label is used to distinguish them from natural philosophers in general). The '*Natur*' indicates their belief that, as human beings, we are
inextricably entangled within the natural world. It is impossible for us to step
outside—we cannot prevent our own minds from constructing in advance how
we are going to analyse and interpret what we see.

Science was, declared the *Naturphilosophen*, heading in the wrong direction. A
disparate group with no collective manifesto, they searched for grand theories that
would unite the Universe by incorporating living beings within its development.
Whereas Newton, Descartes, and the mechanical philosophers thought of creation as a giant astronomical clock, they envisaged a cosmic organism and believed
in an organic, growing nature, a cosmos that is itself alive. If this attempt at a

FIG. 35 Woodcut designed by Johann Goethe to accompany his *Optical Lectures* (1792).

summary sounds vague and obscure, then that reflects the grandiose imprecision of their own writing. But diverse and esoteric as they were, the *Naturphilosophen* exerted an enormous influence, both immediately on nineteenth-century science—Faraday's electromagnetism or Humboldt's terrestrial physics, for instance—and in the long-term, on topics as varied as evolution, quantum mechanics, and environmentalism.

Had the *Naturphilosophen* wanted a corporate logo, they might have chosen Figure 35, taken from a pack of playing cards designed by Johann Wolfgang van Goethe for his early lectures on optics. Yes, Goethe—although nowadays better known as Germany's Shakespeare, Goethe was an active scientific researcher, a minerals expert who owned 18,000 rock samples and also participated in international discussions about biology and optics, especially colour. Goethe's subjective approach towards scientific experimentation deliberately included the observer's own reactions. Here his Masonic eye stares out aggressively, projecting rays of illumination to dispel the dark clouds of ignorance. For Descartes, Newton, and the champions of objectivity, a prism or a lens is used to produce discrete images that can be inspected with detachment, as though the eye were itself an instrument. But according to Goethe, human beings are themselves inevitably involved in the observations they make. He looked directly at his prism, so converting the retina inside his own eye into the projection screen. For Goethe, science belonged out in the world, not just in laboratories—he examined the effects on himself of gazing at a woman's bright clothes or a snow-covered mountainside, and in his novel *Elective Affinities*, modelled marital exchanges on molecular transformations. Rather than denying his imagination, Goethe claimed that as a creative Romantic genius, his emotional intensity and heightened awareness would help him found a more humane type of scientific knowledge.

This picture's rainbow symbolizes Goethe's hostility towards Newtonian optics. Whereas Newton had maintained that colours already exist blended together in sunlight, Goethe insisted that they arise from the conjunction of polar opposites—just look, he said, at the coloured fringes you see against a sharp black/white edge. For British scientists, national honour was at stake, and they defended Newton by ridiculing Goethe's ideas, often in highly charged invective at odds with any notion of scientific decorum. In contrast, German physiologists incorporated Goethe's person-oriented chromatic science into their studies of perception. Many Romantic experimenters (including Samuel Taylor Coleridge, an important conduit for *Naturphilosophie* into Britain) welcomed Goethe's emphasis on polarity, which resonated with their own investigations into magnetic, electrical, and chemical activity—north and south, positive and negative, attractive and repulsive. Just as Goethe used his own eye as a recording instrument, they made their own bodies part of electric circuits (the pain they endured suggests that they, rather than Galileo, deserve to be called 'martyrs of science').

The *Naturphilosophen* have been squeezed out of the history books because—to put it extremely crudely—the ideology of objectivity won during the nineteenth century. Scientists claimed to show the world as it really is, although this goal proved well-nigh impossible to attain. For one thing, in order to avoid ending up with a record of the Universe as big as the Universe itself, selections and summaries must be made—an obvious entry point for subjectivity. Once you have picked a particular plant, crystal, or magnet to examine, you have no guarantee of how typical it is, however meticulously you document its appearance and behaviour.

One way of tackling this dilemma is to depict an ideal version, the finest possible example imaginable, an unrealizable amalgam or distillation of perfect components. This was the solution adopted during the Enlightenment, when artists deliberately improved the specific subject in front of them. The portraits of Joseph Banks and his scientific contemporaries do not depict them brutally as they really were, because painters selected appropriate poses and exaggerated certain features to present their human subjects as ideal types—explorers, administrators, surgeons. Even anatomists who prided themselves on accurate representation converted *Homo sapiens* into *Homo perfectus* by drawing originals who conformed to expectations. They liked to choose male skeletons with big heads and long legs, whereas female ones can seem unusually small, their narrow ribs and wide hips perhaps reflecting deformations due to wearing tight corsets. Similarly, although Pieter Camper measured his skulls accurately (Figure 24), the squared grid lines merely lend an aura of objectivity.

Victorian scientists were appalled to think that subjectivity might lie at the very heart of science. One strategy they adopted was to reject the Enlightenment concept of ideal, universal forms, and to insist that naturalists draw the actual specimens in front of them, warts and all. Scientists were exhorted to exert self-discipline and behave as if they were recording instruments producing an objective view. Any professional artists they employed were closely supervised to prevent them from introducing personal flourishes that might make their images aesthetically pleasing rather than scientifically accurate. The next logical step was to eliminate human observers altogether and replace them with machines, which could be controlled more easily than people. Recording devices would, their inventors promised, provide a direct transcript of the world with no human intervention. For instance, doctors could access the body directly with thermometers and stethoscopes, while photographs would settle once and for all the question of life on the Moon.

Yet whatever defensive steps scientists took to ensure objectivity, personal assessments kept sneaking back in again. Even self-recording instruments raised problems. For example, one ingenious medical device monitored a patient's pulse with a sensitive needle that traced out a zigzag pattern on a piece of glass. But although faithful, this undulating pattern was in itself of little value, and did not immediately reveal its owner's state of health. Doctors complained that 'the record is written in a language which we are only beginning to understand…the oscillations of the lever are quite as meaningless as the vibrations of the telegraphic needle to one who is not furnished with a proper alphabet'.[4] Deciphering the diagnostic graph entailed learning how to relate marks on the glass to the physical events that had caused them—a process of interpretation that demands expertise, experience, and individual judgement. As data collection machines became more sophisticated during the twentieth century, the problem became still more acute. Magnetic maps, X-rays, and cloud-chamber photographs are packed with detailed information—but only for experts who have learnt how to translate them. And because experts are people, they do not always reach the same conclusions.

Even the camera that never lies revealed different versions of the truth. Faraday was an early convert. 'No human hand has hitherto traced such lines as these drawings displayed,' he enthused; 'what man may hereafter do, now that dame Nature has become his drawing mistress, it is impossible to predict'.[5] Yet not everyone was convinced that photography provided unmediated access to the natural world. For one thing, there were technical problems to be surmounted—long exposure times, fragile plates, paper that shrank or stretched and so precluded accurate measurement. And then there was photography's reputation to consider. Spiritualists were exposing visitors from the other side, while money-making opportunists were touting stereoscopic views of the Moon and naked women. How could such a source of entertainment be a legitimate scientific tool?

The word 'photograph' was invented in 1839 by the astronomer John Herschel, and astronomy became the stellar photographic science. Astronomical photographs of distant planets and swirling nebulae captivated Victorian newspaper readers, while Herschel himself was photographed as an inspired genius with a halo of wild white hair. But this was no overnight scientific success story. For one thing, innovators disagreed about the best use of photography in astronomy. Commercial entrepreneurs boasted that the new technology revealed new phenomena, such as the flares visible around the Sun during an eclipse, and they promoted photography as an exciting exploratory tool that would uncover hidden secrets of the cosmos. In contrast, professional astronomers were more interested in accuracy. Responsible for supervising teams of inconsistent observers, they hoped that photography would replace people with precision.

For both camps, human intervention proved essential, and objectivity remained elusive. Because printing photographs involved lengthy and expensive processes, originals were hand-copied for mass reproduction, so that most people saw drawings rather than direct transcripts from nature. Rather than replicating photographs with painstaking accuracy, engravers enhanced images to make important features stand out more clearly. Near the end of the nineteenth century, continuous automatic scanning of the heavens did become possible, and it seemed that perhaps—at last—human fallibility could be eliminated. But now there was a new problem—covering the entire heavens would yield an unmanageable pile of astronomical photographs thirty feet high. Faraday's dream of accessing 'dame Nature' directly seemed impossible to realize.

Photography made its first impact not on science, but on portraiture. Scientists soon took advantage of the new recording medium to advertise themselves, posing stiffly in studios clutching skulls or geological samples. Later, they turned to capturing other people. Although their cameras were supposedly neutral observers, the finished photographs were as artificial as their own publicity shots. Supposedly dispassionate data collection lent itself to social control. For example, doctors in mental asylums produced poignant images of derangement that reinforced existing prejudices and justified keeping patients locked away. Scientists started to catalogue different types of human beings, emphasizing the objectivity of their photographic gaze by reducing their subjects to specimens—colonized subjects forced by anthropologists to stand naked against a measuring grid, convicts reduced to full-face and profile shots that could be quantitatively compared.

By photographing people they judged to be abnormal—mental patients, other races, criminals—scientists were defining what it means to be normal. They used photography, which was supposedly an objective classification tool, to back up personal assessments of who should be fully accepted as members of society. This meant grappling with the problem of representing a group—the insane, respectable citizens, Africans—by somehow encapsulating the characteristics of the individuals within it. The eighteenth-century solution had been to depict an ideal type. The Victorians adopted a different approach—they developed statistical methods for finding an average. To strengthen the illusion of detachment, they gave normality a new numerical foundation.

Statistical thought pervaded nineteenth-century life. Rather than being an esoteric mathematical speciality, statistics not only dominated scientific research but also became a vital weapon for social reformers—Florence Nightingale was an early campaigner, using her data to argue that hospital hygiene cut costs as well as deaths. Statisticians prided themselves on practising the ultimate object-ive science, in which only facts counted and opinions were banned. 'The dryer the better,' intoned a cholera expert—'Statistics should be the dryest of all reading.'[6]

One expert who tried to give numerical data an immediate impact was Francis Galton, Charles Darwin's cousin, an obsessive data collector who devised various clever ways of displaying statistical information. As well as condensing masses of meteorological figures into weather maps, Galton also invented a photographic machine to perform what he called 'pictorial statistics'. First he photographed several members of a category—murderers, sisters, men with syphilis—and then he superimposed their individual pictures to generate a composite image. The top row of Figure 36 illustrates his attempts to identify criminals by this process of mechanical accumulation, betrayed here by ghostly outlines fringing the clearer central faces. The four lower pictures reveal Galton's snobbish assumption that the class difference between officers and privates will be stamped onto their appearance.

Galton's composite photographs portray visually the normal distribution, the bell-shaped curve that falls off symmetrically on either side of the central aver-age, or mean. It is often called the Gaussian distribution, because it was intro-duced by the German mathematician Karl Gauss to estimate the errors involved in astronomy. Gauss showed that if the same measurement is repeated several times, the readings will cluster in a narrow curve around the average: the wider the curve, the more likely that any one particular reading will be incorrect. By enabling scientists to calculate the reliability that could be attached to their results, Gauss's techniques added mathematical conviction to claims that measurements were objectively accurate, uncontaminated by human mistakes.

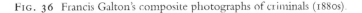

Comparison of Criminal and Normal Populations.

FIG. 36 Francis Galton's composite photographs of criminals (1880s).

Nevertheless, this numerical concept of normality enabled subjective judgements to creep right back in again. It was only a short step from describing to prescribing, from social mapping to social engineering. Galton was just one of many Victorian scientists who believed that measuring physical characteristics would yield unbiased knowledge of people's mental abilities, psychological tendencies, and racial origins. Criminals, for instance, allegedly lay at the tail ends of normal distributions, because relative to the average, they had receding chins and long arms, suggesting that they were somehow degenerate beings of a lower grade than Victorian gentlemen. Conversely, geniuses were said to be unusually thin with bulging brows—no coincidence that Sherlock Holmes matched that description.

Looking back, it seems obvious that, like Camper's project, these arguments were laced with presuppositions and circular loops. In the name of objectivity, scientists emphasized the importance of photographic evidence and precise measurement, but these two discriminatory tools were also deployed in the twentieth century by political parties waging campaigns for what they claimed was social purification. For example, in Germany Nazi activists distributed small cards comparing Jewish and Aryan eye colours, and in Sweden, where people were still being forcibly sterilized in the 1960s, doctors compiled photographic galleries of different racial and psychological types. Clearly, Galton was not to blame for the Holocaust—but perhaps the *Naturphilosophen* were right to question the ideal of objectivity.

4 God

It was the Lord's design he made apparent—
These bands and blocks of azure, umber, gilt,
Set in their flexing contours, solid flow
That had composed itself in its own frame:
Red garnet neighbouring mica, silver white;
A slice of agate like an inland sea...

Clive Wilmer, 'Minerals from the
Collection of John Ruskin' (1992)

In December 1871, the Prince of Wales was hovering near death, desperately ill with typhoid. The Archbishop of Canterbury and his allies moved into action, flashing out orders through the electric telegraph system for special prayers to be read in churches all over the kingdom. The Prince soon recovered, but the nation was divided—had God intervened, or was modern medicine responsible for this apparently miraculous cure? An eminent surgeon suggested that the issue be resolved statistically, making one particular hospital ward the target of prayers for a few years to see if its success rate improved. Although this holy trial was never carried out, the Prayer Gauge Debate continued for years—was disease a punishment under divine law, or could it be prevented by obeying scientific laws of health?

These arguments about praying might look like a direct conflict between science and religion, but they were not so much about who was right, but more about who could be trusted to decide what was right. Traditionally, authority lay in the hands of the Anglican Church, but during the nineteenth century British scientists started to claim power for their new rational priesthood. Ambitious scientists struggling to consolidate their reputation as elite experts squeezed out anyone they thought inappropriate. One move was to establish themselves as professionals by marginalizing those without full educational credentials. By adopting the pejorative label of amateurs, they set aside a large group of knowledgeable people—women, collectors, home-based astronomers.

Another tactic was to establish for the first time a sharp distinction between science and religion. Francis Galton deployed some strategic statistics, crunching through carefully selected samples to expose a supposed dearth of religious leaders on the councils of scientific societies. After a few logical leaps, he concluded that clergymen were no good at science—holding a theological vocation was, he declared, incompatible with being a competent scientist. The most eloquent spokesman attacking the Church was Thomas Huxley, champion of Darwinian evolution and inventor of the word 'agnostic'. Huxley landed his most celebrated coup during a public debate at Oxford, sneering that he would rather have an ape for an ancestor than the bigoted bishop opposing him. Although that may well be an apocryphal tale, Huxley definitely did ferociously condemn anyone who imagined that 'he is, or can be, both a true son of the Church and a loyal soldier of science'.[7]

By parodying the religious opposition, Huxley made Darwin's ideas sound better. Even so, his aggressiveness does indicate how deeply theological issues were entrenched within scientific research during the middle decades of the nineteenth century. Broadly speaking, there were two major themes. One strand of arguments related specifically to biblical theology. Evidence from fossils and rock formations suggested that the Earth was far, far older than suggested in the Bible; more controversially, theories of evolution contradicted traditional beliefs that life has remained unchanged since its creation by God. However, for many Victorians, the scriptural accounts represented powerful metaphors rather than literal reality, so that any contradiction of biblical details was not a prime concern. Instead, science's critics were more worried about the philosophical implications of recent ideas. Christians believed in a teleological cosmos, one created by an omniscient God, a Grand Designer, for a specific purpose. This comforting view was threatened by the new statistical methods in physics, and also by Darwin's theory of evolution, which assumes that chance may intervene between generations to introduce new characteristics.

God had been forcefully excluded from astronomy during the French Revolution, when Pierre-Simon Laplace rewrote Newton's ideas to create his deterministic cosmos, in which scientific laws govern every movement of every planet with no need for divine intervention. Inspired by this success, a Belgian astronomer called Alphonse Quetelet decided that human societies are also controlled by laws. Each country has its own statistical patterns that remain constant from year to year—suicide and crime rates, for instance—and so Quetelet suggested that an 'average man' can consistently encapsulate a nation's characteristics. Politicians should, Quetelet prescribed, operate like social physicists and try to improve average behaviour rather than worry about extreme anomalies. For him, variations from the statistical mean were—like planetary wobbles—imperfections to be smoothed out so that overall progress could be ensured.

Quetelet had introduced a radically new way of thinking about human beings. As one of his admirers put it, 'Man is seen to be an enigma only as an individual, in mass, he is a mathematical problem.'[8] Quetelet's successors took his ideas in many different directions. For one thing, his work was valuable politically because it could be interpreted in different ways. While conservatives insisted that little could be done to alter the current system, radicals accused governments of impeding the natural course of progress, and utopians—such as Karl Marx—envisaged harmonious societies governed by nature's own laws guaranteeing improvement. Data collection projects proliferated, and statisticians searched for laws governing every aspect of life, ranging from the weather to the growth of civilization, from stock market fluctuations to the incidence of disease. Many scientists took their ideas from Quetelet rather than from abstract textbooks—but they added their own twist. Whereas Quetelet regarded individual deviations from the norm as errors to be eliminated, scientists set out to study how variations occur.

In physics, the most important application of statistics was to gases, a topic of major interest in steam-driven Europe, because studying thermodynamics—the links between heat, movement, and power—promised to make industrial factories more efficient. In 1873, James Clerk Maxwell, a Scottish scientist at Cambridge, stunned his audience at the annual BAAS meeting by explaining how, like human populations, the overall behaviour of a gas can be described by looking at the average velocity of its molecules, which he assumed to be moving randomly. Maxwell maintained that it is impossible to establish absolute, comprehensive knowledge of gases and other microscopic systems—only statistical certainty, with its fuzzy edges, is ever possible.

Far from being simply a neat mathematical trick, this move posed fundamental questions about determinism. Quetelet's equivalent bid to describe human societies statistically had already generated intense debate. If on average ten people commit suicide every year, is their destiny decided in advance, or does each individual retain free will? The majority view was that free will is basic to human life—but extrapolating to physics was a different matter. Giving molecules the ability to make decisions would remove the distinction between spiritual beings and inert matter, introducing a philosophy of materialism completely counter to Christian theology. Another problem was Maxwell's assumption that gas molecules move randomly, which smacked of a cosmos arising from chance rather than one planned by God. Statistical techniques flourished in science because they worked, but intractable theological problems continued to plague Maxwell as well as his critics.

Geology raised questions more directly related to Christianity. Genesis, the first book of the Bible, describes how God fashioned the Universe out of nothingness in six days. Some believers clung resolutely to this account, although they were less concerned with God's speed than with the notion that He had created people, animals, and plants exactly as they are now—no room in their views either for extinction or for new species. Their other major preoccupation was the age of the Earth. Bibles often carried a small marginal note explaining that the world originated on Sunday 23 October 4004 BC, a carefully calculated date more or less agreed on by ecclesiastical chronologists (BC, Before Christ, is now generally written as BCE—Before the Current Era).

During the eighteenth century, challengers of this orthodoxy had proceeded cautiously, often preferring accommodation to confrontation. For instance, following the French Newtonian naturalist Georges Buffon, they ingeniously constructed terrestrial histories divided into six epochs, corresponding to the six days of creation in the scriptures. The central problem confronting geologists was to explain how rocks that have clearly been built up from piles of sediment accumulating under water are now on dry land—a transformation conveniently accounted for in the Bible by Noah's flood. One school of geologists—the Neptunists, named after the Roman god of the sea—believed that a vast ocean once covered the Earth's surface, but has now dried up (they avoided explaining where all the water has vanished to). The Neptunist movement was particularly strong in German mining academies, where classifying rocks was a major practical concern. Their opponents were called the Plutonists, after the god of the underworld. They argued that violent earthquakes, caused by the Earth's high internal temperatures, had pushed up areas of dry land from below the surface.

The most influential Plutonist was James Hutton, who moved in the intellectual Edinburgh circle of the engineer James Watt and the economist Adam Smith. After getting bored first as an industrial chemist and then as a gentleman farmer, Hutton turned to geology, viewing the landscape with an eye to its agricultural potential. Geology was, he discovered, hard work—saddle-sore during a field trip to Wales, he moaned 'Lord pity the arse that's clagged to a head that will hunt stones.' Most importantly—and this was a radical innovation at the end of the eighteenth century—Hutton imagined a continuous and extraordinarily slow cycle of change: eroded by wind and water, rock particles are deposited on the sea floor where they harden into sediment before being thrust up into mountains again. This steady-state system demanded unimaginable aeons of time. As Hutton put it dramatically and controversially, 'we find no vestige of a beginning,—no prospect of an end'.[9]

Theologically, this scheme was interpreted in different ways. Although Hutton himself was unconcerned about Genesis, God was central to his cosmos, which he envisaged as a giant perpetual motion machine devised expressly to make its inhabitants happy for all eternity. Some critics were scandalized. Hutton had blatantly contradicted the 6,000 years laid down in the Bible, and they reinstated the Flood. Rejecting Hutton's gradually changing system, they focused on dramatic upheavals, catastrophes such as earthquakes and floods. In contrast, French and German geologists ignored the Bible but had other reasons for favouring catastrophic schemes. How else to explain the presence of gigantic rocks strewn across the landscape?

When the French anatomist Georges Cuvier examined the rocks around Paris, he found that they lay in distinct bands or strata, each with its own characteristic type of fossil. It seemed clear to Cuvier that a series of violent convulsions had separated one era from another—and many geologists agreed. Converts set out to garner evidence of catastrophic events from all over the globe. An Oxford academic found a muddy hyena den in Yorkshire (clear testimony of the Flood, he concluded), Alexander von Humboldt reported on volcanoes in South America, and a myopic British lawyer—Charles Lyell—trained up his wife, Mary, and took her with him to Italy. The sights of Italy convinced Lyell to change his mind, and with Mary's help, he produced *Principles of Geology* (1830–3), three enormously influential volumes that rejected catastrophism and revived Hutton's views.

Change, declared Lyell, occurs slowly and uniformly, over long (seriously long) periods of time, and at the same rate in the past as now. As usual with new ideas, there was no overnight conversion. Like Galton and Huxley, Lyell was determined to separate science and religion—to free geology from Moses, as he pithily expressed it. By this, he did not mean that only non-believers could be scientists—after all, many of his geological allies were Christians—but rather that science could only gain prestige by divesting itself of men who clung to older ways of thought. The Victorian scientists who fought for social authority were not trying to eradicate faith, but rather to prune their community from within by excluding anyone who based scientific hypotheses on religious arguments.

Lyell was extremely proud of his frontispiece (Figure 37), embossed in gold on the cover of de luxe editions. The temple at Serapis, near Naples, was a famous monument of classical civilization, but for Lyell, it was also a monument to geological events that had occurred far earlier. The dark bands on the pillars have been left by marine molluscs, showing that the original building first subsided below sea level and was later pushed up again. Serapis encapsulates the major tenet of Lyell's theory—that small-scale changes, occurring at a steady slow rate, are responsible for even the most spectacular features of the Earth's surface. Contrary to Christian theology, this is a world that continues in a steady state, with no in-built pattern of progress, with no arrow of time pointing inexorably from the past to the future.

Human and geological history are intertwined in Lyell's image. A modern man stands next to the columns, while the seated figure contemplates this double testament of the past. This integrated vision represented a radical shift in people's

FIG. 37 *The Temple of Serapis.* Frontispiece of Charles Lyell, *Principles of Geology* (1830).

perceptions of themselves and their relationship to the cosmos. According to the biblical account, there is only one type of history—human affairs during the past six thousand years since the Earth was created as it is now. Geologists forced Victorians to contemplate a deeper version of time, an uninhabited world that extended back unimaginably far into the past, long before life existed. Huxley made the point eloquently in a lecture to some workers in East Anglia—'A great chapter in the history of the world is written in the chalk.'[10] Examining the chalk beneath their feet or in a carpenter's pocket would, Huxley proclaimed, make them more knowledgeable about the past than learned scholars buried in their books.

Time and space were both stretched out during the nineteenth century. Powerful telescopes revealed vast expanses of space containing apparently never-ending vistas of stars, nebulae, and other planetary systems. Looking outwards towards the sky involves looking backwards in time, because although light travels fast, it does not arrive instantaneously—the further away you can see, the longer ago the light started out on its journey. Boring down into the Earth also meant travelling back into the past (which is why Jules Verne—who kept up with the latest scientific ideas—put prehistoric monsters at the centre of the Earth). Although geologists hesitated about how many zeros they should add to the age of the Universe, it was clear that even the most ancient human kingdoms were merely seconds away on the geological clock.

This expansion of geological time not only rocked science but was also a major transformation in European thought. Like placing the Sun instead of the Earth at the centre of the Universe, it radically diminished the significance of human life. No coincidence that the nineteenth century's most widely read poem was *In Memoriam*, a meditation on God, nature, and life prompted by a young man's premature death. In this famous elegy, Alfred Tennyson—who had read Lyell and kept up with the latest scientific debates—grappled with the anguish of an uncertain, non-teleological cosmos. How, he agonized, could he relinquish his Christian trust 'That nothing walks with aimless feet'? Like many literary men of the nineteenth century, Tennyson was scientifically well read, and he adapted Lyell's own description of the perpetually shifting landscape to emphasize that civilization is but a recent event in the Earth's long history:

> There rolls the deep where grew the tree.
> O earth, what changes hast thou seen!
> There where the long street roars, hath been
> The stillness of the central sea.[11]

5 Evolution

He begins to sense a religious content as its significance swells...
[Darwin's] five hundred pages deserved only one conclusion: endless
and beautiful forms of life, such as you see in a common hedgerow,
including exalted beings like ourselves, arose from physical laws, from
war of nature, famine and death. This is the grandeur. And a bracing
kind of consolation in the brief privilege of consciousness.

Ian McEwan, *Saturday*, 2005

'Nature, red in tooth and claw'—Tennyson's *In Memoriam* famously evokes
the cruel competitiveness underlying Charles Darwin's theory of evolution. But, but, but...although focusing on dates is generally a boring way of
thinking about history, it can be revealing. Tennyson published his elegy nine years
before Darwin's *On the Origin of Species* appeared in 1859. Darwin and evolution
have become virtually synonymous; yet the basic concept of evolutionary change
had been around since before his grandfather's time. Darwin's model was rejected
for decades and was never fully accepted—even the so-called Darwinian synthesis
of the twentieth century was very different from his original formulation.

Darwin now rivals Newton as Britain's major scientific genius; yet in his own
lifetime he was a controversial celebrity. The invective was virulent and public—
one reason why Darwin spent much of his life as a virtual recluse in his country
home. His Victorian critics found it hard to accept that human beings were not
created separately, but were descended from other animals. Although most people
agreed with some form of evolution by the time that Darwin published, countless
caricatures—including Figure 38—depicted him as a monkey, a supreme insult
in nineteenth-century Britain. Here his heavy eyebrows have been exaggerated
to give him an apelike appearance, and his prehensile tail is far longer than his
philosophical beard. This simian Darwin holds his left hand in a warning gesture,
a parody of a Pope's blessing that mocks his sacrilegious theories.

Debates about evolution were ferocious because far more than a scientific
hypothesis was at stake. Even the bitter arguments about the Bible were to some
extent only a colourful cover concealing deeper rifts, acting as steam vents for

FUN.—November 16, 1872.

THAT TROUBLES OUR MONKEY AGAIN.

Female descendant of Marine Ascidian:—"REALLY, MR. DARWIN, SAY WHAT YOU LIKE ABOUT MAN; BUT I WISH YOU WOULD LEAVE MY EMOTIONS ALONE!"

FIG. 38 Caricature of Charles Darwin. *Fun* magazine, 16 November 1872.

fundamental passions. People's opinions about evolution reflected their basic essence, how they felt about themselves and their relationship to the world. The Christianized version of Aristotle's Great Chain of Being had enabled Europeans to envisage themselves safely ensconced at the top of an unchanging hierarchy, under instructions from God to superintend the world and use it for their own advantage. Rich hereditary landowners clung on to this reassuring vision, content to contemplate themselves and their descendants enjoying the privileges they regarded as their God-given rights. In contrast, political radicals welcomed the idea of change—if the natural world had evolved, they argued, then society could also be transformed to break away from tradition and redistribute the nation's wealth.

Evolutionary ideas originated in pre-Revolutionary France, but it was Darwin's grandfather Erasmus who first expressed them coherently—well, as long as you don't examine the details too closely. His central notion was that God had designed creatures who could improve themselves over time, a clear parallel to his own ambitions as a successful provincial doctor, a member of the rising middle classes gaining money and power through their own efforts as Britain industrialized. According to Erasmus Darwin, new organs gradually developed as parents handed down their own small gains to the next generation—just as he and his Lunar Society colleague Josiah Wedgwood, self-made factory owner, would later pass on their combined wealth to their grandchildren, Charles and Emma Darwin. In other words, characteristics acquired during an individual's lifetime could be inherited.

The classic example of this inheritance through life experience is giraffes, whose necks have supposedly stretched higher and higher over many, many generations. However, the man notorious for this view is not Darwin, but his younger contemporary, a French naturalist called Jean Lamarck, who worked in Paris's Natural History Museum. Lamarck adopted the suggestion of Georges Buffon that as the Universe gradually cooled down from its original molten state, life forms were spontaneously generated. Adding his own twist, Lamarck envisaged life forms that constantly improve themselves rather than staying fixed—although organisms originate from many different spontaneous creations, they steadily move upwards along preordained tracks. Lamarck would have been mortified to realize that he is now remembered for a relatively trivial part of his grand theory. For him, the crucial point was not acquired characteristics, but that life continually progresses.

Unfortunately for Lamarck, his major rival also worked at the Natural History Museum. An expert string-puller, Georges Cuvier made sure that he got promotion, that his opinions were listened to, and that Lamarck was eclipsed. Politically conservative, Cuvier consolidated his position by insisting on stability and rejecting Lamarck's principle of change. But although Cuvier himself was an anti-evolutionist, his work profoundly influenced later evolutionary theories because he used anatomy to classify animals into groups. Instead of focusing on a creature's external appearance, Cuvier studied its internal structure. For example, elephants, fish, and snakes might not resemble each other superficially, but Cuvier bracketed them together as vertebrates, because he felt the similarities between their skeletons sharply distinguished them from creatures without a backbone.

One of Cuvier's big innovations was to divide the animal kingdom into four basic types. This was a crucial move because he eliminated hierarchical order. Although he believed that the vertebrates are fundamentally different from the other three groups, which can be represented by oysters, spiders, and starfish, they are not intrinsically superior. Many naturalists found that hard to accept, but Cuvier had at least established the possibility of thinking in a non-linear, non-Chain-like way. He even scorned internal ranking—fishes and mammals were, he claimed, all vertebrates but were adapted, fine-tuned, to live in various habitats. English naturalists imposed a theological spin, using Cuvier's concept of adaptation to back up arguments for God as a loving designer. Ingenuity helped them justify everything. Even Tennyson's 'Nature, red in tooth and claw' could be explained away—predators compassionately rescue their prey from starving to death.

An opportunistic career-builder, Cuvier liked to boast that he could deduce an animal's entire skeleton from a single bone. An exaggeration—but by applying his anatomical expertise to fossils, Cuvier did bring palaeontological arguments into the debates about evolution. Fossils had long been valued as collectors' curiosities, but it was hard to fit them in with the biblical story of unchanging creation. Why should God have put them there—or were they perhaps the outcome of some hidden powers operating within the rocks? And what were those massive elephant-like skeletons that explorers in Siberia and America had started to unearth at the end of the eighteenth century?

By systematically examining fossils, Cuvier proved anatomically that the Earth was once home to species now extinct. Collectors sent him specimens and drawings from all over the world, and he demonstrated beyond doubt that mammoths and mastodons were different from modern elephants. Cuvier also carried out his own excavations near Paris. By reconstructing vertebrates from the fossil bones in deeper and deeper layers, he discovered that the older the rock, the more unfamiliar the creatures it contained. As a staunch conservative, Cuvier refused to accept change—successive catastrophic events had, he maintained, eliminated the species of each era (he never did explain satisfactorily where the next ones came from). Yet as fossil evidence continued to accumulate—including some dramatic dinosaurs—pro-evolutionists took over Cuvier's results to support their own views.

Political and religious attitudes were central to debates about evolution. Many people favoured Lamarck's concept of progress, but rejected spontaneous generation, with its implication that life could be created from matter—a materialist view transgressing Christian beliefs. To make progressive evolution palatable, an enterprising Scot called Robert Chambers presented change as part of a divine plan for the entire Universe. A self-made middle-class publisher, Chambers was politically committed to progress. He wanted not only to make money for himself, but also to use his steam presses for providing cheap yet instructive reading matter that would improve the working classes.

Cleverly, Chambers opened his *Vestiges of the Natural History of Creation* (1844) with relatively uncontroversial accounts of astronomy, but he quickly moved onto progressive geology and evolution, envisaging a law-governed process of progress towards higher and higher life forms, so that fishes, reptiles, and birds were followed by mammals and ultimately by human beings (unsurprisingly, he placed men and Caucasians higher than women and other races). At first, reviewers raved about the book's democratic message and eloquent style. Tennyson was thrilled to find ideas corresponding so closely to his own, and he incorporated much of Chambers's vision into *In Memoriam*. But soon *Vestiges* was being slated—scientists picked holes in the facts, conservatives detested its materialist accounts of life and intelligence, and almost everybody was horrified at the idea of being descended from animals.

Savaged so critically that sales rocketed, Chambers's book caused an international sensation. By the time Charles Darwin went into press fifteen years later, much of the passion against evolution had been defused. Darwin had read *Vestiges* only a few months after finishing a long draft for his own book, and he followed the controversy closely. Chambers was trying to establish a universal law of life, and Darwin was relieved to discover that he had already anticipated most of the criticisms likely to be levelled against his own theory—he approached a particularly harsh review of Chambers 'with fear & trembling, but was well pleased to find, that I had not overlooked any of the arguments, though I had put them to myself as feebly as milk & water'.[12] Nevertheless, Darwin watched and waited, avoiding public laceration by postponing publication, and constantly working on his grand theory. Meticulous to the point of obsession, he embarked on an eight-year study of barnacles.

Darwin had never been an obvious candidate for the accolade of genius. A mediocre student, he left university with a passion for bugs and geology, but few ambitions except to escape the clergyman's life prescribed by his father. Inspired by Humboldt, Darwin sailed round the world aboard the *Beagle*, and with Lyell's *Principles of Geology* as his travel guide, he interpreted the phenomena he saw in terms of change. For example, he concluded that coral reefs had risen and fallen like Lyell's Temple at Serapis, with plenty of time for slow transformations. In South America, Darwin discovered fossils similar to its unique living fauna, and he watched European colonialists who had survived by adapting to their alien environment. Did animals perhaps alter in reaction to their surroundings?

Generally a meticulous collector, Darwin unfortunately missed some important clues in the Galapagos Islands, where he indiscriminately stuffed bags full of specimens from different locations without listening closely enough to the locals. Too late, he learnt from them that tortoises in each island could be distinguished by the shapes of their shells, and he realized that he should have been more careful with his labelling. The muddle with his birds was only completely sorted out (by somebody else) after Darwin got back to England, when the contrasting beaks of finches on neighbouring islands provided valuable evidence in support of his theory.

For quarter of a century, Darwin observed, read, and observed some more, eventually arriving at his theory of natural selection. His lyrical prose reveals how scrupulously he studied, and how emotionally he responded. 'There is grandeur in this view of life,' he wrote in the famous last sentence of his book on evolution; 'whilst this planet has gone cycling on according to the fixed law of gravity, from so simple a beginning endless forms most beautiful and wonderful have been, and are being, evolved'.[13] From talking to pigeon fanciers and farmers, Darwin learnt how they could breed new varieties, picking out particular features to modify birds and animals for human requirements. In addition to these products of artificial selection, Darwin accumulated countless examples of natural adaptation—clover flowers that matched different bees, dandelion seeds light enough to be scattered far away, water beetles with fringed legs. Sometimes he got swept up in wishful thinking—can he really have believed that whales evolved from bears swimming with wide open mouths to trawl for water insects?

Darwin also thought about human societies. The work of Thomas Malthus, an eighteenth-century economist, was an eye-opener for him. Malthus opposed social reform—if you improve conditions and encourage people to breed, he argued, the population will rapidly outstrip the available food supply. By Darwin's time, Malthus's dire predictions seemed to be coming true. City populations were escalating, and although the economy was booming, industrial capitalism had countless casualties. Darwin lived comfortably off his inherited wealth, but he saw death and struggle all around him. Desperately poor workers were emigrating to Australia and Africa where many of them died—and where they decimated indigenous populations, who succumbed to imported European diseases.

Personal competitiveness finally nudged Darwin into publication. When a letter arrived from an unknown collector in Malaysia, Darwin realized that other people were thinking along similar lines. Protected by his allies, he warded off this potential rival and rushed into print with *On the Origin of Species*. Darwin argued not only for change—by then, widely accepted but also presented his original notion of natural selection based on the competitive struggle to survive. In a hostile environment, claimed Darwin, any organism with even the slightest advantage will thrive better; over the aeons of time established by Lyell, beneficial characteristics will be handed down through the generations, eventually yielding a new species better adapted—better fitted—to suit its surroundings.

The first edition was snapped up by booksellers, but the long-feared hostility soon started. Rallying round him, Darwin's prestigious friends ensured that natural selection was taken seriously—without their support, Darwin's book might well have been criticized into obscurity. For many Christians the major objection was the absence of God. Instead of a teleological universe planned by a divine designer, Darwin had substituted one ruled by chance, with no moral guidance to ensure spiritual progress. He scarcely mentioned the human race in *On the Origin of Species*, at this stage remaining prudently quiet about the possibility that people and apes might be close relatives.

Although a clergyman apparently pointed out Darwin as the most danger-ous man in England, the attacks were not solely on religious grounds. Darwin described his book as one long argument from beginning to end, and he was right. Sceptics accused him of piling up example upon example to support his theory, rather than providing definitive proof. Darwin kept diplomatically quiet about several tricky points, but scientists posed some tough questions. Where, they asked, was his explanation of how changes originated? Was it feasible that an organ as complicated as the human eye could have emerged without advance planning? And how did the first life forms appear? Aware that spontaneous generation was a dangerous topic, Darwin avoided tackling that problem.

A reclusive semi-invalid, Darwin left others to defend his theory for him—even the catchphrase 'survival of the fittest' was invented by somebody else. Years later, when the furore had died down, he dared to go public with his views about people. The cartoons were merciless, but they do indicate how well informed ordinary people were about scientific controversies. Figure 38 appeared in a cheap popular journal; yet understanding it required readers to have a good grasp of the latest details. Marine ascidians (familiar to Victorian seaside collectors as sea squirts) are sedentary rock-dwellers descended—according to Darwin himself—from virile swimming organisms. Such degeneration was the flip-side of evolu-tion as progress. Victorians were terrified that regression to previous stages was also possible, that civilization might decline. These fears underpinned bestselling novels such as *Dracula* and *Dr Jekyll and Mr Hyde*—and they subsequently fuelled Nazi purification propaganda.

This caricature also comments on Darwin's attitude towards women. This emotional fashion-plate turns away to avoid the stare of 'our monkey' because she is blushing, and Darwin is feeling her pulse to indicate that her behaviour is caused physically, and so is essential to her female nature. Darwin regarded women as inferior, asserting that the traditional feminine attributes—intuition, imagin-ation—were 'characteristic of the lower races, and therefore of a past and lower civilisation'.[14] From studying birds such as peacocks, Darwin concluded that flashy feathers give males an advantage because they attract superficial females looking for suitable mates. Victorians disliked such arguments abut sexual selection not because they disagreed with his verdict on feminine taste, but because Darwin had given a key role to women, allowing their choices of partner to guide the path of evolution. As his arch-critic John Ruskin sneered, what would the human race have been like if blushing young maidens had held a baboon-like predilection for blue noses when choosing their men?

Like Linnaeus, Darwin was affected by prejudices of the time. Taking for granted that women are vain and shallow, he interpreted his observations to build up an argument that inevitably confirmed his original assumptions. Because his pronouncements carried the prestige of science, they could be used to justify discrimination by allowing people to say that women are just made that way—no point fighting against nature. Similarly, politicians used Darwinian evolution to support their laissez-faire approach, arguing that nothing should be done to alleviate the misery of the workers because this would contravene nature's ruthless struggle. Citing the law of natural selection, American industrialists got rich fast by unscrupulously stamping out their rivals, and German Darwinists backed political campaigns to establish a superior race and dominate Europe.

Yet by 1900, although evolution was accepted, Darwin's explanation seemed on its way out. His supporters had failed to come up with any mechanism for change, and rival theories proliferated. Some scientists—including Darwin himself—revived Lamarck's opinion that acquired characteristics can be inherited. Psychologically, Lamarckism is appealing because it can be given a hopeful teleological twist—if parents are passing down their advantages, they are in a sense choosing how they deal with their environment rather than leaving everything to chance.

Other opponents came across some experiments performed years earlier in a mountain garden by an obscure Austrian monk, Gregor Mendel, and they adapted his results to challenge natural selection. Mendelian genetics is now a vital component of modern Darwinism, but for twenty years, their supporters wrangled bitterly. Progress was the clarion call of Victorian scientists, but evolution's history makes it clear that science does not advance inexorably in straight lines. Mendel wrote nothing about genes, and modern Darwinism is very different from Darwin's Darwinism.

6   Power

When man wanted to make a machine that would walk he created the wheel, which does not resemble a leg.

—Guillaume Apollinaire, *Les Mamelles de Tirésias* (1918)

When the traveller in H. G. Wells' *Time Machine* lands millions of years ahead in the future, he witnesses a lifeless planet: 'The darkness grew apace...All the sounds of man, the bleating of sheep, the cries of birds, the hum of insects, the stir that makes the background of our lives—all that was over.'[15] A fictional narrative—but one based on solid nineteenth-century science. According to British physicists, the Earth was inexorably winding down, irreversibly moving towards its end. Its lifespan was, they declared, too short for Darwin's protracted processes of natural selection.

The evidence for this inevitable death of the Universe came from steam engines. Leaping from engines to evolution might seem an odd line of argument, but the missing link was energy, Victorian scientists' favourite concept for analysing power and productivity. The laws governing energy emerged from the new science of thermodynamics, which addressed the two most pressing problems confronting Europe as it industrialized—how to make machines more efficient, and how to make factories more profitable. By providing answers to these commercial questions, British physicists gained power, establishing themselves as national experts qualified to pronounce on every major scientific issue. Power was buried deep inside the Victorian laws of physics.

This close fusion of physics and industry took place exceptionally early in Britain. In France, Napoleon had poured money into technical education, so that science became mathematized well before it did in Britain. Nevertheless, industry stagnated, the victim of rigid centralization and an educational system so compartmentalized that advanced engineering institutes had little impact on practical development. Instead, theoretical research carried out by French engineers was picked up by British physicists, who explored what happens when the mathematical principles of thermodynamics are applied to real equipment in factories, rather than to ideal machines on pieces of paper. One of their conclusions became known as the Second Law of Thermodynamics—and it was this Law that predicted Wells' bleak scene of finality.

The Second Law of Thermodynamics has acquired the reputation of being dauntingly incomprehensible, but it rests on two common sense observations. Heat cannot move on its own from a colder to a hotter body, but needs to be pushed—a fridge needs an engine to cool down the air inside it. And however well tuned a machine, it is never 100% efficient, so that small amounts of energy are constantly being wasted as friction or heat. On the basis of these everyday realities, physicists depicted a gloom-laden scenario. Once lost, energy can never be retrieved, but becomes permanently unavailable to carry out useful work. Eventually, everything will cool down to the same temperature, and molecules will stop moving. In the cosmos of uniformity, organization will disappear and information will cease to flow.

Predicting the end of the world brought physics into line with the Bible, and gave life on Earth a direction missing from Darwin's evolution by natural selection. The major champion of heat death was Victorian Britain's most celebrated physicist, William Thomson, who achieved the then-novel feat of using science to become rich as well as famous. This Scottish professor is best known for laying the trans-Atlantic telegraph cable, but that was just one of the profitable engineering projects that led Queen Victorian to honour him as Baron Kelvin of Largs. A powerful advocate for thermodynamic arguments, Kelvin forced geologists to incorporate methods taken from physics when they calculated the age of the Earth. For him, the Sun is the Earth's powerhouse, and he opposed evolution by insisting that there had been insufficient solar energy to keep the Earth going during the long aeons needed for Darwin's theory. Obdurate to the end of his long life, Kelvin never admitted defeat, refusing to recognize that the hidden power of radioactivity provided an additional energy source.

By bringing together science, engineering, and economics, Kelvin made himself enormously wealthy, but he also perpetuated the thrifty values of northern industrialists—energy must be conserved, waste avoided, self-discipline exerted. Kelvin carried this Christian work ethic beyond the shop floor and into the laboratory, making energy a basic tool of Victorian physics. Previously, British physicists had followed Newton's lead by focusing on the forces between individual objects, but during the nineteenth century, they started to think about the possibilities for work and movement embedded within an entire system. For instance, in electricity, instead of describing how charged particles attract and repel one another, Faraday developed his field theory by envisaging contour lines of energy spreading out through space, so that electricity flows away from peaks with high potential energy to other points in the field at lower levels.

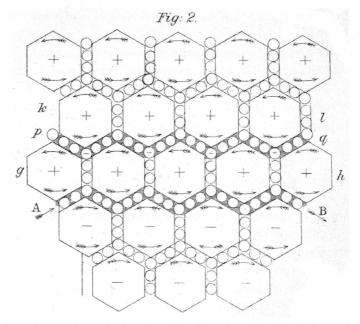

FIG. 39 James Clerk Maxwell's physical model of the ether. *The Scientific Papers of James Clerk Maxwell* (1890).

Theoretical scientists were determined to show that they knew more about efficient telegraph systems than practical engineers. Whereas it made pragmatic sense to think of electricity as water running down a pipe, Kelvin adopted Faraday's more conceptual approach, analysing how electromagnetic fields spread out invisibly around the wires packed together inside the cable casing. The most influential field theory physicist was another professor from Scotland, James Clerk Maxwell, who went south to found the Cavendish laboratory in Cambridge, which rapidly became a leading centre. Whereas down-to-earth Kelvin only believed in things he could measure, Maxwell operated on a more abstract plane, constructing imaginary mathematical analogies. With his mathematician's approach, Maxwell was more interested in developing equations that worked than in pinning them firmly to physical reality.

Maxwell asserted that the whole of space is filled with an invisible electromagnetic aether, which he visualized in Figure 39, a honeycomb cross section showing lines of force as tubes filled with an incompressible fluid. Hexagonal whirlpools of fluid are separated by rotating idle-wheels, which represent tiny particles thrust sideways when a current flows. The small arrows indicate how the fluid rotates anti-clockwise inside the tubes; the role of the idle-wheels is to turn in the opposite direction, and so coordinate the fluid's movement (although there are mistakes in the drawing below the line AB, where several arrows are pointing in the wrong direction). Confronted by gasps of disbelief, Maxwell conceded that his diagram did not exactly represent the aether itself, but maintained that it did provide an invaluable conceptual model—after all, he pointed out, an orrery is not identical to the real Solar System.

For Maxwell, the electromagnetic energy in his field was interchangeable with the mechanical energy powering industrial equipment. Foreign critics were scathing about the British fashion for pulleys and pumps. 'We thought we were entering the tranquil and neatly ordered abode of reason,' sneered one French physicist; 'but we find ourselves in a factory'.[16] But that was exactly the intention—one and the same set of physical laws could be used equally well to describe the aether and machines. At least, it could in Britain. In principle, during the second half of the nineteenth century, experimenters all over Europe had plenty of opportunities to learn about each other's results—more and more journals were being produced, communications were improving, and there were no major wars. Nevertheless, researchers often ignored foreign theories or else tried to refute them. Despite rhetorical claims that knowledge flowed freely within the scientific community, scientists developed their own distinctive national styles.

The most striking differences were between Britain and Germany, two major industrial powers with contrasting opinions about how and where science should be performed. While British investigations into energy were promoted by economical engineers, in Germany, the laws of energy were developed by physiologists. By the middle of the nineteenth century, German scientists wanted to dissociate themselves from the earlier *Naturphilosophen*, and they set out to eliminate fuzziness from physiology by measuring working bodies with the same precision as factory equipment. As Kelvin's counterpart Hermann Helmholtz pointed out, 'The idea of work is evidently transferred to machines from comparing their performances with those of men and animals…We still reckon the work of steam-engines according to horse-power.'[17]

Like Kelvin, Helmholtz was his country's leading scientist, but there the similarities stop. Educated in German philosophy and French mathematical physics, Helmholtz envisaged an orderly cosmos governed by Newtonian forces rather than a vibrating aether or an energy-packed field. Whereas Kelvin had taken academic physics into industry, Helmholtz was originally trained as a doctor (financed by the army), and only later became a university experimenter. As a quantitative physiologist, Helmholtz brought numerical precision into his mechanical universe, insisting that inputs and outputs must balance—an early version of the law that energy can never be created or destroyed, but is always conserved. Drawing on Germany's chemical expertise, Helmholtz analysed how food is processed by human engines, regarding nature as a giant warehouse of energy available for powering people as well as windmills and steam engines.

Following the German model for teaching, Helmholtz attracted a group of admiring acolytes who turned to him for inspiration and worked under his close supervision. His prize student was Heinrich Hertz, an engineer turned physicist who inherited Helmholtz's theoretical allegiance to Newtonian forces. In Hertz's opinion, Maxwell was clearly wrong. Hertz believed that electrical effects somehow leap over empty space rather than being carried by an aether—but after a few years of careful experiments, he started to change his mind. In a dramatic series of demonstrations, Hertz showed that electricity behaves as if waves are travelling through a fluid, even reflecting and refracting his waves to prove their close resemblance to light. Writing with the enthusiasm of a convert, Hertz boasted that there could no longer be any doubt about Maxwell's mathematical theories. His—Hertz's—experimental work had provided incontrovertible verification.

In Britain, the theoretical physicists were delighted. Hertz had vindicated their hypotheses, and—just as significantly—had justified their claims to superiority over practical engineers. In contrast with Hertz, who dedicated himself to revising Maxwell's electromagnetic theories, British scientists enthused about the commercial potentials of electric waves. One of Maxwell's colleagues pointed out that whereas light rays are blocked by walls and London fogs, radio waves (as they are now known) can travel right through them. Might it be possible to establish a new form of telegraphy, one that needed no expensive subterranean cables? Keen to make science profitable, the British government backed an Italian inventor called Gugliemo Marconi. This turned out to be a worthwhile investment. In 1901, Marconi sent an air-borne message from Cornwall to Newfoundland, and for the first time, the two sides of the Atlantic were in virtually instantaneous contact. As the twentieth century opened, radio had effectively shrunk the globe.

Another crucial difference between German and British science was organization. Britain remained the land of free enterprise, where science was closely intertwined with commercial interests. In Germany, on the other hand, the state decided to invest in scientific education. Reformers were inspired by Justus von Liebig, an enterprising chemist who, during the second quarter of the nineteenth century, converted his small provincial training school for pharmacists into a large international centre for organic chemistry. Liebig's great innovation was to combine research with education. As well as teaching his students how to use his precise measuring instruments, he also assigned them individual projects that would supplement his own investigations. By administrating and educating this collective work force, Liebig became Europe's most influential chemist. His laboratory became an efficient factory for generating chemical knowledge, so that Liebig's ideas were carried by his many graduates into pharmacy, industry, and agriculture.

Funded by the state, German universities started to build laboratories along Liebig's model, with a dominant professor perpetuating his particular research style by encouraging students to participate in projects and seminars. Special institutes for experimental physics were founded, dedicated to producing graduates who could boost Germany's manufacturing industries. Secondary education was also improved by introducing a new tier of technical schools which focused on practical skills—science, technology, modern languages. By the 1870s, the results were obvious. The general level of scientific knowledge was far higher in Germany than the rest of Europe, and industry was booming. German capitalists had packaged knowledge, making systematic training into a commodity as vital for economic power as modern equipment and a large labour force.

Science is now so ubiquitous and so powerful that, in retrospect, its worldwide spread seems to have been inevitable. But although many other countries aspired to Germany's success, the routes they followed varied enormously In Britain, for instance, the government continued to give comparatively little financial support, and there was no effective technical education until well into the twentieth century. Some scientists did set up specialized laboratories—such as Maxwell's Cavendish at Cambridge—but lacking any centralized organization, research schools depended on the initiative of particular individuals (in Liverpool, one professor set up a physics department in a former mental hospital, even incorporating the padded cell).

Germany influenced the United States of America more directly. But even there, instead of importing the German system wholesale, universities grafted it onto existing Anglo-American institutions. They created a new entity—the Graduate School, where students were taught by a group of experts rather than clustering tightly round a single influential figure. These American Graduate Schools offered a carefully structured approach to advanced education, providing their students not only with specialized research skills, but also with the prospect of climbing up a systematic career ladder to reach the peak of their profession. By the end of the nineteenth century, America had become a major scientific power in its own right.

Technological science did not spread evenly around the globe, but was modified and adopted to fit local requirements. In Japan, for instance, the political regime changed dramatically in 1868, when the Meiji emperors came to power and began opening the country up to the outside world. Suddenly, Japan started importing expertise in science and technology, and transformed its scientific organization. The Ministry of Education launched intensive campaigns to advertise European achievements, printing translated books and distributing posters showing legendary moments of scientific inspiration, such as Watt and his kettle, or Franklin and his kite. These scientific heroes were used to promote the importance of patient, dedicated work—the traditional meaning of 'industry'. Under the centuries-old feudal system, Japanese people had loyally served local overlords, and the new professional scientists transferred their allegiance to the nation. By the early twentieth century they were carrying out world-class research, although this abrupt transition from isolation left a long-lasting legacy—well after World War II, Japanese engineers and scientists were still accused of relying too heavily on Western innovations rather than originating their own ideas.

Even within a single country, there was no uniform response to the introduction of science. In China, for example, European astronomy had first arrived with Jesuit missionaries in the seventeenth century. Since their main concern was to recruit Roman Catholic converts, they avoided discussing the controversial theories of Copernicus and Galileo, but did try—often unsuccessfully—to impress their hosts with elaborate instruments. Some Chinese astronomers preferred to resurrect the ancient techniques of their own predecessors, and it was only in the second half of the nineteenth century, when imperial powers were interfering with internal Chinese politics, that a new wave of Protestant missionaries imposed European education on large sectors of the population and the government reluctantly began teaching modern science.

Similarly, inhabitants of the British colonies did not necessarily welcome the arrival of heavy machinery and the imposition of new school curricula. In nineteenth-century India, although ambitious middle-class people decided that it was in their interests to collude with British settlers and secure their own power, other groups were less receptive. Farmers resented mechanization, because they found that their traditional techniques of ploughing and sowing were more effective for coping with local agricultural conditions; under pressure from them, British scientists revised their own practices and incorporated this indigenous expertise. Many Indians found that imported medicine was ineffective, and they preferred to rely on ancient remedies rather than pollute their bodies with foreign chemicals. By the twentieth century, Indian nationalists seeking independence had converted European science into a symbol of British oppression.

From the perspective of European capitalists, scientific progress brought power both at home and abroad, as new technologies such as steam transport and the electric telegraph network enabled them to control large areas of the world. Many imperialists genuinely believed that they were improving the lives of those they had conquered, and found it hard to understand the unenthusiastic responses they encountered. Nowadays, politicians are more aware of science's potential for damage. Scientists should perhaps have listened more closely to Mahatma Gandhi in 1928, when he encouraged Indians to demand self-rule. 'God forbid that India should ever take to industrialization after the manner of the West,' he proclaimed—'If an entire nation of 300 million took to similar economic exploitation, it would strip the world bare like locusts.'[18]

7 Time

Keeping time entails exerting control. When mechanical clocks were introduced at the end of the thirteenth century, they imposed regular church hours on traditional village activities; six hundred years later, more accurate timepieces were disciplining society ever more strictly. In Paris during the 1880s, clocks throughout the city were coordinated by puffs of compressed air sent through underground pipes from a central machine room. Citizens wanting to tap into this pneumatic system could visit a luxurious showroom, lit by a Statue of Liberty lamp as though unaware of the restrictions incurred by centralized automation. These wealthy customers had been swept up in the cult of precision.

During the nineteenth century, measuring accurately—and still more and more and more accurately—became an obsession. When an American inventor displayed freshly minted optical instrument ruled with a stunning 43,000 lines to the inch, German scientists were appalled at losing their lead in the race for high precision. British soldiers were instructed to count how many telegraph poles whizzed past their train window every mile, a lone meteorologist made hourly observations for twelve years, and the astronomer John Herschel routinely waited in the cold night air for two hours until he had reached the same temperature as his telescope.

Victorians valued precision as a hallmark of modern science. Nevertheless, like time, precision is not an abstract absolute, but involves reaching agreement. When American psychologists introduced intelligence tests to vet would-be immigrants, they produced numerical scores to back up their claims that northern Europeans are superior to Jews, Italians, and blacks. However, knowing whether someone rates 105 or 106 or 107 is irrelevant unless everyone accepts that this is a meaningful thing to do. Similarly, scientists must decide that the value shown on an instrument's scale is valid, that it is accurately recording something useful.

Being sure that an instrument shows the right reading is more complicated than it might seem. One precaution is to carry out practical checks—make sure that an instrument isn't worn out, that it can withstand climatic extremes, that it gives the same result for every observer (well, properly trained ones, anyway). But there are more fundamental issues to be resolved. No piece of apparatus shows exactly the same value every time—but if you want to overturn an old theory with a new reading, do you have to be 99% or 99.99999% sure that it's accurate? And if two highly skilled scientists end up with conflicting results, how do you decide which to choose? Resolving such questions demands consensus.

Scientists were forced to confront these problems by the growth of the telegraph network, which demanded international coordination. If British engineers wanted to control the Empire electrically, they had to make sure that what worked at home could be perfectly replicated in India or Africa. As James Clerk Maxwell explained in the *Encyclopaedia Britannica*, 'The equations at which we arrive must be such that a person of any nation, by substituting the numerical values of the quantities as measured by his own national units, would obtain a true result.'[19] When telegraph experts compiled conversion tables comparing the results obtained by experimenters in different countries, they discovered that although all the readings had been recorded to several significant figures, they failed to match.

Establishing universal standards aroused huge controversy. National pride was at stake, and British chauvinists opposed French plans to bring back the metric system that had been abandoned a few years after its introduction during the French Revolution. Some British archaeologists insisted that measuring an Egyptian pyramid in yards revealed how its ancient builders had been divinely inspired to create perfect proportions. More influentially, the astronomer John Herschel described how military surveyors in India had proved that the metre is not an exact fraction of the Earth's dimensions. In any case, he continued, since Britain's empire dominates the world, everybody else should comply with us. Such arguments might not sound scientific, but they proved powerful—Maxwell persuaded the rest of the world to accept the electromagnetic standards established in his Cambridge laboratory.

Of all the quantities being argued about, time was the most vital. Until the mid-nineteenth century, towns operated on their own local time, and residents set their clocks by the stars and the Sun. When trains started linking places hundreds of miles apart, it became essential to coordinate clocks throughout the network. In Britain, the railway companies agreed to use London time, measured from the stars every day at Greenwich and flashed out all over the country by telegraph signals along the tracks. The situation in other countries was more complicated. In

France, the trains ran on Rouen time, which was five minutes behind Paris, where the clocks inside stations were set five minutes behind those outside. The United States was so vast that they decided to set up internal time zones, a compromise later adopted all over the world.

As international electrical communication networks were established, countries were forced to collaborate more closely than ever before, agreeing on standardized systems of measurement so that the entire world would tick with the same clockwork rhythm. Telegraph and radio signals linked far-flung regions into one single system, effectively shrinking the globe so that information could be distributed immediately rather than taking weeks or even months to arrive. As had happened with trains, introducing these new technologies not only opened up opportunities for centralized control, but also raised new demands for coordination—this time on a global scale.

International collaboration was essential for map-making, which depended on telling time accurately. The big problem was longitude—surveyors wanted to measure the time gap between two points on the same latitude so that they could work out the distance between them. Prizes for a reliable technique had been on offer since the early eighteenth century, when many disasters occurred at sea through ships inadvertently wandering off their course. Money-hungry inventors submitted some ingenious suggestions, and once the stranger ones had been weeded out, it seemed clear that a clock was the best answer. Unfortunately, even the sturdiest instruments failed to keep time sufficiently reliably during the long stormy trip across the Atlantic.

Eventually, telegraphy promised a solution. Since electric signals travel virtually instantaneously, local times hundreds of miles apart can be compared almost simultaneously. Cartographers had to abandon the habit of placing their own capital city at the centre of their world maps, and instead agree on a universal numbering system for longitude. And once again, Britain prevailed—in 1884 an International Committee decided to draw the zero line for the entire globe through Greenwich (although the French clung to Paris until 1911).

As accuracy escalated, people started worrying about being even slightly out of step. Whereas before the nineteenth century, telling the time exactly was still so unimportant (and difficult) that many clocks had no minute hands, by the 1880s, citizens hooked into Paris's underground pneumatic network were complaining that it took several seconds for air puffs to travel from the central control room to various parts of the city. By the turn of the century, major metropolises were installing networks of city clocks linked together electrically to make sure they all showed the identical time.

A particularly acute form of this problem confronted telegraphic surveyors, who had to make allowances for the minute length of time—fractions of a

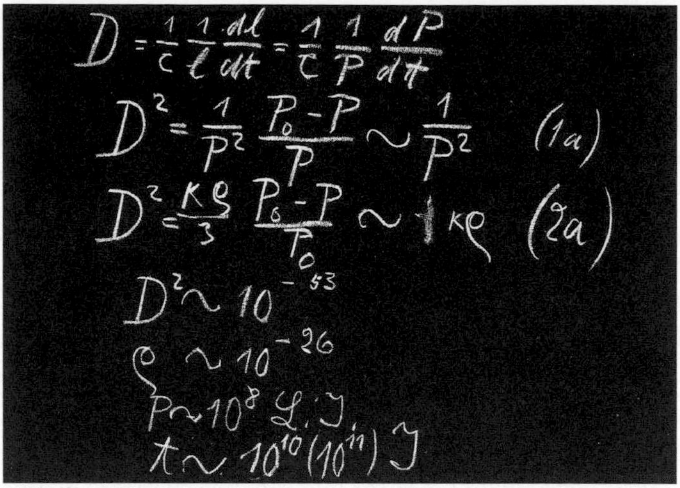

$$D = \frac{1}{c}\frac{1}{\ell}\frac{d\ell}{dt} = \frac{1}{c}\frac{1}{P}\frac{dP}{dt}$$

$$D^2 = \frac{1}{P^2}\frac{P_0 - P}{P} \sim \frac{1}{P^2} \quad (1a)$$

$$D^2 = \frac{\kappa \varrho}{3}\frac{P_0 - P}{P_0} \sim \frac{1}{3}\kappa\varrho \quad (2a)$$

$$D^2 \sim 10^{-53}$$

$$\varrho \sim 10^{-26}$$

$$P \sim 10^8 \, L.J.$$

$$t \sim 10^{10}\,(10^{11})\,J$$

FIG. 40   A blackboard from Albert Einstein's lecture of May 1931 in Oxford.

second—needed for an electric signal to reach a destination on the other side of the world. When ultra-high precision was demanded, even small errors in matching up times could lead to significant discrepancies in measured distances. The longitude race was reborn in a twentieth-century version, as optimistic inventors designed devices to synchronize timepieces all over the world. Aiming to protect the fortunes they envisaged reaping, they applied for patents in Switzerland, centre of the clock-making trade. And many of their designs landed on the desk of a philosophical physicist who was originally more interested in thermodynamics than in time—Patent Officer Albert Einstein.

Einstein has become an icon of theoretical incomprehensibility. For most people, the equations of Figure 40 are meaningless squiggles; yet visitors flock to admire this blackboard lovingly preserved in Oxford since 1931 (its Cambridge counterpart was thrown out years ago). When Einstein chalked up these equations, the audience gradually slipped away from his supposedly introductory lecture on relativity. 'I don't blame them,' commented one scientist; 'If their maths are good enough to follow him their German certainly is not.'[20] These particular calculations conclude with Einstein's estimate of the Earth's age—he was worried about recent evidence showing that he'd got it wrong. By then, he was no longer vetting Swiss patents, and had transformed his experience with the practical problems of keeping time into mathematical models of the entire cosmos.

'Why is it', Einstein asked a *New York Times* journalist in 1944, 'that nobody understands me and everybody likes me?' He was not, of course, expecting an answer, but it is an intriguing question. How did the obscure creator of an arcane cosmological theory become world famous? Until he was forty years old, nobody outside a small circle of mathematic physicists had even heard of Einstein. He first hit the international headlines in 1919, when a British expedition investigating a solar eclipse confirmed his theory of general relativity. Soon Einstein's trade mark Brillo-pad hair and droopy mustache immediately identified the hero who had dared to challenge Isaac Newton—and had won the intellectual contest.

Like many scientific heroes, Einstein was an expert self-publicist who lectured enthusiastically and relished explaining how his theory of relativity had revolutionized time. 'An hour sitting with a pretty girl on a park bench passes like a minute,' he quipped, 'but a minute sitting on a hot stove seems like an hour.' Einstein was not alone in his fascination with time. Many avant-garde artists, musicians, and writers also wanted to find new ways of representing the world, and they claimed to be stimulated by his physics. To his disgust, 'Everything is relative' became a popular catchphrase that was devoid of meaning but reinforced the notion that Einstein was a supernatural genius who had created a theory incomprehensible to normal mortals.

Einstein's first article on relativity, published in 1905, had no references or footnotes, as though it were a patent application submitted by an inventor claiming originality. Although its appearance is now heralded as a momentous event, there was no immediate revolution. Relativity theory was developed and tested over many years, some aspects not being experimentally confirmed for half a century—even the famous equation $E = mc^2$, which ties together energy, mass, and the speed of light, was not in Einstein's initial paper.

Relativity overturns the common sense ways of thinking about time and space established by Newton three centuries earlier. For Newton, space is fixed and time flows inexorably past us at a steady rate. For Einstein, time depends on where you are and how fast you're moving, so that it only makes sense to define your own personal time with respect to—or relative to—something else. In Einstein's relativistic cosmos, just one quantity seems the same to everybody: the speed of light. By making this basic postulate, Einstein eliminated conjectures about Maxwell's aether. Scientists had tried to detect how light might be slowed down if it was moving in the opposite direction to an aether swirling around the Earth (or pushed faster by an aether behind it) but relativity dispensed with the need for such hypotheses.

Einstein produced two versions of his theory. The Special Theory of 1905 is comparatively straightforward mathematically, but ten years later Einstein published his more comprehensive General Theory, which takes account of gravity and makes some weird-sounding predictions—light curves when it goes near the Sun, and space travellers will return to find themselves younger than the friends they left behind on Earth. Satirists latched on to these bizarre phenomena. A *Punch* cartoon showed policemen trapping a burglar with torches whose rays bent round corners, while a witty poet coined this limerick:

> There was a young lady named Bright,
> Who travelled much faster than light.
> She started one day
> In the relative way,
> And returned on the previous night.

There are at least three perversions of science in this verse: Ms Bright would die long before she travelled fast enough to be affected by relativity; nothing can travel faster than light; and the order of events can never be changed, only the interval between them.

But however confused and confusing the jokes, they worked because the inventor of relativity was famous—and they had the effect not of making him seem ridiculous, but of making Einstein a household name. Yet when inspected in detail, it seems less clear that he deserved such accolades. For one thing, as he developed his theories, Einstein relied heavily on the work of other mathematicians. Even the experimental confirmation of the General Theory looks suspect under close examination. Einstein suddenly became a media star in 1919, when a British eclipse expedition led by a Cambridge astronomer, Arthur Eddington, supposedly proved that he was right and Newton was wrong. The planning had started a couple of years earlier, when Eddington was searching for a way to avoid detention as a conscientious objector during the war. To justify the expenditure on a non-military project, Eddington was committed in advance to vindicating Einstein, and he made the trials sound simple.

According to Einstein's theory, light from a distant star is bent by gravity as it travels past the Sun. This means that for people looking up from Earth, the star will appear to be in a different place than if its light were travelling in a straight line. So if you measure the star's position near the Sun during an eclipse, and compare it with other recordings when it is in distant parts of the sky, you can see if the results match up to Einstein's or to Newton's prediction. But what sounds like a crucial test to distinguish between two competing theories was far more complicated in practice. Most annoyingly, no suitable stars were close to the Sun during the eclipse period. This increased still further the demands on accuracy, so that the observers faced roughly the same task as trying to measure a small coin from a mile away. And in addition to many technical difficulties, on the big day itself, the weather was cloudy.

When the results proved inconclusive, Eddington set about massaging the data, summarily discarding photographs that failed to confirm his pre-established views and suppressing contradictory evidence gathered by other eclipse teams. Many scientists were already converted to Einstein's ideas, and Eddington drew on his influential contacts to convince them that his expedition demonstrated the truth they wanted to hear. Nevertheless, some experts remained sceptical, and two decades went by before American astronomers were converted.

Einstein liked to think of himself as Newton's successor in the pantheon of great geniuses, almost as if some immaterial numinous power could be passed on from one extraordinary intellect to another. Even so, his arcane theory was rooted in the practical problems of clock coordination under the nineteenth-century regime of precision. Scientific icons are worshipped as other-worldly beings who float above the realities of daily life. Einstein illustrates how even the most apparently abstract thinkers do not conform to such idealized visions.

VI

Invisibles

During the nineteenth and twentieth centuries, scientists developed increasingly precise instruments, yet they repeatedly summoned up undetectable entities to explain natural phenomena. It seemed that however closely they searched, ultimate causes always lay elusively beyond their grasp. Although some physicists confidently predicted that the laws of nature were nearly sewn up, their complacency was shattered by radioactivity, which revealed an unexplored and unsuspected micro-universe whose behaviour could only be described by the anti-intuitive laws of quantum mechanics. Uncertainty appeared to be an integral part of the Universe, making it theoretically as well as practically impossible for atomic scientists to know everything about everything. In unsuccessful quests for the material basis of life, biologists also struggled to make the invisible visible, relying on powerful optical and chemical tools to probe deep inside cells and reveal concealed agents affecting human existence. Launched in the name of progress, research programmes claiming to further science and improve humanity carried political and commercial implications that prompted deep ethical reservations.

I *Life*

Propaganda is a soft weapon: hold it in your hands too long, and it will
move about like a snake, and strike the other way.

—Jean Anouilh, *The Lark* (1953)

'It was on a dreary night of November,' recalled Victor Frankenstein. 'I collected the instruments of life around me, that I might infuse a spark of being
into the lifeless thing that lay at my feet…I saw the dull yellow eye of the creature
open; it breathed hard, and a convulsive motion agitated its limbs.'[1] Frankenstein
has become a semi-mythical creature of nightmares, a scientific monster often
conflated with the innocent creature he electrified into life. Yet he alarmed readers not because he performed impossible feats, but because his activities hovered
on the brink of feasibility.

Frankenstein's own creator, Mary Shelley, had steeped herself in the latest scientific research. Her book may be fictional, but it provides an acute commentary
on the scientific scene of 1818, when *Frankenstein* was first published. Then, resur-
recting the dead seemed a real possibility. Drowned corpses had been successfully
revived, and anatomists performed public experiments on freshly hung criminals,
jolting them with massive electric currents until their limbs convulsed, their backs
arched, and their eyes leered open. Research continued, and within twenty years,
scientists were eagerly debating widespread reports that a geologist had created
insects by electrifying a stone.

Biology was being promoted as a new science (first identified in an obscure
German footnote of 1800), and its pioneers were pursuing the hardest question of
all—what is the nature of life? British opinion was divided into two major camps.
According to traditional views, life arises when a soul or spirit is infused by God. At
the opposite extreme lay materialists, scientific reductionists who insisted that life
lies in matter itself, somehow springing from a rearrangement of fundamental units.
Their ideas seemed particularly threatening because they originated in France, perceived by chauvinists as the source of revolution and atheism. The arguments were
bitter. Trying to mediate, an eminent London surgeon diplomatically proposed

that life is imparted to inert substances by some sort of vitalizing force, perhaps a superfine fluid analogous to electricity. Shelley was aware of these debates when she portrayed Frankenstein as a blundering experimenter, an alchemical mystic whose attempts to generate life electrically proved disastrous.

Shelley was familiar with the controversial line taken by her husband's doctor, William Lawrence, who poured scorn on any links with electricity. The only thing we know about life, he declared, is that it is passed on from one generation to the next—its secrets clearly lie in the fabric of the bodies themselves. The new biologists preferred to ally themselves with physicists and chemists rather than natural historians, declaring that instead of simply collecting and classifying living organisms, they would use precise experiments to search for life itself. This type of laboratory-based research expanded rapidly after the 1830s, when high-quality microscopes were introduced. Although microscopes had been invented a couple of centuries earlier, the images they produced were too blurred for precision observation. After the optical components had been redesigned, biologists could identify minute micro-organisms and look deep inside the cells of living creatures.

Nevertheless, even when armed with powerful instruments, biologists were unable to agree, and debates about life remained acrimonious throughout the nineteenth century. Far more was at stake than simple facts. Devout Christians found it sacrilegious to suggest that the gift of life was not unique to God—one reason why *Frankenstein* was greeted with such horror. This religious opposition to materialism became more outspoken once the possibility of evolution started to be taken seriously. Although theories (including Darwin's) skirted round the issue of how life got started in the first place, they did imply that spontaneous generation—the independent creation of life from matter—had taken place at least once in the long distant past. Many people found this notion hard to accept.

Religion and science were not automatically at loggerheads. Across Europe, battle lines and alliances were drawn up differently depending on local political allegiances. In France, the Church had joined forces with the Emperor (Napoleon's nephew) to maintain an authoritarian regime, and the main critic of spontaneous generation, Louis Pasteur, was a staunch Catholic committed to defending God's role in creation. The national controversies were inflamed still further by the French translation of Darwin's *Origin of Species*, which carried a provocative preface denouncing Catholicism as a damaging religion imposed by corrupt priests. Being a good patriot meant rejecting spontaneous generation—a loyalty that Pasteur turned to his advantage as he manoeuvred to become France's most famous scientist.

A provincial expert in brewing wine and beer, during the 1860s Pasteur set about converting himself into a national hero by tackling the biggest scientific question of the day—can life emerge from dead plants and animals? His chief opponent was Félix-Archimède Pouchet, already an eminent naturalist, who had devised an elegant apparatus that seemed to produce micro-organisms. Pouchet's basic technique was to boil hay with water and create a sterile infusion which he isolated from air with a trough of mercury—and then wait for minute signs of life to appear.

Pasteur approached his research not with ideological neutrality, but with preconceived ideas. His boast that 'chance favours only the prepared mind' has become one of science's most famous maxims, because approaching nature with a blank mental slate is not generally very productive—success often depends on recognizing the significance of some tiny effect that has previously been ignored. Interpreted less charitably, it implies that Pasteur aimed not so much to find out what was right, but more to find evidence that would prove he was right. An ambitious, intolerant man, Pasteur was notorious for appropriating his assistants' results, and his notebooks reveal that when he went into battle against Pouchet, he was already convinced of the answers he would get.

Keeping quiet about some early failures, Pasteur eventually worked out how to test spontaneous generation with what he claimed was a crucial experiment. Using special flasks whose long contorted necks were designed to keep out pollution from the atmosphere, Pasteur compared the fates of two types of sugared yeast-water, boiled and unboiled. In his experiments, only the unboiled samples went mouldy—no living organisms appeared in sterile solutions. According to Pasteur, he had conclusively demolished spontaneous generation. Although Pouchet and his allies raised many objections, Pasteur managed to navigate his way around them; for instance, he wriggled out of several awkward situations by taking a risk and accusing Pouchet of using contaminated mercury. Pasteur did perform fine experiments—even climbing with his flasks to the top of a mountain glacier in search of pure air—but he also concealed counter-evidence and ignored his rivals' conclusions when it suited him, feeling justified not by the data but by his own convictions. Logic evaporated as he claimed to have proved a negative—that spontaneous generation can never happen anywhere.

Pasteur became a hero, his preliminary assumption became an established fact, and Pouchet was forgotten. The two scientists had been trapped in an experimental double-bind, because they were using similar procedures to obtain conflicting results. How could anyone know which of them was right? This is a common research problem, similar to the electrical dilemma of resolving incompatible readings given by different types of instrument. As with the British and the telegraph system, such discrepancies can enable the strongest team to decree what counts as truth. Several years later, it turned out that Pouchet's results had indeed been valid, because some of the micro-organisms in hay can withstand boiling—everybody had wrongly assumed that his infusions were sterile. If Pasteur had followed strict scientific protocol, he would not have been able to eliminate spontaneous generation so effectively.

Science and religion were not always allies in controversies about life. After Otto Bismarck became Chancellor in 1871, the newly unified Germany became Europe's leading scientific nation, and many scientists backed the government's campaign against the Catholic Church. Ernst Haeckel, Germany's leading defender of Darwin, denounced religion as though he were writing a political manifesto rather than a scientific textbook. In the holy war against 'intellectual servitude and falsehood', he proclaimed flamboyantly, 'embryology is the *heavy artillery in the "struggle for truth"*'.[2] Embryology was a major science in Germany, and Haeckel hoped that by comparing the development of different animals, he could uncover the mysterious forces that convert a few tiny cells into an independent living creature.

Science needs instruments as well as theories. Without dramatic improvements in microscope technology, Pasteur could not have examined so closely the minute organisms that cause disease and fermentation, and embryology would never have become so important in Germany. Whereas English producers concentrated on providing expensive instruments for natural historians, German manufacturers were producing excellent cheap microscopes that even students could afford. By using these instruments to compare the internal structures of plants and animals, German scientists decided that cells must be the basic building-blocks of life, even though they vary enormously in appearance.

Cell theory dominated German research during the latter two-thirds of the nineteenth century. After several decades of systematic investigation, involving the sacrifice of several tentative theories, biologists agreed on a general description that still stands today—all cells have a nucleus, which contains chromosomes and is suspended inside a jelly-like cytoplasm. But although these components could be seen through a microscope, their functions remained unclear. And although animals developed from eggs and sperm into fully formed embryos, the processes involved were mysterious. Scientists were still frustratingly far from revealing the secrets of life.

When Haeckel investigated embryos, he benefited by belonging to this cell-oriented research community of microscope experts. He also inherited specifically German beliefs that some sort of guiding force gives life a direction. The *Naturphilosophen* had envisaged development taking place in distinct stages towards goals that have been decided in advance, so that mind emerges from matter in some sort of preordained process that they never specified very clearly. For the *Naturphilosophen*, the same fundamental laws of progress govern the micro- and the macro-worlds. To illustrate this parallelism, they pointed to embryos, which seemed to grow along predetermined guidelines until they reached their final

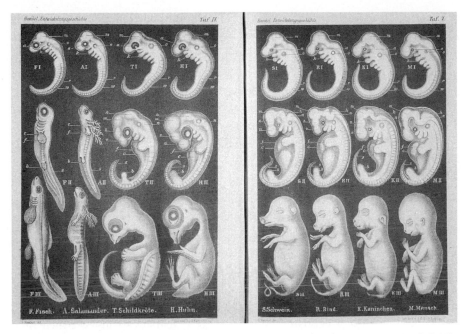

FIG. 41 Comparisons of embryos in three different stages of evolution. Ernst Haeckel, *The Evolution of Man: A Popular Exposition of the Principal Points of Human Ontogeny and Phylogeny* (1883).

form, almost as if they were repeating—recapitulating—the entire ancestry of the species inside the womb.

These idealistic visions were at complete odds with materialist theories of evolution. In its original form, natural selection held out no promises of progress, although some of Darwin's closest British supporters—including Huxley, renowned as his bulldog—did later insist that there must be some sort of preordained direction from lower to higher organisms. Making a similar move, Haeckel fused together natural selection and innate progress, so that he became known as 'the German Darwin', despite expressing views that differed from Darwin's own. An expert publicist, Haeckel formalized his personal brand of evolution by coining a memorable summary—'Ontogeny is the brief and rapid recapitulation of phylogeny.' This pithy slogan is a smart way of saying that the development of an individual (ontogeny) recapitulates the progressive stages travelled through during the evolution of its species (phylogeny), so that embryos of more recent species pass through forms of older ones.

As Figure 41 illustrates, Haeckel aimed to short-circuit the long words by making recapitulation self-evident. The diagram's eight columns compare the appearance of eight animals at three stages before birth, ranging in complexity from a fish at the left to a human at the right. In the top row, the young embryos closely resemble each other, whereas the older ones at the bottom have diverged apart. Haeckel used such visual tables to make it clear that vertebrates share a common pattern of development, and also to explain how embryos pass successively through forms lower down in the evolutionary tree. For instance, the undeveloped human foetus at the top right bears features that resemble gill slits and fins belonging to the more mature form of its fishy ancestor on the bottom left.

Haeckel's powerful diagrams, colourful prose, and anti-religious rhetoric ensured that his version of recapitulation became enormously influential all over Europe and beyond. Even in the mid-twentieth century, long after recapitulation had been discredited, Haeckel's striking pictures were still being reproduced to back up Darwinian arguments that although animals may look very different, they share a common ancestor. Haeckel's images look convincing—but do they tell the right story? Like Pasteur, Haeckel believed that science involves persuasion as well as investigation, and he wanted to present his evidence in the most advantageous light. Tactics that he thought admissible—such as simplifying originals to accentuate particular features—were denounced as fraudulent by his critics. These accusations were later exploited by right-wing Christians to support their own political positions.

Haeckel did not embark on outright forgery, but he did manipulate his data to stress the similarities that he was sure existed. Rivals with different preconceptions from Haeckel also examined embryos, and, unsurprisingly, arrived at different theories. Because they wanted to undermine recapitulation, they downplayed likenesses, and highlighted distinguishing features instead. To emphasize the influence of local environments over ancient ancestry, some scientists started intervening in embryonic development to isolate the physical and chemical causes of specific changes. Gradually, Haeckel's appealing recapitulation theory became discredited, although he was still hailed as Germany's evolutionary pioneer.

Haeckel was unlucky rather than corrupt. Many scientists have pursued their prior convictions with equal determination—including Pasteur, who failed to follow the strict rules of correct scientific procedure, but is celebrated as a hero because his rejection of spontaneous generation was subsequently confirmed. Unfortunately for Haeckel, he backed a hypothesis that scientists now declare to be false. Evolution remains a contentious topic, and modern Creationists pounce on old accusations of Haeckelian forgery to attack the entire edifice of Darwinian science. In singling out this one example to bolster their own case, they are—similarly to Haeckel—engaging in self-serving yet genuinely intentioned exaggeration.

2 Germs

> But we would gladly know more than we see, and there's the difficulty:
> for if we could see, we should know enough; but we see most Objects
> quite otherwise than they are; so that the true Philosophers spend their
> time in not believing what they see, and in endeavouring to guess at
> the knowledge of what they see not.
>
> —Bernard de la Fontenelle, *A Discovery of New Worlds* (1686)

Victor Frankenstein tried to create life, whereas doctors set themselves the less ambitious goal of prolonging it. However, eighteenth-century medicine was not very effective. Even the best-trained physicians could often do little more than help their patients die comfortably, while high fees and dubious remedies made some of them seem disturbingly close to unscrupulous quacks—when Mary Wortley Montagu brought back small pox inoculation from Turkey, she despaired of finding any doctors with 'virtue enough to destroy such a considerable branch of their revenue for the good of mankind'.[3]

Doctors diagnosed patients on a personal basis, trying to cure them by rebalancing their humours and taking account of their inherent constitution. At least, that's what happened for rich patients—the poor often died, or consulted an apothecary or a village wise-woman. After the French Revolution, medicine became more egalitarian, and hospitals were established to care for large numbers of sick people. Patients with similar symptoms were grouped together in wards, so that doctors started to recognize and treat illnesses as entities in themselves rather than regarding complaints as being inseparable from individuals.

Hospitals were dangerous places—one eminent Victorian surgeon commented that a patient 'laid on an operating-table…is exposed to more chances of death than the English soldier on the field of Waterloo'.[4] Sick people who were rich enough preferred to hire a family physician, who would dispense individual care in the quiet comfort of a patient's own home. Although new drugs were being developed in laboratories, general practitioners were reluctant to see their authority being eroded by science-based medicine. To reinforce their status as experts equipped with an all-seeing gaze, they introduced new diagnostic techniques based on intimate physical examination and accurate measurement.

Infectious diseases such as smallpox regularly wiped out large numbers, but nobody had any clear idea of how they spread. Many people believed that epidemics were caused by poisonous miasmas, invisible yet all-pervasive atmospheres of toxic matter. Unlike germs, which cause specific illnesses, these miasmas were said to affect victims differently, giving them influenza, cholera, or some other affliction, depending on their inherent constitution and state of health. When Charles Dickens and other middle-class campaigners protested against filthy overcrowded slums, they were also interested in protecting their own health by ridding cities of virulent miasmic clouds.

Looking back towards the nineteenth century, historians can tell the past as a continuous stream of success stories. Doctors introduced vaccines to prevent fatal illnesses, micro-biologists isolated individual germs responsible for specific diseases, surgeons drastically reduced hospital infections with antiseptics, and laboratory chemists started producing powerful drugs. Death rates were slashed (or at least, they were in Europe and North America), and doctors could at last offer realistic hopes of curing their patients. On the other hand, not all research projects proved fruitful and the causes of major killer diseases—malaria, flu, beriberi—remained unidentified. The progress that now seems so clear was at the time played out in confusion.

Take smallpox. Doctors did adopt the Turkish inoculation recommended by Montagu, but it was a painful, risky procedure that killed off many healthy children. In the 1790s, the country surgeon Edward Jenner took the sensible step of listening to local farmers and dairy-maids, who maintained that after people had recovered from cowpox, they rarely succumbed to smallpox. Although Jenner is one of medicine's great heroes, nowadays he would be banned for his testing procedures. After selecting an eight-year-old boy, Jenner first injected him with cowpox, and then tried to infect him with smallpox. Luckily for them both, his volunteer/victim survived, and Jenner was rewarded by the government with £30,000 for inventing the smallpox vaccination.

Like so many other innovations, vaccination seems momentous in retrospect, but was hotly contested at the time. The denunciations included a colourful but well-informed caricature by James Gillray satirizing one of the free clinics set up in London (Figure 42). Dressed in a formal jacket (probably dirty), Jenner is wielding the blades of his vaccinating instrument, helped by a charity boy holding a tub of 'VACCINE POCK hot from yᵉ COW' and an assistant ladling out 'OPENING MIXTURE'. This fantasized scene represented very real fears. A woman on the right is simultaneously vomiting and giving birth to cows, while several patients have grown cow-shaped excrescences, a reference to the renowned case of the

The Cow Pock — or — the Wonderful Effects of the New Inoculation!— vide the Publications of y Anti Vaccine Society

FIG. 42 James Gillray, *The Cow-Pock—or—The Wonderful Effects of the New Inoculation* (1802)

so-called ox-faced boy whose glands swelled up dramatically after vaccination. Protests against vaccination continued right through the nineteenth century, long after its efficacy had been demonstrated.

Doctors and patients had good grounds for caution. The procedures were painful, disfiguring, and administered so unhygienically that infections spread rapidly. Lacking any strong theoretical justification, Jenner's explanations of anomalous results limped—and as a mere surgeon, he was easily associated with butchery and grave-robbing. Above all, it seemed sacrilegious to contaminate a human body with matter taken from a sick animal lower down the scale of life. Although vaccination did appear to work, powerful campaigners—including Florence Nightingale—protested that the state had no right to interfere in people's lives by making vaccination compulsory. This opposition to government control meant that smallpox epidemics continued to break out in Britain even in the twentieth century.

International trade and travel spread diseases around the globe. Smallpox decimated indigenous American populations who lacked resistance to this European disease; similarly, British and French explorers exported sexually transmitted infections to the Pacific islands they praised as paradise. By the early nineteenth century, germs were travelling in the opposite direction. When cholera started to spread out of Asia and across Europe, doctors panicked, realizing they had no idea how to stem its flow. It first reached Britain in 1831, inducing similar feelings of doom and helplessness as prevailed in the early years of AIDS. The crowded hospitals were soon overflowing with patients who inevitably died in agony within a few hours. Although that epidemic receded, others arrived, goading reluctant officials into action.

The government half-heartedly tried to improve basic living conditions, although many British people resented this state interference. Victorians excelled at recording facts and figures, and by analysing the statistics, doctors eventually worked out how cholera is transmitted. The conventional hero of this story is John Snow, a London surgeon who supposedly solved the problem single-handedly by logical deduction. In fact, he spent several years reading up on previous research, and concluded in advance of his own investigations that cholera is carried by microbes in water. Armed with his preconceived convictions, Snow set out to prove that he was right. By systematically mapping out the cholera cases in his own neighbourhood, Snow argued that one particular pump was the source of the infection. In dramatic versions of this myth, he is even credited with taking away the pump handle to stem the outbreak, although it was actually removed by a committee after the danger was already over.

Like Pasteur, Snow was subsequently vindicated, even though his evidence did not provide cast-iron certainty. As his opponents pointed out, Snow's data could equally well be explained by assuming that cholera had been carried through the air from a pile of dirt next to the pump. And, they added, even if cholera does travel through water, he had no right to claim that *no* illnesses are caused by miasmas. In a second neat study, Snow showed that the rates of infection in a house depended on which company was supplying the water. This result provided useful ammunition for campaigners insisting that London should be cleaned up, but medical experts remained unconvinced—although cholera appeared repeatedly on the agendas of international sanitation committees, Snow's research was not discussed. Far from being a lone genius, Snow was just one of many talented contributors to the collective project of saving lives and reducing deaths.

Snow was lucky—he backed the water-borne theory of cholera, the one that survives today. Other medical heroes managed to establish their posthumous reputation on the basis of theories later abandoned. One prime example is the surgeon Joseph Lister, now world-famous for his campaigns to eliminate germs from hospitals. A handsome, charismatic self-publicist, Lister presented himself as a surgical saviour whose team of white-robed assistants ritually sprayed his operating theatre with carbolic acid like incense-swingers in a church. Sceptics sniped, 'And now let us spray.'

Lister was a skilled surgeon who did much to improve operating conditions. On the other hand, he borrowed his rivals' techniques, clung on to old-fashioned ideas, and rewrote his career to make himself appear a leading advocate of modern germ theory. For one thing, he was far from the first to stress hygiene. With some postoperative death rates as high as 65%, public health reformers were already whitewashing walls, separating patients, and improving ventilation. Although designed to tackle miasmas, these steps were effective—and they included using carbolic acid. So surgeons were initially unimpressed when Lister insisted that soaking instruments and dousing open wounds with carbolic would reduce mortality, especially as he still wore blood-spattered clothes, never scrubbed his hands, and was unconcerned about his patients' soiled sheets (to say nothing of the pain his acid must have caused to raw flesh). When Lister later looked back over his career, he astutely revised his opinions to keep up with current thinking. Originally, he had regarded germs as general sources of infection, but in the 1880s, he replaced these all-purpose germs with specific ones, shifting his theoretical position so adroitly that admiring fans still celebrate him as a pioneer of modern germ theory.

The grand hero of germ theory is Robert Koch, the German bacteriologist who shot to fame after he identified the organism responsible for industrial Europe's biggest killer—tuberculosis (then often called consumption or phthisis). Koch's success depended on devising new methods and knowing how to advertise his discoveries. By inventing Petri dishes (named after one of his assistants), Koch grew solid cultures in carefully controlled environments, and he developed techniques for photographing microscopic images. These innovations meant that he could compare his procedures quantitatively, and could communicate his findings effectively. By setting out a clear set of experimental procedures (called Koch's postulates), Koch established cause and effect with 100% certainty, demonstrating beyond doubt that particular bacteria are responsible for particular diseases.

Laboratory researchers claimed great progress, emphasizing that as Koch and his successors identified more and more micro-organisms, death rates from infectious diseases plummeted. However, not everyone concerned with public health was convinced. There were less romantic causes of the falling figures—people were eating better food, sanitation was improving, and governments were putting money into hospitals and health education. But those important measures made boring reading, and people preferred to hear that heroic doctors were battling against enemy diseases. According to the germ propagandists, bodies are not governed by balanced humours which may go out of kilter, but instead are collections of cells vulnerable to attack by invisible yet deadly micro-organisms: rather than being in harmony with the Universe, humans are independent entities who need to protect themselves against microbial intruders.

Corresponding to this concept of enemy invasion, medicine acquired a military vocabulary of breakthroughs, defeats, and destruction. Like other scientific

A.—Going to the attack.

FIG. 43 David Wilson, 'Going to the attack' (1899). Pen and ink based sketch, based on photographs of bacteria by Elie Metchnikoff.

metaphors, these images operated both ways, not only reflecting how illness was conceived, but also affecting how foreigners should be treated. These associations are illustrated by Figure 43, which is called 'Going to the attack' and is one of a series about 'The Army of the Interior'. The two inset drawings display the microscope's revelations of internal warfare, as warrior white corpuscles pass through the walls of a vein to attack alien bacteria among the large flat red blood cells. To hammer the message home, the cartoon shows white British soldiers braving a hostile river to capture the interior of Africa or Asia. Up on the cliff top, menacing black bacterial demons conjure up caricatures of colonized Africans threatening European explorers.

Just as bodies had to be protected against intrusive microbes, so wealthy nations tried to defend themselves against infectious immigrants. Diseases had often been blamed on foreigners, but germ theory provided new grounds for rationally explaining old fears. Prejudices about race and cleanliness could now be given a scientific label. When a Chinese resident of San Francisco died from plague, the whole of Chinatown was quarantined, a discriminatory step justified as medical precaution. The USA was attracting millions of settlers, and in the early twentieth century the government set up a screening programme to assess their health. Although this was advertised as medical security, in reality the patchy vetting procedures meant that rich Europeans found it far easier to enter than other applicants.

European researchers were also invading Asian countries, trying to stem the massive epidemics that threatened trading profits. At the end of the nineteenth century, India effectively became a giant laboratory, as Koch and other experts arrived to study in detail how bacteria transmitted plague, and which treatments worked best. Indians bitterly resented this intervention, which implied that they were incapable of looking after themselves. When Europeans ruthlessly enforced health regulations, they often transgressed local proprieties—placing members of different castes in the same hospitals, preventing Muslims from embarking on their Haj to Mecca, subjecting women to physical examinations carried out by male doctors. Controlling diseases entailed curtailing individual freedom, the same objection levelled by anti-vaccination campaigners.

Despite the medical breakthroughs they claimed, scientists did not achieve total victory over the enemy of disease. Often what seemed straightforward in the laboratory proved complicated outside it, and such discrepancies lent power to germ theory's critics. Even though Koch had proved that nobody gets consumption without being exposed to the tuberculosis bacillus, he was unable to explain why only about ten percent of people became infected; still worse, the cure rate proved lower than had been hoped. Victorian statisticians set about compiling data, but although the disease was clearly most prevalent in poor industrial areas, they could find no simple cause and effect. The enemy agent had been identified, but it seemed to leave many potential victims unscathed, suggesting that the sick were somehow tainted in advance. The stereotypes shifted: in the late eighteenth century, consumption had been regarded as a romantic affliction sustained by poets and artists with inherently delicate constitutions; a hundred years later, TB gained a new identity as a contagious disease that circulated in squalid city slums, a mark of inferiority rather than of aesthetic vulnerability.

Patients were isolated in large sanatoria, regimented not only for their own good but also to protect society. Acquiring TB became a matter of shame, as though patients had been picked out as culpable rather than being innocent victims of neutral microbes. Frightening and mysterious, the illness became a tabooed topic: 'in discussing tuberculosis', the Prague novelist Franz Kafka reported two months before he died of the illness, 'everybody drops into a shy, evasive, glassy-eyed manner of speech'.[5] This stigma was shared by other contagious diseases, especially syphilis, which was often blamed on women. The name syphilis comes from a poem describing how the evil seductress Luxury lured Hercules into sin, a Greek version of Eve tempting Adam in the Garden of Eden. In the corresponding medical mythology, prostitutes, not their male customers, were castigated as the corrupt sources of pollution.

Such emotional attitudes still prevail. Towards the end of the twentieth century, cancer became the new TB, the big C that could not be mentioned by name, and

AIDS transmitted to male homosexuals the condemnation previously directed at female prostitutes. In his play *Ghosts*, which explores the irrational guilt induced by inherited syphilis, the Norwegian playwright Henrik Ibsen wrote: 'It's not only what we've inherited from our parents that haunts us. It's all the old obsolete ideas…they stick to us and we can't shake free of them.'[6] Ghosts from the past continue to plague the present, but perhaps exorcism can come from understanding their origins.

3　Rays

> Vague and insignificant forms of speech, and abuse of language, have so
> long passed for mysteries of science; and hard or misapplied words, with
> little or no meaning, have, by prescription, such a right to be mistaken
> for deep learning and height of speculation, that it will not be easy to
> persuade either those who speak or those who hear them, that they are
> but the covers of ignorance, and hindrance of true knowledge.
>
> —John Locke, *Essay concerning Human Understanding* (1690)

By photographing bacteria, biologists showed that they existed. And by transmitting radio messages, physicists proved that the aether was real. Or did
they? Seeing was not always believing; conversely, when hypothetical entities per-
sistently refused to make themselves visible, they remained shrouded in doubt.

Distinguishing between truth, falsehood, and self-deception was not always
straightforward. Enthusiasts claimed that their cameras never lied, but many people found it hard to accept that photographs of blurry faces or women in trailing
robes genuinely revealed visitors from the spirit world. On the other hand, if radio
waves could vibrate round the globe through an undetectable aether, then perhaps
the notion of supra-telegraphic communication with the dead was not so bizarre
after all? Some sensitive subjects could detect luminous auras flaming around
people's bodies, while others readily succumbed to the mesmeric forces generated
by animal magnetizers. Mediums, seances, and telepathic communication were
later denounced as fraudulent—but during the second half of the nineteenth
century, scientists were not so sure.

In this atmosphere of uncertainty, strange effects were regarded with suspicion,
and some atomic phenomena initially appeared no more authentic than spirit
manifestations. When the British physicist William Crookes announced that he
had discovered radiant matter, his colleagues remained unconvinced. But Crookes
was a clever experimenter, and he devised clear demonstrations that something
strange was happening inside his glass discharge tubes. Resembling small neon
lights, these instruments contained gas at a low pressure, and were extremely
popular with electrical performers because they glowed when a current passed

through them. Crooke ingeniously customized a tube by inserting a miniature paddle wheel on rails inside it. When the power was switched on, the wheel trundled towards one end, pushed—according to Crookes—by particles of radiant matter streaming out from the metal plate (the cathode) at the other.

Crookes's evidence was persuasive, and he was later vindicated when his mysterious cathode rays were shown to exist, even though they were given a new label—electrons. Nevertheless, Crookes also applied his undoubted experimental skills to verifying spiritualism, a phenomenon now discredited. As an electrical expert, he felt well qualified to assess claims about long-distance communication, and after several prominent mediums survived his rigorous tests without being caught cheating, many Victorians were convinced that it really was possible to contact the dead. Sceptics sneered that a gullible physicist was being duped by charlatans (or perhaps was one himself), but Crookes and his colleagues wanted to provide a material explanation for these supposedly spiritual effects. They suggested that radio might have a human analogy, so that people with especially sensitive organs can tune in to unsuspected aetherial vibrations.

Crookes accused his fellow scientists of betraying their calling. By refusing to examine spiritualism, he charged, they were judging it in advance, which 'appears like reasoning in a circle: we are to investigate nothing till we know it to be *possible*, whilst we cannot say what is *impossible*, outside pure mathematics, till we know everything'.[7] His self-defence went to the heart of what it means to be a scientist. If you automatically reject the unfamiliar, and refuse to investigate it, then nothing new will ever be revealed. Making major discoveries entails going outside the box by eliminating preconceptions and challenging previous knowledge. Although it would be counter-productive to check every previous experiment, it is not necessarily eccentric to challenge accepted wisdom.

Some of the other bizarre phenomena unexpectedly turning up in scientific laboratories initially seemed as inexplicable as table-tapping at seances. In 1896, an unknown German professor called Wilhelm Röntgen stunned the world with an X-ray photograph of the bones in his wife's hand, her wedding ring eerily floating around her finger. He had stumbled across this mysterious radiation accidentally when he was carrying out some experiments with discharge tubes. Other scientists, including Crookes, had already noticed that wrapped photographic plates kept getting fogged over, but Röntgen decided to find out what was happening.

In contrast with Pasteur's maxim that 'Chance favours the prepared mind,' Röntgen tried to avoid making any theoretical presuppositions. As he put it, instead of thinking, he investigated. By exploring the properties of his new rays experimentally, Röntgen showed that they were different from the cathode rays also being produced in discharge tubes. Whereas X-rays are neutral, weightless radiation like light or radio waves, cathode rays carry an electrical charge (negative) and weight (although not much). Within a few years, X-ray machines became standard hospital equipment, Röntgen won the first Nobel Prize for physics, and X-rays entered the repertoire of fairground performers:

> I'm full of daze
> Shock and amaze
> For now-a-days
> I hear they'll gaze
> Thru' cloak and gown—and even stays
> These naughty, naughty Roentgen Rays.[8]

Only a few months later, a similarly modest scientist in Paris made another momentous chance discovery—Henri Becquerel, the first to detect radioactivity. Although he too became an early Nobel Prize winner, Becquerel self-effacingly credited his achievement to his distinguished family's tradition of investigating phosphorescence. During a project to examine X-rays systematically, Becquerel was routinely working his way through phosphorescent substances when he decided to test a uranium salt (one that his father had made earlier). Becquerel struck lucky because he neglected to follow strict scientific protocol—bored with waiting for the Sun to shine, he developed some photographic plates long before he expected them to show anything.

Instead of the faint images he had anticipated, Becquerel found himself gazing at clear black and white silhouettes. He was mystified—but rather than throwing them away as a mistake, he continued investigating. Soon he discovered that even without sunlight, his father's uranium crystals produced a photographic image. Eventually, after further systematic experiments, he realized that phosphorescence—his starting point—was irrelevant. What mattered was the uranium. But why? And was uranium unique, or did other substances behave in this extraordinary way? The cause and the extent of the phenomenon remained mysterious.

Over the next few years, bewildered scientists invented new words—including radioactivity—for the inexplicable effects that kept turning up. Tentatively, they added three types of other ray to the mysterious X-rays, naming them after the first three letters of the Greek alphabet: alpha (α), beta (β) and gamma (γ). (Some scientists have a quirky sense of humour—hence 'quarks' in modern subatomic physics—and when Ralph Alpher wrote an article with George Gamow, they recruited a third author, Hans Bethe, to make their combined surnames sound like Alpha, Beta, Gamma.) More precise labelling emerged only sporadically, and it eventually turned out that only gamma-rays are radiation (like electricity or X-rays)—the other two rays were later found to be streams of particles. Alpha rays are composed of positively charged helium nuclei (two protons and two neutrons), while beta rays (also called cathode rays) are far lighter, negatively charged electrons.

Yes, it's complicated. And at the time, confusion reigned. Radioactivity, spirit communications, cathode rays—these curious phenomena had no discernible cause, and they apparently belonged together in confounding the stable, law-governed Universe that nineteenth-century physicists aspired to establish. As the twentieth century opened, it was hard to predict their future significance. Some scientists were still deeply committed to spiritualism, and Becquerel's radioactivity seemed to be an obscure anomaly of little interest—which is why an obscure, anomalous researcher called Marie Curie was able to pick it up as a research topic. It was only later that scientists recognized how crucial Becquerel's discovery had been.

The big subjects in electricity, those pursued by mainstream physicists, were cathode rays and X-rays. Scientists first went back to Crookes's discharge tube, modifying it in order to examine cathode rays more closely, still unclear what this radiation might be. The most famous of these researchers was Joseph John Thomson (a name that his colleagues would not have recognized: his friends called him J-J, but modern publishing conventions demand that full first names be provided). Head of the Cavendish at Cambridge, Thomson made his discharge tube behave similarly to a television, bending the cathode rays with electric and magnetic fields. This ability to manipulate the rays proved that they must be different from X-rays, and Thomson claimed that they were streams of tiny particles, which he called electrons.

In retrospect, Thomson's identification of electrons is celebrated as a path-breaking experiment, but many of his contemporaries remained unconvinced by his evidence. Aether enthusiasts refused to change their minds, and they had some good arguments. Most obviously, nobody had ever seen an electron—unlike the spirits revealed by cameras. Even measuring the electron's charge was proving incredibly difficult, and possible values spread over a wide range. This lack of precision enabled Thomson's critics to use the same results for supporting the opposite point of view—that electricity is not packaged into discrete particles, but is continuous, as though waves are travelling through the ether.

Although Thomson was so clumsy that his assistant kept him at a safe distance from the delicate apparatus, his research students adored him, reporting that when they asked for intellectual rather than manual help, Thomson intuitively diagnosed their problems and suggested solutions. *Intuitively*—that might not sound like an authentic way to approach research, but intuition and instinct were vital for persuading recalcitrant, unsophisticated equipment to yield the right results. But what are the 'right' results? This is the same circular problem that keeps cropping up. If you're measuring something new, how can you be sure that your apparatus is working properly? Scientists are often faced with the problem of deciding what to do with a result that contradicts expectations. If they take it seriously—as

Crookes did spiritualism—then they run the risk of sabotaging their reputation or of skewing a numerical average by including a mistake. If they ignore it, their results look more consistent—but Röntgen and Becquerel succeeded precisely by refusing to overlook the unexpected.

This dilemma has no 'right' answer. Although scientists are meant to be impartial, some major achievements have resulted from being selective with the data—including the experiment that supposedly provided a definitive measurement of an electron's charge. It was devised by Robert Millikan, an anti-aether American physicist, who ingeniously suspended electrified drops of oil between electric plates. By finding out the electric force needed to hold them stationary against the downward pull of their own weight, Millikan calculated the charge they carried. At least, that was the principle. But his delicate apparatus was easily disturbed, and—armed with his conviction that electrons really do exist—Millikan discarded around two-thirds of his readings. According to his laboratory notebooks, he knew precisely what he was looking for. 'This is almost exactly right & the best one I ever had,' he exulted in December 1911; by April 1912, he was still more confident. 'Too high by 1.5%', he noted by one; 'Perfect Publish', by another.[9] It's tempting to denounce Millikan's high-handed behaviour, yet scientists need to benefit from experience as well as being inexorably logical. Sensitively attuned to the vagaries of his apparatus, Millikan produced a value for the charge on an electron very close to the one accepted today.

Convinced in advance that electrons exist, Millikan gambled with his future and won a Nobel Prize. In contrast, a group of French scientists behaved similarly, and ended up looking foolish. While investigating X-rays, they claimed to have found yet another mysterious form of radiation, which emanated from living creatures as well as ordinary matter, and could—like spirits—only be seen by super-sensitive observers. Just as visible light can be split into a spectrum, so these N-rays (named after their home university of Nancy) could allegedly pass through an aluminium prism to form a pattern on a special screen. Rather than being a minor blip, N-rays featured in around 100 serious scientific papers between 1903 and 1906, and many observers genuinely believed they could detect their effects. International scepticism mounted, but the French scientific community rallied round the Nancy experimenters, defending them particularly vigorously against German critics.

They were eventually proved wrong by a devious American visitor who sneakily removed the vital prism while his hosts proudly continued making measurements. Easy to laugh—but initially N-rays seemed no more implausible than radioactivity. Easy to criticize—but were the Nancy enthusiasts acting so much more unethically than Becquerel, who abandoned his schedule and developed some plates early? Or than Millikan, who sacrificed neutrality by jettisoning inconvenient readings? When the scandal erupted, France was reorganizing its scientific structure, aiming to revive research and rescue its international status. Is it so surprising, or so reprehensible, that high-flying French scientists should close ranks against hostile foreigners and try to establish Nancy as a prominent research centre?

In 1903, two French discoveries hit the headlines—N-rays in Nancy and radium in Paris. Ironically, the country's scientific saviour did not originate from the privileged elite, but was a doubly marginalized outsider who married into it—a Polish woman called Manya Skłodowska, later world famous as Marie Curie. Lacking funds or even a doctorate, Curie had decided to pursue Becquerel's work on uranium salts, and find out if any other substances behaved similarly. Recruiting her husband Pierre's assistance, she spent six years hunting down minute traces of radioactivity, manipulating delicate apparatus with great skill, yet also performing the hard physical labour needed to process vast masses of experimental material.

Eventually Curie isolated two new radioactive elements—first polonium (named after Poland), and then radium. Radioactivity was rapturously welcomed. Figure 44, taken from a British series called 'Men [sic] of the Day', shows the proud couple gazing in awe at a shining radium crystal, blazing out like a miraculous star. Curie herself would die of leukaemia caused by her work, but the initial universal enthusiasm was untarnished by any knowledge of Hiroshima or Chernobyl. As France's first female professor, Curie set up a research centre to study the new phenomenon, and during World War I, she demonstrated the practical value of radioactivity by running mobile X-ray units for wounded soldiers.

The first scientist to win two Nobel Prizes, Curie has become a role model for aspiring girls. This caricature reveals how double-edged the mythology can be. Here a diminutive figure, she peers from behind her husband's shoulder as he triumphantly holds up the test tube radiating the glory of genius onto his exaggerated forehead. Although she published theoretical papers independently, here he clasps a book while she—the practical subordinate—leans on the table of apparatus, her plain dress indicating that she has renounced conventional feminine interests. The couple's frugal lifestyle fueled romanticized horror stories about their sordid working conditions, and in many accounts, Marie features as the domestic drudge who devotedly sifted through tons of dirty pitchblende, the laboratory equivalent of routine cooking.

The message seems clear: scientific women are trapped inside a special category, neither top-rate scientists nor normal females. This prejudice survived for decades (and still does, according to many women), although in the early twentieth century, some men were also experiencing discrimination. Like Curie, Ernest Rutherford became world famous but was initially looked down on as an outsider. A farmer's son from a state school in New Zealand, Rutherford had the wrong

FIG. 44 Caricature of Marie and Pierre Curie by Julius Mendes Price ('Imp'), *Vanity Fair* (1904).

clothes and the wrong accent when he arrived at Cambridge University. Even so, he eventually succeeded J-J Thomson as head of the Cavendish laboratory, and became a major pioneer in nuclear physics. Rutherford remained proud of his New Zealand origins, and when he was made a Baron, concocted an elaborate coat-of-arms featuring a kiwi, a Maori warrior, and Hermes Trismegistus, the patron saint of alchemy.

According to Rutherford, the most amazing event in his whole life took place in Manchester in 1909. By then, he had been working on radioactivity for over ten years, and was already notorious for his heretical suggestion that some atoms are unstable, emitting rays and particles as they disintegrate from one element to another. One day Rutherford had a hunch. Working with Hans Geiger (whose counters are still used), he had already realized that a beam of alpha particles does not travel straight through a thin sheet of metal foil, but is scattered in different directions. What would happen, Rutherford wondered, if he put counters on *both* sides of the foil? The result, he reported, 'was almost as incredible as if you fired a 15-inch shell at a piece of tissue paper and it came back and hit you'—some alpha particles were being reflected back by the wafer-thin screen.[10]

Sixteen months went by before Rutherford was ready to go public with his explanation. His experiments showed that metals are not made of atoms packed closely together like oranges in a box, but have small heavy nuclei separated by distances enormous on the subatomic scale. If a relatively light alpha particle just happens to hit one of those nuclei, it bounces back. Rutherford explored atoms and nuclei for the next two decades, often working with collaborators but continuing on his own during the First World War, when his younger colleagues were at the Front (where some of them died).

Under Rutherford's guidance, atomic structures became clearer—if no less extraordinary. By the time he died in 1937, scientists had discovered protons as well as neutrons; they could split nuclei artificially by bombarding them with neutrons; and the first linear accelerators had been built to produce high-speed beams of particles. Yet in retrospect, Rutherford seems to have been over-concerned to play down the potentials opened up by his research. In one of last public lectures, Rutherford warned his audience that the 'outlook for gaining useful energy from the atoms by artificial processes of transformation does not look promising'. War broke out only a couple of years later, and his prediction soon sounded surprisingly naive—the first atomic bombs were dropped in 1945.[11]

4 Particles

I have found in Dame Nature not indeed an unkind, but a very coy Mistress: Watchful nights, anxious days, slender meals, and endless labours, must be the lot of all who pursue her, through her labyrinths and meanders.

—Alexander Pope, *Memoirs of the Extraordinary Life,*
Works and Discoveries of Martin Scriblerus (1741)

For chemists who value order, the Periodic Table represents the supreme testament of human ingenuity. Celebrated as the decoding key to the cosmos, this logical sequencing condenses the Universe's myriad substances into a straightforward pattern resembling a giant chemical Sudoku puzzle. The elements proceed steadily across the rows, ranked numerically, uniformly increasing by a single proton from one atom to the next. In contrast, when the Table is read vertically, the elements in each column are united by having the same number of free electrons. But this taxonomical beauty was far from self-evident, and only became apparent after decades of investigation. When the Table was being drawn up, in the second half of the nineteenth century, scientists knew nothing about the internal structure of atoms, but regarded them as indivisible particles, the basic building-blocks of matter. Early attempts to marshal the elements into a mathematical arrangement were disfigured by anomalies, gaps, and misfits.

A more cynical browse through the Table's terse symbols suggest that political decisions are also deeply embedded within it. Traditionally, an element's name hinted at its properties—Argon the Inactive, Oxygen the Acidifier (a mistaken legacy from Antoine Lavoisier), or fast-flowing Mercury, Greek messenger of the gods. Later ones reflected their discoverers' glory, such as Polonium for Marie Curie, or Europium for the Englishman William Crookes (also notorious for his spiritualism). After World War II, when heavy unstable substances were being flashed briefly into existence inside accelerators, science became a battlefront in the Cold War—no coincidence that Americium, Berkelium, and Californium lie close together in a single row, that the Soviet Kurchatovium was diplomatically renamed Rutherfordium, and that American researchers decided against christening an element after the Table's creator, the Russian chemist Dmitrii Mendeleev.

Or did Mendeleev invent the Periodic Table? Outside Russia, some historians have been reluctant to award Mendeleev this accolade. In their versions of the past, Mendeleev features as just one amongst six scientists who gradually developed similar schemes at around the same time. But for Russians, there is no question—Mendeleev, chemistry professor at St Petersburg and government consultant, is the nation's greatest scientific hero. According to them, after fifteen years of careful study, the concept of periodicity sprang into Mendeleev's mind one day in 1869; however, despite his brilliance, he was forced to spend the next few decades fighting ignorant, hostile critics. During the Soviet era, Mendeleev's status as a national figurehead grew still further, because he was politically committed to improving industrial production and preached that society can be explained scientifically—doctrines that meshed well with Marxist ideology.

The arguments about Mendeleev's status as an inspired genius follow predictable lines. Pro-Russianists retort that Westerners are unable to read the original sources, while sceptics emphasize the widespread and long-lasting opposition to Mendeleev's ideas. Such debates about heroic discoverers recur again and again when analysing science's past, because neither theories nor inventions are born fully fledged, but instead are developed for years, often long after the originator's death. Nevertheless, perhaps Mendeleev has merited a special mention by working exceptionally hard and long to promote his Table. As Sigmund Freud quipped, although other scientists were acquainted with the periodic law, Mendeleev married it.

Mendeleev initially lined up the elements according to their chemical properties, arranging them like cards in a game of scientific patience. Some relationships now seem obvious. For instance, copper, silver, and gold—vertically aligned in his Table—had been bracketed together for centuries; similarly, the recently discovered gases fluorine, chlorine, and iodine also resemble one another. Next, Mendeleev thought it would also make sense to arrange the elements in the order of their atomic weights—although chemists had never seen a single atom, they had discovered ways of measuring how heavy they are. But this was where things seemed to go wrong. However often he shuffled his pack of chemical cards, Mendeleev failed to make the elements obey his simple rules.

To resolve the impasse, Mendeleev made a bold and controversial move—instead of changing his hypothesis, he attacked his data. Mendeleev decreed that some atomic weights had been wrongly measured. He also predicted that in the future, new elements would be discovered to fill awkward gaps in his Table. Even in Russia, conventional chemists derided this theoretical approach, while far away in Western Europe, his bizarre suggestions were ignored. But evidence was piling up to support them, and eventually, Mendeleev scored a public victory over a French scientist, who had recently isolated a new element. The Frenchman patriotically named his discovery gallium, but it was the Russian chemist who successfully predicted its properties from his Periodic Table.

Even though Mendeleev always denied that electrons exist, they later turned out to be vital for ordering the elements in his Table. An element's position is determined not by its atomic weight, but by its atomic number—the number of positively charged protons in the nucleus, balanced electrically by the same number of negative electrons. In the early twentieth century, many of the investigations into atomic structure were carried out in the Cavendish laboratories at Cambridge, where Thomson had identified electrons and Rutherford's scattering experiments showed that atoms have tiny nuclei. Cambridge scientists soon realized that nuclei must themselves be built up from still smaller particles, and their first suggestion was that each nucleus contains large positive protons as well as tiny negative electrons, the entire assembly surrounded by a fuzzy cloud of extra electrons. Nevertheless, there were some perplexing puzzles—why did similar elements have very different numbers of external electrons?

Like Mendeleev and his contemporaries, atomic scientists searched for numerical patterns. If you believe that the Universe is arranged in a logical way, then it makes sense to seek out mathematical simplicity —the sort of aesthetic law-like beauty in which Einstein believed. Faced with a mass of bewildering observations, British researchers took the reassuring step of modelling the unknown on the familiar. They visualized each atom as a miniature planetary system, with the Sun/nucleus at the centre, and planets/electrons revolving around it. But analogies are never perfect—which is where the Danish physicist Niels Bohr stepped in.

Bohr is often said to be second only to Einstein amongst twentieth-century physicists, and the Cavendish group nicknamed him 'The Great Dane' to reflect both his nationality and his lugubrious demeanour. It was Bohr who imposed order on atomic electrons. As one of his colleagues put it, Bohr decreed that electrons should behave like trams rather than buses, travelling round on tracks rather than roaming where they fancy. In Bohr's model, electrons are confined to specific orbits, which each hold a maximum number of electrons; when one orbit is full, electrons start to fill up the next available shell out. This provided a reasonable explanation for the similar behaviour of elements in each column of the Periodic Table—they each have the same number of electrons in their outermost orbits.

Bohr's explanation made the layout of the Periodic Table gratifyingly clear, although it was still hard to understand how charged particles could remain crammed together inside a tiny nucleus. The solution made the situation both less and more complicated—in 1932, James Chadwick discovered a third subatomic particle, the neutron. (In one of those notorious 'What if?' questions of history, perhaps Chadwick would have achieved this earlier if he had not been interned in Germany during World War I, chronically ill and forced to live in a cramped stable.) Chadwick's experiments showed that neutrons are heavy, like protons, but carry no charge—hence their name. Scientists set about reorganizing their atomic models, banishing electrons from the nucleus to make it consist only of protons and neutrons. But answering that question generated several more. There were already three subatomic particles (electrons, protons, neutrons)—how many more might there be? What sort of glue held the nucleus together, preventing all those positive protons from repelling each other and exploding apart? And what about the Russian doll principle—if atoms contain nuclei, and nuclei contain protons and neutrons, could there be still tinier particles waiting to be revealed?

Yes, there were. By 1959, there were already thirty, even though research had slowed down during World War II when many scientists were diverted to military work. The most important instrument for bringing these subatomic mites into visibility was the cloud chamber, prototype of modern apparatus and source of countless photographs like Figure 45. Criss-crossed by enigmatic white tracks, these images reveal permanent vestiges recording the swift flight and brief existence of otherwise invisible particles.

Like other pieces of scientific equipment, cloud chambers had originally been designed for a different purpose—to reproduce the weather. These devices that now belong to modern high-energy physics were invented by Charles Wilson, a Victorian meteorologist working on the fringes of Cambridge science. Whereas the mainstream Cavendish approach was to analyse and experiment, Wilson wanted to understand natural phenomena through replicating them, and he tried to generate inside his laboratory the glorious effects of sun-drenched clouds in the Scottish mountains. At first, Wilson produced artificial mist by condensing water droplets on to floating specks of dust; after years of research, he ended up with an electrical instrument that photographed the trails of drops created by charged particles. Wilson had changed meteorology by importing techniques from physics, but he had also expanded the vista of physicists by enabling them to view the submicroscopic universe of cosmic rays and atoms.

Analysing the results demanded skill and also patience. For non-experts, there was little to distinguish one set of splodges and streaks from another. And although many, many thousands of pictures were produced by cloud chamber apparatus, only very occasionally did a photograph turn up that had captured a rare collision or explosion to expose the existence of yet another particle. Scrutinizing the masses of snapshots provided underpaid employment for American housewives, that rarely acknowledged yet vital labour force of atomic physics, their eyes and brains attuned to detect the slightest anomaly in the black and white patterns.

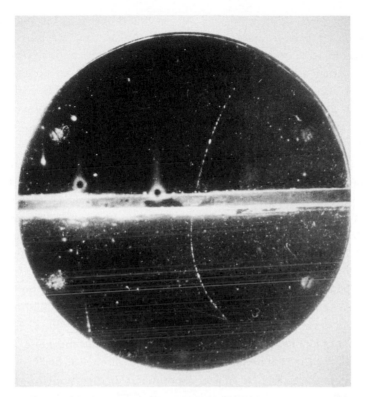

FIG. 45 Carl Anderson's first photograph of a positron (1932).

In 1932, the same year that Chadwick discovered neutrons, the photograph of Figure 45 was developed in California. Although scientists were extremely excited, they disagreed fundamentally about what it reveals. Ignore the thick horizontal bar—that's a lead plate. The picture's crucial feature is the spidery arc stretching vertically across it. Four different explanations were offered, but only one of them survived the ensuing arguments—two particles have been created simultaneously, a negative electron curved downwards by a magnetic field, and its mirror image, a minute positive particle steered in the opposite direction. Despite the initial doubts, and despite this photograph's opaqueness to untrained viewers, scientists now accept it as unequivocal, visible proof that positrons exist as the positive partners of electrons.

There's a certain irony about visual information, which supposedly lays bare the secrets of nature for everyone to view. When Rutherford was questioned about the reality of alpha particles, he retorted testily that he could see them as clearly as the spoon in his hand. Well, yes—seeing is believing, but only if you're a fully trained expert. Instinctively, it would seem that using machines to collect data should increase objectivity by eliminating fallible human observers. In practice, as Victorian scientists came to realize, the converse is often true, because so much interpretation is needed to decipher the results generated by technological equipment. Photographs of subatomic particles do provide definitive evidence that exists independently of any human being, but—like body scans or thermal maps—they are unintelligible to all but a few.

Victorian chemists imposed order on the elements, arranging them into the neat boxes of the Periodic Table. Similarly, during the second half of the twentieth century, physicists organized subatomic particles into an array called the Standard Model. It made the subatomic world appear neat and tidy, ranked in neat rows of particles with unfamiliar names—fermions and bosons, leptons and gluons. Yet, as with the Periodic Table, achieving this taxonomic simplicity was far from straightforward. One high-energy physicist entered a cautionary tale into his laboratory's logbook: 'The man was tired, for he had diligently worked the area for weeks. He stooped low over the pan at the creek and saw two small glittering lumps. "Eureka!" he cried, and stood up to examine the pan's content more carefully. Others rushed to see, and in the confusion the pan and its content fell into the creek. Were those lumps gold or pyrite? He began to sift through the silt once again.'[12]

Scientists need theories as well as experiments. If they want to avoid being deceived by fool's gold, they have to work out in advance the best place to look for the real thing. Mendeleev's counterpart was Murray Gell-Mann, an American scientist who—like Mendeleev—confidently left a gap in his Model, predicting (correctly, as it turned out) that another particle would be found to fill it. Gell-Mann also imagined into existence a new family of still smaller particles, the minute internal components of protons and neutrons—quarks, a name he borrowed from James Joyce's *Finnegan's Wake*. Research teams adopted various tactics to confirm Gell-Mann's ideas, and eventually, in 1974, they found the proof they needed to vindicate his speculations.

Since then, more quarks have been predicted, and more quarks have been found, categorized into whimsically named flavours, such as charm or strangeness. But they remain invisible: nobody has ever seen a quark. So how can we be sure they exist? Scientists had their answer. Quarks must be real, they maintained, because they explain so accurately what happens in experiments. Time after time, in diverse situations, researchers predicted theoretical results by assuming that quarks exist, and then found that these matched perfectly the actual results obtained experimentally. This was true for many different laboratories working with many different approaches, not just for one team set on confirming its own particular convictions. As the evidence accumulated, it became harder and harder to conceive that any rival explanation might be possible.

Nevertheless, mysteries remain. Take mass, which intuitively seems easier to grasp than esoteric quantum forces and spins. Scientists have neatly grouped all their particles into three families, light, medium, and heavy. The difference between the largest and the smallest member parallels that between an elephant and a tiny ant—but why should their masses vary so much? In any case, what is mass? It turns out that although there are three possible explanations, two of them depend on assuming that yet another conjectured particle—the Higgs boson—will eventually be detected. Even proving that these bosons exist would still leave another question unresolved—why is the world we see composed only of particles in one family, including Higgs bosons?

When Mendeleev brazenly left gaps in his Table, scientists accused him of overriding observation with theory. Looking for Higgs bosons means building bigger and bigger instruments which are tailor-made in advance to search for something specific. Ideologically, scientific research means testing predictions. In practice, as the theoretical and financial stakes get higher, it can also mean confirming them.

5 Genes

Ah, love, let us be true
To one another! for the world, which seems
To lie before us like a land of dreams,
So various, so beautiful, so new,
Hath really neither joy, nor love, nor light,
Nor certitude, nor peace, nor help for pain.

—Matthew Arnold, *Dover Beach* (1867)

A doting father, Charles Darwin was wracked with guilt about the implications of his own theories, accusing himself of having reduced his children's fitness to survive by marrying his first cousin. He was also concerned that the entire nation might be degenerating, warning his affluent Victorian readers that 'the reckless, degraded and often vicious members of society [his not-so-subtle dig at the Irish] tend to increase at a quicker rate than the provident and generally virtuous members [subtext: people like him].'[13]

Politicians shared his worries, seizing on Darwin's ideas about evolution as scientific justification for their attempts to purify the race by controlling childbirth and eliminating anyone seen as defective. 'This savage breed', thundered one twentieth-century demagogue, must be 'as far as possible wiped out…by intelligent artificial selection, and the nation which produces the finest, noblest and most intellectual race will win in the long run'. This goal of racial improvement was pronounced not by Adolf Hitler, but by one of Britain's most eminent doctors as he railed against Germany.[14] Hitler now epitomizes evil, but when the Nazis set out to purify the Aryan race, they were following the example of social cleansers all over Europe and the United States of America.

From the 1880s and well into the twentieth century, campaigners for artificial selection adopted two complementary approaches. One group, the positive eugenicists, encouraged educated, wealthy people to have more children. However, they were less influential than their allies, the negative eugenicists, who imposed draconian restrictive measures, recommending compulsory sterilization of the poor and incarceration of single mothers. In Sweden, around 60,000 people underwent compulsory sterilization as part of a state-run welfare package designed to release funds by eliminating the 'biologically unfit' (generally interpreted as low scorers in IQ tests); this project was abandoned only in 1967. For the Nazis, America provided a particularly potent role model. Across the Atlantic, wealthy donors were contributing millions of dollars for 'better breeding' research. Several states had passed laws permitting those judged mentally and physically inferior to be sterilized, and immigration regulations effectively excluded people of non-Nordic origins. Hitler carried the American system to an extreme conclusion, killing off anyone he deemed undesirable—not only Jews, but also gays, Romanies, and the mentally disabled.

Far from being an unfortunate blind alley, eugenics is buried deep within the modern life sciences. Eugenic science originated in Britain, founded by another of Darwin's cousins, Francis Galton, and led in the twentieth century by Darwin's son Leonard. All these family relationships are significant, because Galton compiled an influential survey of Britain's leading intellects to back his claim that genius is passed down from father to son (although he did recognize that mothers play some part in this process). Not everyone was immediately convinced that intelligence is due only to inheritance. The nature/nurture debates were as fervent in Victorian times as now, and discerning critics pointed out that rich children clearly benefited from their better education. To counter such objections, Galton defended his eugenic programme by developing new statistical techniques—the same quantitative, mathematical methods that later helped to found modern Darwinism.

When the twentieth century opened, Darwin's ideas were being slated in books with dramatic titles, such as *At the Deathbed of Darwinism*. This virulent opposition was voiced not only on religious grounds, but also on a range of scientific ones. That evolution had occurred was no longer in doubt—the problem was to decide how. Paleontologists emphasized the large gaps in the fossil record, which implied that instead of occurring evenly and gradually, evolution had taken place in sudden leaps at different parts of the globe. Critics also accused Darwin of failing to devise any convincing explanation of how organisms might change between one generation and the next. Some scientists (including Darwin himself) echoed the Lamarckian idea that children inherit characteristics acquired by their parents; others suggested that a child's development might be determined—somehow—by mixing together each parent's germ plasm, defined vaguely as an immutable substance holding sets of coded instructions. Many scientists hoped to find a model of evolution that did not rely on the ruthless struggle for survival they found so morally repellent.

One of the leading anti-Darwinian movements was Mendelism, a surprising contender considering its present centrality in evolutionary theories. Confounding any image of straight-line scientific progress, modern genetics is based on experiments initially used to argue *against* Darwin's evolution by natural selection—unlike Darwin's model of gradual continuous change, Gregor Mendel's research supported the notion of abrupt transformations. This Central European monk has acquired a romantic aura which is intensified by the paucity of solid information. An impoverished scholar, Mendel was a contemporary of Darwin's, yet conducted his research in a remote monastery and published his key results in an obscure local journal. Over three decades went by before several biologists (the precise number is a matter of academic debate) independently rediscovered Mendel's work and incorporated it within their own projects.

'Darwin + Mendel = Modern Darwinism'—if only it were so straightforward! Unfortunately for those who like their history in neat patterns, this simple equation simply does not work. A revised version of evolution by natural selection was developed only gradually and sporadically over the first half of the twentieth century, as the insights of disparate research groups were contested, altered, and moulded into a new pattern. Mendel himself knew nothing of genes, a concept introduced long after his death, and was more interested in breeding plants than in arguing about the origins of human beings. He believed that he could produce a new species of plant by hybridization (combining two different species, the plant equivalent of creating mules from horses and donkeys), and he set out to describe the process mathematically.

Like many scientists, before Mendel started experimenting he already had a good idea of what he wanted to find. In addition, his results are suspiciously precise. For his research subject, Mendel chose ordinary garden peas. After cross-fertilizing contrasting strains, he counted up how the parents' characteristics were passed down to the offspring. For example, when he mixed tall and short peas, Mendel found that the first generation of his new plants were all tall; however, when he interbred those, the next generation contained (exactly!) three tall ones for every short one. The concealed shortness had reappeared, as though linked to some sort of recessive factor outweighed by the dominant tallness.

When Mendel's work was rediscovered in 1900, his mathematical approach appealed to a new generation of researchers who wanted to establish biology as a modern science based on solid evidence. Mendel's greatest champion was William Bateson, a Cambridge evolutionist who invented the word 'genetics' and was convinced that heredity takes place in small discontinuous steps. After Bateson's experiments confirmed Mendel's numerical ratios, professional plant and animal breeders were delighted, because being able to predict offspring more accurately enabled them to boost their profits. In contrast, scientists were slower to be converted. Antagonists included not only the British pro-Darwinian statisticians, Galton's successors who insisted that evolution is continuous, but also Thomas Hunt Morgan, an American embryologist.

Fɪɢ. 7.—Mutants of Drosophila melanogaster, arranged in order of size of wings; *a*, cut,; *b*, beaded; *c*, stumpy; *d*, another stumpy; *e*, vestigial; *f*, apterous.

Fɪɢ. 46 Some wing mutants of the fruit fly, *Drosophila*. Thomas Hunt Morgan, *Evolution and Genetics* (1925).

Morgan is another counter-intuitive hero in Darwinian evolution. Although he eventually won the Nobel Prize for demonstrating that Mendelism works, Morgan originally opposed this new theory. Mendel's dominant and recessive 'factors' were, he said, too vague to be taken seriously, and in any case, probably only affected peas and a few other plants. Morgan's own research convinced him otherwise. He transformed genetic science by moving it out of natural environments—gardens, forests, farmyards—and into laboratories, where carefully controlled experiments could be carried out. Operating with only a paltry budget, Morgan and his team related Mendel's woolly 'factors' to physical entities inside living cells.

Crucially, Morgan chose a highly suitable organism—*Drosophila*, a fast-breeding fruit fly with clearly visible mutations. Figure 46 shows six of his experimental subjects, bred with many thousand others inside washed-out milk bottles stacked up on old wooden tables in Morgan's homely 'Fly Room'. These artificial mutants have different wings, described with prosaic labels, including 'cut' (top left) and 'stumpy' (top right and bottom left). By examining how characteristics such as wing shape and eye colour are inherited, Morgan's group identified the vital role played by chromosomes, tiny threads inside the cell nucleus that they could see through their microscopes. They went still further, tentatively mapping out the chromosomes into separate parts called genes, at that time invisible hypothetical units responsible for transmitting specific features—sex, white eyes, beaded wings—from one generation to the next. Morgan made Mendelism convincing by showing that its numerical laws have physical origins inside cells.

Yet although genes lie at the heart of modern Darwinism, laboratory genetics had anti-Darwinian origins—Morgan published his fly pictures in a book titled *A critique of the theory of evolution*. Even though he came round to agreeing that tiny mutations can eventually affect an entire population, Morgan could never accept that evolution operates by ruthlessly annihilating the weak. Instead, argued Morgan and his followers, a race is gradually improved by incorporating small, beneficial changes. These debates continued right up to the middle of the twentieth century, so that almost a hundred years after Darwin published *On the Origin of Species*, there was still no scientific consensus about natural selection.

Between around 1920 and 1950, a fresh form of Darwinism was synthesized by combining different approaches from all over the world, and melding them together into a new theoretical model. In addition to the contributions made by American laboratory experimenters, another important strand came from theoretical mathematicians in London. Instead of looking at individual organisms, they studied large populations statistically, drawing smooth curves to suggest that characteristics are inherited continuously, rather than in the discrete steps indicated by Morgan's work. To reconcile these views, the mathematical population geneticists carried out some nifty (and very complicated) calculations, showing that although small individual alterations may take place abruptly, large overall transformations can appear gradual.

Yet another set of insights came from field naturalists, scientists who studied wild plants and animals out in the countryside. Although unfamiliar with the abstruse formulae produced by statisticians, they had reached the same conclusion—it was essential to study populations. During the early twentieth century, this pro-Darwinian approach was particularly important in Russia, where evolutionary biologists investigated what happened to local fruit flies when they introduced some of Morgan's laboratory-reared specimens. They explained their results by assuming that variations float around unused and often unsuspected in a population's gene pool until they are needed to cope with a change. For instance, when the trees of industrial cities were dirtied by smoke in the nineteenth century, the grey peppered moth became vulnerable to predators, easily visible against the darkened trunks, until in a process of rapid adaptation, the previously rare black form started to predominate.

Over several decades, representatives of different research traditions came into contact with each other and exchanged ideas, gradually building up the modern synthesis that underpins current Darwinism. As just one example of this coalescence, in 1927 a Russian scientist called Theodosius Dobzhanksy went over to America and joined Morgan's team, taking with him his own national approach towards populations. During his personal investigations into insects, Dobzhansky combined his practical experience as a naturalist with the abstract formulations of mathematicians and the laboratory statistics of geneticists. By the second half of the twentieth century, such collaborative styles of investigation had become standard, and scientists agreed that genes are the secret mechanism of heredity that Darwin had never known about.

At least, most of them did—but not in Stalinist Russia. Despite the importance of earlier Russian research, by 1940 the country's eminent geneticists had been dispatched to Siberia, and a Ukrainian agriculturalist called Trofim Lysenko was in charge. In a new twist on Lamarckian ideas, Lysenko claimed that permanent, heritable effects could be produced by changing the environment. Like many Russian scientists, Lysenko was committed to relieving food shortages, and his early trials in growing wheat by subjecting the seeds to low temperatures happened to yield precisely the impressive results that politicians wanted to hear. Adeptly manoeuvring his way up through scientific and political hierarchies, Lysenko seized power by proclaiming that Josef Stalin supported his plans to remodel Soviet agriculture and outlaw Western genetics. Under their agricultural regime, farming output dwindled and people starved, until Lysenko was finally ousted in 1965.

Lysenkoism is—like mesmerism—often denigrated as a pseudo-science. But at the time, the situation was not so clear-cut, and the battle was waged with words, not ideas. To demote Lysenko, critics both inside and outside Russia whipped up public hysteria by abusing him as a charlatan. Arguing on the other side, Lysenko's allies adopted complementary tactics, denouncing Mendelism for its links with fascism, imperialism, and corrupt bourgeois culture. Although Lysenkoism itself has long been discredited, its supporters did level cogent criticisms of the majority view. The caricatures in Figure 47 appeared in a popular weekly Soviet news magazine, drawn to accompany a savage article denouncing the followers of Mendel and Morgan. Called 'Fly-lovers—Man-haters', it reviled Anglo-Americans for using science to sanction racist ideologies.[15]

Marxists believed that rearing fruit flies represented a fruitless search for invisible genes. Moreover, from their perspective, genetics was inseparable from politics and big business. The author of 'Fly-lovers—Man-haters' was neither a journalist nor a politician, but a well-informed Soviet professor of biology. Focusing on American eugenicists and their recent suggestions for sterilization, he evoked for his readers the twisted face of a Ku Klux Klan vigilante peering out from behind a geneticist's shoulder—in the sketch on the left, the three men linked arm-in-arm personify racism, deadly science, and the state. In the central cartoon,

FIG. 47 Caricatures by Boris Efimov in *Ogonek (Little Flame, 1949)*.

a Nazi pamphlet protrudes from the pocket of a microscopist pointlessly searching for immaterial entities. And on the right, a gross American capitalist clasps
the reins of puny scientists claiming independence, but waving a 'Banner of pure
science' branded with a large dollar sign.

The label 'Darwinism' has been around since the middle of the nineteenth
century, yet its meaning has changed enormously. For people who link science with progress, the modern synthesis is indubitably a great improvement
on Darwin's original version because it is based on complicated statistics, solid
laboratory research, and countless experimental confirmations. But do maths and
microscopes inevitably lead to better science? As Lysenko and his Soviet allies
pointed out, it depends on what you mean by 'better'. Their agricultural ideology
may have devastated the national economy, but their disparagement of the ties
between genetics and eugenics were well founded—and many of the ethical issues
they raised still cloud present-day research. Galtonian statistics and Mendelian
genes gave Darwin's unproven theories a sound quantitative basis, but they also
rationalized the prejudices of eugenic reformers. Science and politics were tightly
entangled on both sides of the former Iron Curtain.

6 Chemicals

What is matter?—Never mind.
What is mind?—No matter.

—Thomas Hewitt Key,
Punch (1855)

Denigrating women has been a popular sport ever since Eve was accused of falling prey to a serpent's wiles in the Garden of Eden. When Darwin claimed that natural selection had resulted in men being superior to women, even his critics approved. Like them, Darwin found it hard to escape from his Victorian convictions. He had been brought up to believe that the English are more civilized than the Irish, that white Europeans should rule over Africans, and that men are both stronger and smarter than women. Building such prejudices into his theory of natural selection, Darwin consolidated them by justifying them scientifically.

Endorsing discrimination through science was not peculiar to Darwin. Enlightenment anatomists exaggerated differences between male and female skeletons to conform with ideal body shapes, and ranked primate skulls in an order conveniently confirming European supremacy (Figure 24). During the twentieth century, the mask of objective rationality still concealed old prejudices. According to conventional accounts of conception, a female egg lies dormant like Sleeping Beauty awaiting her prince, the heroic sperm who surpasses his rivals and braves hostile vaginal fluids to awaken his quiescent target. From the 1980s, post-feminist scientists twisted anthropomorphization in the other direction, visualizing eggs as femmes fatales who entice errant sperm with attractive chemicals and trap them in finger-like tendrils.

Even the most scrupulous laboratory experimenters approach their research from different angles. In the middle of the nineteenth century, some German physiologists conceptualized people as chemical machines, oiled by internal fluids to keep mind and body operating smoothly and in unison. 'The brain secretes thought,' pronounced the controversial scientist Karl Vogt, 'as the stomach secretes gastric juice, the liver bile, and the kidneys urine'.[16] In principle, this group argued, if they could work out exactly how the human engine operates, then they could develop chemical remedies to make it function perfectly. Critics loathed this reduction of living beings to complex molecules. Nevertheless, this materialist approach did produce results, such as drugs to defeat infections and fine-tune internal mechanisms. By detecting concealed chemical differences between men and women, or between Africans and Europeans, it also provided new scientific rationales for discrimination.

Physiologists probed ever deeper into the body's internal tissues, methodically searching out gender distinctions in brains, hair, and arteries until only the eye remained unisex. In the early twentieth century, masculinity and femininity were attributed to hormones, blood-borne chemical messengers produced by the glands to control behaviour and physique. At the same time, physicians were diagnosing previously unsuspected conditions said to be race-specific. Some were transparently biased, such as the syndrome of Ethiopian insensibility to pain, invented to provide a convenient let-out for plantation slave owners. Others were based on solid laboratory evidence, yet were used to confirm existing bigotry. Sickle-cell anaemia, for instance, was traced to a type of deformed red blood cell then found more often in black Africans (and African Americans) than in white Americans. Black blood became denigrated as bad blood, a dangerous carrier of diseases that threatened white survival through intermarriage.

Laboratory science grew dramatically and altered medicine for ever in the second half of the nineteenth century. In broad brush terms, diseases and their cures became defined not by doctors examining surface signs, but by research chemists exposing invisible entities. Powerful instruments made it possible to identify microscopic organisms responsible for infectious illnesses—tuberculosis, cholera, anthrax—and also to investigate ways of destroying them. Techniques developed inside laboratory walls could then be tested outside, so that sick people became incorporated within investigations orchestrated by laboratory researchers.

Diagnosis moved away from patients' bedsides into hospital laboratories, where illnesses could be identified through applying standard tests to samples that originated from individual bodies, yet were rendered anonymous for chemical examination. Traditional family practitioners protested that although medicine might be becoming more efficient and effective, attention was focused on the disease rather than on the individual sufferer. Some doctors apparently became so determined to analyse diseases that they overlooked their primary goal of preventing suffering. This happened not only in Nazi Germany, but also in the USA. For example, one research team deliberately stopped treating their syphilis patients—crucially, a group of black men—coolly watching them disintegrate in order to track the illness's terminal stages.

Chemical physiologists started to regard patients themselves as mini-laboratories. For them, illnesses were a type of experiment. In real life, they explained, a sick body's normal operations have been disturbed naturally yet unintentionally; in a laboratory, normality is also disrupted, but artificially and deliberately. Some experimenters tested their chemical remedies on themselves (including the elderly man who claimed to feel miraculously rejuvenated after injecting himself with extracts of dog testicles). Others picked out involuntary volunteers. After a teenager was bitten by a dog, Louis Pasteur injected him with an untested rabies vaccine, running the risk of killing him even though it was unclear whether he had been infected. Other people were made sick under controlled conditions. For instance, suspicions about the transmission of yellow fever were confirmed by exposing healthy victims to mosquitoes and watching them become ill, while similar trials were devised to explore the effects of chemicals on the brain and nervous system.

When illnesses could not be attributed to an external invader, such as a TB bacillus or a parasite, researchers searched for internal causes. To understand what was going wrong, they needed to find out what happens when people's bodies are working properly, and they invented all sorts of instruments to measure normal body functions numerically. Some of these were mechanical, designed to record patterns of heart beats, breathing or temperature; others were chemical, measuring concentrations of acid in the stomach or minerals in bones. Some were painless—taking blood pressure, for instance—while others were more invasive, demanding blood samples or even exploratory surgery, often without any anaesthetic. Unsurprisingly, experimental subjects were generally poor, black, or confined to mental institutions.

Being ill came to mean deviating from normality—not the patient's habitual personal equilibrium of humours, but the statistical normality of the whole population. Starting in early nineteenth-century Paris, patients with similar symptoms were grouped together in large hospital wards, where they could be treated collectively. By counting up treatment successes and failures, illnesses could be systematically classified, the effects of new drugs closely monitored, and records kept to compare physical characteristics of the sick and the healthy. As statistical approaches gained strength, doctors tackled illness by measuring a patient's body with scientific instruments—thermometers, stethoscopes, blood-pressure cuffs—and then describing it in terms of numbers. It was the quantitative aspect of this approach that was novel. For instance, William Harvey had demonstrated the pumping action of the heart in the early seventeenth century, but over two hundred years went by before physicians started routinely recording pulse rates accurately.

Chemical investigations of normality enabled doctors to alleviate diseases that had been threatening human survival for millennia. Take diabetes. Since antiquity,

FIG. 48 Maggi Hambling, *Dorothy Hodgkin* (1985).

successive generations of physicians had gradually refined its diagnosis, and by the mid-nineteenth century they had narrowed down its origin to the pancreas. The rate of discoveries then accelerated, as first researchers fascinated by chemical hormones isolated insulin, and then clinicians tested it on patients in hospital wards. The next stage of this story is unsavoury, clouded by invidious wrangling about priorities. Nevertheless, the outcome dramatically confirmed the power of physiological and biochemical research—after the spectacular recovery of some early experimental subjects, insulin went into mass production during the 1920s, and diabetes became a condition to live with rather than an illness to die from.

Such cumulative progress is a gratifying concept, implicitly conveyed in many renditions of science's history. In Figure 48, the model of an insulin molecule sits amidst the cluttered papers on the desk of Dorothy Hodgkin, who won a Nobel Prize for her research in crystallography, a discipline with an unusually high number of eminent women. Light from the window is channelled on to a ball-and-stick replica which will, like the knowledge it represents, endure for ever as evidence of increasingly successful scientific research. In contrast, the half-eaten sandwich is destined to disintegrate, as is Hodgkin herself, whose human transience is emphasized by her gnarled hands, the painful consequence of chronic rheumatoid arthritis.

Here symbolically engaged in multiple activities—writing in her notes with doubled-up hands, inspecting a drawing, peering through a duplicated magnifying glass—Hodgkin was also in reality involved in several different research projects. As well as analysing insulin, she studied penicillin and Vitamin B_{12} —three major chemicals that transformed medicine, but in very different ways. Whereas insulin was eventually isolated to treat one particular illness, penicillin was World War II's wonder drug that eliminates a wide range of infections, while Vitamin B_{12} is essential for curing pernicious anaemia, a disease that had not previously been identified. Their stories illustrate the diverse impacts of personal interests on scientific research. Even when invisible phenomena are revealed by laboratory equipment, medical researchers do not necessarily interpret them with clinical objectivity.

The accidental discovery of penicillin has become one of modern medicine's most powerful myths. The plot runs like this: 'Alexander Fleming, a Scottish research scientist, noticed that some mould had destroyed the bacteria he was cultivating; immediately spotting the significance of this chance occurrence, Fleming developed penicillin, a powerful new drug that revolutionized the treatment of disease.' British chauvinists loved that version of events, and during the War, Fleming's small-scale initiative in grasping an unexpected opportunity came to symbolize the courage of an independent island people. But for historians who like accuracy, this account limps because it glosses over many less glamorous aspects of what happened—Fleming's failure to report his experimental procedures conscientiously, a gap of fifteen years when he scarcely mentioned penicillin, the committed research of an Oxford team (minus Fleming), and the copious financial investment of American manufacturers.

Vitamin B_{12} also has a complicated past. Before medicine became technological, young women who seemed pale and apathetic were often told by their doctors that they suffered from an illness called chlorosis. The alleged causes varied—virginity, corsets, too much freedom, not enough freedom—because chlorosis was fashioned not only by its symptoms, but also by public opinion about how women should behave. Cures generally involved being forced to conform with expectations by adopting a different lifestyle—in other words, what looked like medical prescriptions were effectively moral injunctions. Chlorosis sufferers were held to have brought the problems on themselves, rather like late-onset diabetes patients are no longer seen as helpless victims of an inherited affliction, but are judged guilty of precipitating ill-health by eating the wrong foods.

In the early twentieth century, chlorosis virtually disappeared—not because the symptoms no longer existed, but because laboratory researchers were making blood the fashionable site for diagnoses. First came iron-deficiency anaemia, confirmed under the microscope but also identified as a female disease. Regarded differently from chlorosis, anaemia was depicted as an inevitable burden, one of several 'female ailments' unfairly borne by overworked women struggling to earn a living and care for a family. Pharmaceutical companies welcomed the chemical approach to a female illness, marketing a gallimaufry of iron tonics and pills aimed at women. In contrast, doctors objected, seeing their role being eroded by technicians using invisible entities to diagnose a disease that hadn't previously existed. The contest intensified after laboratory experts identified yet another disease—pernicious anaemia, rarer and fatal, affecting men as well as women. Pharmacists raked in the profits by selling liver extract, which seemed to work—but only sometimes, so that physicians were reluctant to prescribe questionable commercial products for an illness diagnosed outside their own surgery. Although B_{12} was eventually pinpointed as the chemical culprit, its discovery involved gendered preconceptions and bitter conflicts between medical groups with different interests. Similar prejudiced assumptions, but based on ethnicity, surrounded the laboratory identification of sickle-cell anaemia, which acquired the reputation of being a 'black blood' disease.

During the twentieth century, the expanding armament of chemical drugs meant that attitudes towards health changed. Instead of contemplating the likelihood of chronic debilitating sickness, people began to expect and demand longer, fitter lives. Doctors assumed an unprecedented role—they aimed to preserve well-being rather than to help patients die comfortably. And as part of that new mission, during the 1960s an extended drug trial was carried out on a large, healthy population, when the first contraceptive pills were marketed on the basis of only sketchy trials. Wealthy women in Europe and the USA welcomed this chemical innovation, but unwittingly became experimental subjects in a global testing programme.

Hormones had meant big profits since the 1920s, when some of the same pharmaceutical companies that sold insulin also started to market sex hormones. From the beginning, female sex hormones were a far greater commercial success than male ones. For centuries, women had been defined in terms of their reproductive systems, and emotional as well as physical problems had been attributed to their wandering wombs. In the early era of laboratory medicine, anaemia took some of the blame. And then research chemists devised a still more ingenious explanation for women's behaviour—their hormones, soon said to affect every single aspect of their minds and bodies. By the 1930s, manufacturers of sex hormones were boasting success in treating 'practically all the special ills the female human is heir to'.[17]

Yet even though sex hormone therapy became a near-universal panacea, another twenty years elapsed before it was used for contraception—or expressed in terms of how it was first promoted, for controlling reproduction. Control was the operative word. The initial funding and the inspiration for the contraceptive pill came not from the government or the chemical industry, but from Margaret Sanger, an American feminist who wanted to give women the possibility of controlling their own lives. Despite encountering huge opposition, her campaign eventually succeeded, strengthened by eugenicists and population planners who were trying to improve society and reduce birth rates.

Although deemed acceptable at the time, the testing programme fell far short of modern standards, and can sound almost farcical. To get round the moral and legal antagonism towards contraception, experiments were conducted clandestinely, and ostensibly geared towards gynaecological problems. First, doctors tried out sex hormones on women being treated for infertility, and then performed an unsuccessful trial in a mental hospital, where the male (!) subjects objected that their testicles were shrinking. The project later moved to Puerto Rico, which effectively became an island laboratory until many women dropped out, disillusioned by frequent physical examinations and debilitating side effects. Concerned about small sample numbers, the researchers indulged in some creative thinking—by reporting results as menstrual cycles rather than as individual women, they dramatically increased the apparent size of their database.

The trials may now seem perfunctory, but in 1957 the pill was approved in the USA—although effectively disguised as a treatment for menstrual disorders. Everyone knew what that meant. Only two years later, half a million American women were using the pill as a contraceptive, and demand spread rapidly amongst those rich enough to afford it. At last, the drug was being tested on a large scale, albeit on unwitting human guinea pigs. To ensure that the pill was effective, massive doses were initially prescribed, and several types of side effect started to emerge. Although researchers used these findings to change the formulation, some early purchasers were severely affected or even died, fuelling feminist protests about manipulation by an oppressive medical hierarchy. Even so, the advantages seemed clear, and half a century later, over seventy million women were gratefully swallowing a pill every day.

For the first time, doctors were mass-prescribing drugs for the healthy rather than for the sick. This made the contraceptive pill fundamentally different from hormone treatments intended to compensate for malfunctioning organs. Insulin, for instance, remedies diabetes by compensating for the inability of a patient's pancreas to produce it naturally, whereas the pill is more of a designer drug consumed generally for convenience rather than necessity. Whereas insulin saves lives, some other hormone drugs—those accelerating growth, for instance, or reducing spots—have been developed to make lives better.

Medical therapy can easily slide into cosmetic enhancement. Thought of in this way, hormonal adjustment represents a third form of eugenics, neither positive nor negative, but one that promises to improve the population by transforming individuals into desirable norms. Being able to fulfil your dreams by consuming chemicals depends on where you live and how rich you are. It also depends on your gender. For the female pill, forty years went by between the initial inspiration and its eventual low-key launch. In contrast, in the 1990s, Viagra was rushed through the vetting committees in a few months, funded by major pharmaceutical companies that marketed it aggressively.

7 Uncertainties

It isn't that they can't see the solution. It is that they can't see the problem.

—G. K. Chesterton, *The Scandal of Father Brown* (1935)

Albert Einstein courted publicity, playing to the media by ostentatiously refusing to wear socks and coining enough quotable quotes to fill a personal dictionary. On one unusual occasion, he was reduced to silence at a dinner party when someone asked him if relativity theory, Freudian psychoanalysis, and the League of Nations were linked together as products of a revolutionary era. Eventually he agreed—physics, psychology, and politics were all intertwined, different aspects of contemporary intellectual and social upheavals.

Since then, Freud and Einstein have often been bracketed together as the two most important scientists of their era. Both of them voiced the significance of their Jewish identity, and both were deeply affected by the uncertainties of the twentieth century, which contrasted strongly with the political security and scientific conviction into which they had been born. In the 1920s, when many Germans were predicting economic collapse and social disintegration, physicists banished certainty from the subatomic world, declaring that it was impossible to predict events with 100% confidence. In their new quantum mechanics, only likelihoods were allowed—knowledge was restricted to probabilities.

Einstein never accepted that these edicts represented any ultimate explanation of reality, rather than just being useful mathematical tools. In his most quoted quotation, he rejected physical uncertainty, declaring that 'God does not play dice with the world.' A committed pacifist, Einstein also sought political stability. In his quest for world peace, in 1932 Einstein turned to another world-famous Jew living in Germanic Europe—Sigmund Freud. Freud's prognosis was gloomy, couched in the vocabulary he used for describing the psychological warfare inside people's heads—'there is no likelihood', he wrote, 'of our being able to suppress humanity's aggressive tendencies'.[18] Einstein was horrified when Freud suggested that war might be averted by emphasizing its terrifying consequences. Yet within seven years, Einstein had signed a letter urging the American president to preempt the Germans by building an atomic bomb, and they had both emigrated to avoid Nazi persecution.

To find peace in his own life, Freud opted for marriage and children, abandoning his laboratory research into the structure and function of brains, and instead setting up a more lucrative business as a private doctor specializing in nervous disorders. A traditional family man, Freud governed his wife and his patients, his children and his followers, with the same paternalistic authority. Freud's influence has been enormous. Nevertheless, although he thought of himself as a scientist searching for universal laws governing human psychology, his critics are adamant that he should be excluded from any scientific hall of fame. Feelings run high on both sides of the so-called 'Freud Wars', and it is hard to find impartial accounts of his life and work.

A fully qualified medical scientist apparently destined for a conventional career, Freud left behind familiar laboratory research when in 1886, he converted his Viennese medical practice into a new type of laboratory dedicated to establishing the first science of the mind. Relinquishing his former methods of examining dead brains with scalpels and chemicals, Freud started exploring live psyches with mental instruments that he had invented himself, such as dream analysis and free association. These quasi-anatomical techniques proved effective, mapping out the unconscious by revealing points of resistance when the patient—the experimental subject—felt uncomfortable; these psychological iceberg tips indicated where Freud should probe deeper.

His next step was to construct two types of interdependent theory, one meta-psychological and one therapeutic. By collecting together many observations elicited from his patients and also from himself, Freud developed his model of a dynamic psyche powered by both conscious and unconscious forces—in Freudian minds, the primal instincts of sex and destruction are constantly battling against repressive, rational powers trying to impose conformity. Psychoanalytic techniques heal, Freud concluded, by exposing these hidden conflicts and so alleviating physical disturbances, which are the visible manifestations of this unsuspected inner strife.

Freud changed how people think about themselves. By emphasizing the importance of childhood desires and events, and incorporating them within a scientific theory of development, Freud challenged traditional beliefs that human beings are born with a specific personality. He also removed the unity of an individual's psyche, setting up a model based on ambivalence, in which concealed memories, desires, and feelings of guilt result in contradictory behaviour and conflicting emotions. For Freud's followers, sexuality had become a fundamental component of life, central to the psychological makeup not only of adults, but also of children.

Although Freud is as famous as Einstein, scepticism about his ideas is no less virulent now than it was during their lifetimes. One leading antagonist was the philosopher Karl Popper, who denounced psychoanalysis as a pseudo-science. Although some fifty years younger, Popper resembles Freud in several ways. Both were intellectual Jews from Vienna forced to emigrate during the Nazi regime, and both searched for universal laws—one to define science, the other to describe the mind. Yet for Popper, psychoanalysts are as unscientific as astrologers, because it is impossible to refute their conclusions. Good scientists, declared Popper, constantly seek to test their theories by trying to falsify them, whereas pseudo-scientists only try to support them.

Popper was the twentieth century's most important philosopher of science. After listening to Einstein lecture on General Relativity in 1919, he declared that a clear line can be drawn to demarcate true science from non-science. According to Popper, any hypothesis deemed worthy of being called scientific can be disproved. Einstein passes Popper's test because the eclipse expedition measurements falsified Newton's theory, whereas psychoanalysts fail because they have no objective criteria for deciding between alternative explanations. Freud declared that when a small boy called Hans dreamt about a horse, he was expressing sexual fears about his father—but why is that explanation more likely than Hans having been terrified by a real horse in the street?

On the other hand, Popper continued, scientific theories can never be definitively proved right. There is no way of being absolutely sure that the Sun will rise tomorrow, because there might be some unsuspected super-law decreeing that tomorrow is when the Sun will suddenly appear to stand still. In principle, all the theories about cosmic regularity could be ditched if the Sun did not rise one day. And if some future experiment disagreed with Einstein's predictions, then General Relativity would also have to go. Popper might seem to be undermining the entire edifice of scientific certainty; yet he was convinced that scientific knowledge is special knowledge. Although scientists do not always appreciate outsiders' accounts of their activities, they welcomed Popper's vision of crucial experiments ruthlessly distinguishing between right and wrong.

Amongst Freud's opponents, women have been particularly vocal. Ironically, this forceful authoritarian welcomed female disciples, perhaps anticipating (mistakenly) that they would be more subservient than his rebellious male colleagues. At first, feminists welcomed Freud's unprecedented recognition of female sexuality, but they soon started criticizing his male-centred perspective, claiming that he had neglected the vital nurturing roles played by mothers and fabricated the concept of penis envy. Freud often treated his patients as unreliable witnesses, preferring his own interpretations over their own. As he developed his theories, a key turning point occurred when he decided to reject women's own testimony of having been sexually abused as children. In a move that his critics find offensively patronizing, Freud asserted that they were deluding themselves, gripped by fantasies of seducing their fathers.

For other critics, the clinching objection to Freud stems from his therapeutic claims. Whatever promises psychoanalysts hold out, they can provide no experimental proof that their treatment is more effective than time, drugs, or other therapies—too often, snipe critics, the major beneficiary is the analyst's bank account. In any case, even if Freud was right about his own family and his own patients, does it make sense to generalize from small groups—mainly privileged Viennese Jews—to the whole of humanity?

Freud gathered round him a close group of disciples to spread the psychoanalytic word. Although some of them defected to set up their own sects, they succeeded in making Freud famous in the USA as well as in Europe—although not in Britain, where Freud's emphasis on sexuality rendered his theories totally unacceptable. But after World War I broke out, British military doctors made Freudianism acceptable by producing a sanitized version. Soldiers were returning from the front with inexplicable symptoms. Although their bodies seemed intact, they no longer functioned normally. After concocting a vague diagnosis of 'shell shock', doctors searched for organic causes of these perplexing cases—blindness, paralysis, memory loss. It soon became clear that this physical approach was leading nowhere, and they decided instead that their patients were suffering from hysteria. This was a controversial conclusion, because hysteria (named after the Greek for 'womb') had always been denigrated as a women's illness. To understand what was happening, psychiatrists adapted Freud. Glossing over the sexual aspects they found so distasteful, they retained his theories of repression to explain how feelings of terror and disgust experienced on the battlefield were initially pushed down into the unconscious, but erupted later as physical symptoms.

Psychoanalysis received another boost during World War II, especially in the USA. This time, medical advisers were prepared in advance, and they recruited a miniature army of psychiatrists who boosted troops' morale with propaganda programmes, rapidly returning casualties to the battlefield. Since then, psychiatric medicine has boomed, becoming a standard approach for improving not only military efficiency, but also industrial productivity and individual well-being. Resembling preachers of a new secular faith, psychologists market programmes of self-examination aimed at personal awareness and inner improvement.

Despite all the criticisms levelled against him, Freud has been—and still is—enormously influential. Whatever the limitations of his mental model, Freud made it possible to think about minds and bodies, about families and sexuality, about health and sickness, in totally new ways; in particular, he put unconscious urges and childhood sexuality on the conversational map. But even if his theories are non-scientific, is this a good reason for rejecting them? Perhaps science is not the only route to progress. For anyone who gives priority to understanding personal relationships, whether in real life or literature, Freud's contribution to civilization ranks higher than Einstein's. His therapy may not be a universal panacea, but it has encouraged people to reflect on themselves and their lives, hopefully to improve them.

Freud and Einstein have become twinned icons of modern free thinking, heralded as iconoclasts who overturned old-fashioned common sense. Born into certainty, they followed their nineteenth-century predecessors by searching for universal laws that stamped order on the cosmos. To some extent, they both became alienated from the consequences of their own innovations. Just as Freud favoured an authoritarian, directive approach and disowned the disciples who developed his theories in new directions, so too Einstein never fully accepted the inherent uncertainties of the quantum mechanical world and dissociated himself from conclusions reached by physicists inspired by his own work. Freud dealt with human minds, Einstein with the cosmos and subatomic particles, but they shared a classical preference for deterministic laws that enable the past to be used for predicting the future.

Even when probabilities had been introduced into physics during the nineteenth century, determinism was retained. Although average behaviours were being calculated statistically, this did not undermine the fundamental principle that—in principle if not in practice—it is theoretically possible to predict the path of every single particle. In the 1920s, a small break-away group took the radically different step of declaring that it is fundamentally impossible ever to know everything. Even with the finest imaginable instruments, micro-level measurements are intrinsically fuzzy—not because they are inaccurate, but because they just cannot be made. This disturbing concept was expressed by the German physicist Werner Heisenberg in his famous Uncertainty Principle. If you know where a particle is, claimed Heisenberg, you can never be sure how fast it is moving; if you measure its speed (more strictly, its momentum), then you have no way of pinning down its location—certainty will always slip away.

In the subatomic realm of quantum mechanics, nothing is known for certain in advance—there are only probabilities. Influenced by a school of philosophical thought going back to Goethe and the *Naturphilosophen*, Bohr insisted that physicists are part of the system they are observing. This means that detached observation is inherently impossible, because every time scientists try to measure something, they alter it with their presence. By intervening in a situation, they precipitate one particular outcome, which beforehand had been only one of several possibilities. When physicists look at a particular light beam, the instruments that they introduce influence what they see—how they set about recording the

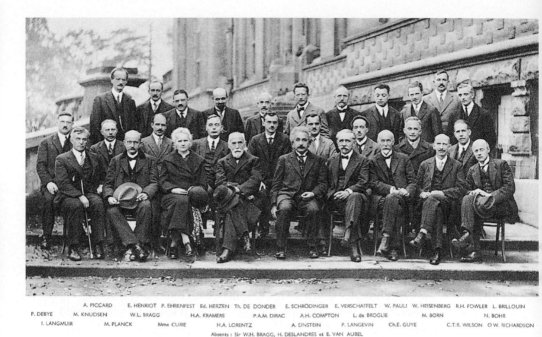

A. PICCARD   E. HENRIOT P. EHRENFEST Ed. HERZEN Th. DE DONDER   E. SCHRÖDINGER E. VERSCHAFFELT   W. PAULI W. HEISENBERG R.H. FOWLER L. BRILLOUIN

P. DEBYE   M. KNUDSEN   W.L. BRAGG   H.A. KRAMERS   P.A.M. DIRAC   A.H. COMPTON   L. de BROGLIE   M. BORN   N. BOHR

I. LANGMUIR   M. PLANCK   Mme CURIE   H.A. LORENTZ   A. EINSTEIN   P. LANGEVIN   Ch.E. GUYE   C.T.R. WILSON O W. RICHARDSON

Absents : Sir W.H. BRAGG, H. DESLANDRES et E. VAN AUBEL

Fig. 49   Scientists at the fifth Solvay Conference, Brussels, 23–29 October 1927.

light affects whether it appears to be waves or particles. Instead of thinking about light either as a wave or as a stream of particles, Bohr maintained that light behaves as *both*.

Counter-intuitive and hard to credit—that's how it still seems, and that's what Einstein thought too. The major showdown between Einstein and Bohr took place in 1927, at a big conference in Brussels. As Figure 49 illustrates, all the world's important physicists were there. Easiest to identify are Marie Curie and Albert Einstein, sitting near each other in the front row. Heisenberg is less immediately recognizable, placed in the back row towards the right. He is standing near his major ally and Einstein's rival—Niels Bohr, the 'Great Dane' from the Cavendish, who sits next to his close colleague, the German Max Born. By this time, Bohr was back in Denmark, where he had established his own Copenhagen School of Physics, which amalgamated his own ideas with those of Heisenberg, Born, and others.

When Bohr arrived at the Solvay conference, he was confident that Einstein would accept his latest Copenhagen interpretation of quantum mechanics. Bohr was wrong. Every morning at breakfast, Einstein produced a new objection; and by dinner time, Bohr had always worked out how to refute it. Instead of carrying out real experiments, Einstein dreamt up imaginary ones that would—he claimed—lead to more definite information than was allowed by Bohr's model, with its inbuilt uncertainties. But every time, Bohr picked up a small detail that Einstein had forgotten to include. This battle of thought experiments went on for years, until Bohr finally defeated Einstein by showing that he had neglected to take his own General Relativity into account. Nevertheless, Einstein spent the rest of his life fruitlessly searching for fundamental laws of certainty that would encapsulate the Universe with mathematical beauty.

The unsettling notions of quantum mechanics were developed during the interwar years, when physicists were themselves subjected to uncertainty, and science was being developed for military purposes. Similar photographs exist for many international conferences, but although they provide tangible evidence of these brief reunions, they reveal no hint that the Jewish scientists among them would soon be dispersed under Hitler's regime. Einstein emigrated to America, Born went to Cambridge, and Bohr escaped to Sweden in a fishing boat before arriving, like Freud, in England. Perhaps some of them later pondered over this carefully posed portrait, an island of certainty in a world clouded by confusion and subterfuge.

Heisenberg, who was not Jewish, remained in Germany, although this turned out to be no safe haven. Einstein's so-called 'Jewish' physics had been officially banned under the Nazi regime, and Heisenberg was attacked for refusing to condemn it. He was also slated from abroad for agreeing to head the German atomic research programme. And that is in itself another uncertainty. Heisenberg claimed that by occupying this key position, he could prevent Hitler from acquiring a bomb. Not everybody believes him.

VII

Decisions

During the twentieth century, scientific research projects were increasingly funded by governments and commercial organizations. At the same time, they dramatically expanded in scale and began to resemble industrial operations. The earliest of these factory-style 'Big Science' enterprises involved physicists, and they focused on atomic research, warfare, and space travel. However, after the double helix of DNA was decoded, money began pouring into the life sciences and genetics became international big business. These massive investments in science, technology, and medicine have yielded rewarding returns. Compared with the Babylonians at the beginning of this book, modern scientists know far, far more not only about the structure of the Universe but also about the mechanisms of living organisms. Nevertheless, they are still unable to answer some of the basic questions about human existence being posed several millennia ago. Celebrated scientific achievements are Janus-faced. Picking apart the atomic nucleus did release untapped energy—but in bombs as well as in power stations. Introducing chemical pesticides improved crop production and reduced starvation—but it also decimated natural food chains. Many people lead healthier, more comfortable lives than ever before—but the world's population is exploding and global warming threatens the planet. Taking advantage of scientific discoveries entails making political decisions about how to use them.

1 Warfare

Now, it must here be understood that *Ink* is the great missive Weapon, in all Battels of the *Learned*, which, convey'd thro' a sort of Engine called a *Quill*, infinite Numbers of these are darted at the Enemy, by the Valiant on each side, with equal Skill and Violence, as if it were an Engagement of *Porcupines*.

—Jonathan Swift, *A Full and True Account of the Battel Fought last Friday, Between the Antient and the Modern Books in St James's Library* (1704)

In his campaign to prevent World War II, Einstein joined forces with his British friend, the mathematical philosopher Bertrand Russell. A more belligerent pacifist than Einstein, Russell had served a couple of prison sentences, sat on pavements, and enlisted Einstein's help for organizing an international pressure group of scientists dedicated to peace. During the first half of the twentieth century, Einstein and Russell witnessed science, government, and industry meld together. Both born in the 1870s, they lived through World War I—the chemists' war of poison gases and explosives—and then on through World War II—the physicists' war of radar, computers, and bombs. Science became entrenched right at the heart of political decisions, a symbiotic relationship illustrated by Figure 50, which shows the quantum physicist Robert Oppenheimer discussing an experimental explosion with his army boss, General Leslie Groves, organizer of the American bomb programme. The two atomic bombs that devastated Japan crystallized growing disillusionment. As Russell remarked, 'Change is one thing, progress is another. "Change" is scientific, "progress" is ethical; change is indubitable, whereas progress is a matter of controversy.'[1]

Warfare may have been a backward step ethically, but it did accelerate the growth of Big Science that characterized the twentieth century. Thinking big was not in itself new—Chinese and Islamic astronomers had built observatories, European Christians had built cathedrals, and Victorian industrialists had built factories. The Big Science that mushroomed in the first half of the twentieth century was different not just in terms of size, but also by being closely linked with the

FIG. 50 Robert Oppenheimer and General Leslie Groves at Ground Zero (1945).

state and with large commercial organizations. It was driven by money, manpower, machines, the military, and the media.

These five Ms worked together to consolidate Big Science. As governments came to recognize the value of scientific involvement in war and defence, they poured money and manpower into building machines for military projects. For example, during World War I, Winston Churchill summoned into the British

Admiralty Chaim Weizmann, a biochemist who had originally escaped from Russia by floating westwards on a log raft. In response to a War Office circular requesting useful discoveries, Weizmann had volunteered that he could produce acetone, vital for gun shells. 'Well', demanded Churchill, 'we need thirty thousand tons of acetone. Can you make it?'[2] Provided with a gin factory for his laboratory, Weizmann scaled up his bench-top chemical processes and went into mass production, eventually overseeing six converted distilleries and a national campaign to collect horse chestnuts as raw material.

Collaboration between scientists and politicians operated in both directions. Weizmann, a committed Zionist, negotiated as his reward Britain's promise to support the establishment of Palestine as a Jewish homeland. Other British scientists also recognized that this unprecedented call for their expertise put them into a strong bargaining position, and they stipulated requirements of their own, insisting that they be involved in policy planning and receive state funds for research. Similar changes had been taking place in other countries since the beginning of the century, and by World War II, scientific, political, industrial, and military interests had become inextricably interwoven.

The fifth M—the media—also helped science to grow. When the eclipse expedition hit the headlines in 1919, Einstein was converted into an international celebrity overnight; similarly, dramatic newspaper accounts of atomic experiments pushed high-energy physics to the front of funding queues. During the 1930s, the media fuelled an international race to discover what lies inside atomic nuclei by breaking them open. At first, scientists used X-rays and neutrons being naturally emitted from radioactive substances. Their next step was to build large accelerators that would speed up subatomic particles artificially until they were energetic enough to split a nucleus apart.

The scientists who raised the most money to finance the biggest accelerators worked in the USA. Most successful of all was Ernest Lawrence at Berkeley, California. He built the first cyclotrons, circular machines that use electric and magnetic fields to make charged particles—such as electrons or protons—race faster and faster around a spiralling circular path. Lawrence started out with a conventional piece of apparatus that fitted on top of a table, but he kept thinking bigger, planning equipment on an unprecedented scale. He realized his ambitions because he persuaded business leaders—especially in the burgeoning electrical industry—that providing finance would be to their mutual advantage. Originally trained as a physicist, Lawrence became a scientific entrepreneur, effectively managing factories dedicated to producing high-energy particles.

Performing a Big Science experiment came to involve many hundreds of scientists, engineers, and technicians cooperating to run an industrial-style operation sponsored by external patrons. Swept up in his own success, as soon as one machine was under construction Lawrence started planning a still larger one, and photographs show his teams dwarfed by massive electromagnets and giant curved tubes. All over Europe and America, physicists recruited experts trained under Lawrence to help them construct their own accelerators. Inspired by his example, they enlisted the support of governments, businessmen, and medical charities to fund their own nuclear research projects.

During the tense years leading up to World War II, scientists in rival laboratories—Rome, Berlin, Cambridge—were competing for first place in the race to understand what lies inside an atomic nucleus. Once military warfare started, the implications of a perplexing experiment on uranium became especially important. This research was based in Munich, although one member of the group, the physicist Lise Meitner, sent in her contributions from Sweden. Like many Jewish scientists, she had fled to escape Nazi persecution, and these forced emigrations strongly affected scientific research. Meitner adamantly refused to be involved in the American bomb project; yet it was she who worked out the physics of nuclear fission that made the bomb possible. To explain her colleagues' strange results, Meitner tentatively concluded that when the nucleus of a uranium atom is hit by a neutron, it splits into two, simultaneously releasing a massive amount of energy, and—just as significantly—emitting more neutrons. When these hit nearby atoms, the process is repeated, each time producing more energy and more neutrons, escalating into an explosive chain reaction that rapidly becomes unstoppable.

As soon as scientists recognized this experiment's significance, any pretence of international collaboration ceased. In scientific journals, the sudden absence of reports on nuclear research made it clear that laboratories in the USA and Britain were exploring its military potential. But what was happening in Germany? Although nobody knew for sure, a group of Jewish émigrés enlisted Einstein's help to convince the American government that a German bomb was a very real possibility. As the war proceeded, this threat provided a convenient justification for continuing, even though there was little evidence of German success. British physicists also became involved, trading in their advanced research on fission for America's expertise in the five Ms essential for a Big Science project. Sponsored by the state, scientists set out to create destruction.

During World War II, US funding for science escalated from $50 million a year to $500 million. Much of it went into the Manhattan bomb project, which operated with military efficiency after General Groves (Figure 50) took over in 1942. He started by establishing a nation-wide network of industrial sites, several of them the size of small cities, whose construction soaked up a large proportion of the budget. Radioactive elements were produced by accelerators and other giant instruments, operated by thousands and thousands of workers with no idea that they were helping to make a bomb. Because Groves imposed a strict need-to-know

FIG. 31 The first nuclear pile, Chicago. Painting by Gary Sheahan (1942).

policy, by 1945 fewer than a hundred people appreciated the full scope of the development programme. Atomic towns were created in deprived areas, experiments in social planning as much as in nuclear physics. Equipped with shopping malls, cinemas, and modern fitted kitchens, their military goals were concealed beneath American normality.

The atmosphere was totally different in experimental stations such as Chicago and Los Alamos, where atomic scientists worked with unprecedented fervour, swept up in their shared enthusiasm to solve problems. Many of them later remarked that these wartime activities had been the best of their lives, and the painting in Figure 51 illustrates how their experiences became romanticized. Dramatically lit, these physicists are poised in expectation, formally dressed in the fashion of the time, and radiating an almost palpable tension as they wait to see whether the Munich discovery of nuclear fission could operate on a larger scale. No indication here that they were miserably cold and dirty, working in sub-zero temperatures and graphite-laden air beneath a Chicago football stand, many of them in pain from accidents incurred during the unplanned building process.

The man in charge, standing on the balcony of this converted squash court and holding a slide-rule in his hand, is Enrico Fermi, who has managed to escape from Fascist Italy. On the right, the stepped brick structure is the experimental nuclear pile containing radioactive materials; three young men sit on top as a suicide squad, ready to dowse the pile with chemicals if it runs out of control. On the basement floor, another scientist manually operates a cadmium rod to control the fission rate. After hours of waiting and an unscheduled lunch break, Fermi eventually tells him to withdraw the rod further. The clicks of the neutron counter merge into a roar, the recording pens go off the scale, and Fermi raises his hand to halt the trial and announce its success.

In some ways more crucial than Hiroshima, this was the decisive day when it became clear that a bomb was possible. The observers reported feeling flat afterwards, forced to contemplate the unknown consequences of what was supposed to be a triumph. They communicated in coded telephone messages, speaking of Fermi as a new Columbus, the Italian navigator who had landed in the new world to find that the natives were friendly. Soon, Fermi moved to the cloistered community of Los Alamos, a self-sufficient industrial township hidden away in the New Mexican desert, where soldiers, scientists, and engineers collaborated to solve the outstanding practical problem. How could nuclear fission be safely packaged inside a transportable bomb?

To run Los Alamos, Groves appointed Oppenheimer, a quantum physicist with no experience of organization. Although apparently an ill-matched pair, the ruthless workaholic general and the nervous intellectual with left-wing leanings made excellent working partners (Figure 50). Smashing their way through decisions, they abandoned normal protocols of pilot schemes and spent lavishly in order to achieve their goal. After Germany surrendered in May 1945, the original claim of needing the bomb as a European deterrent finally lost all validity. But for those involved, it was hard to stop when the goal was so close. In any case, the Americans were still at war with Japan—even Fermi's five-year-old son had learnt to chant 'We'll wipe the Japs / Out of the maps.'[3]

Oppenheimer set up a full-scale test that he code-named Trinity, his idiosyncratic interpretation of the Christian concept that through death comes redemption. As in Chicago, the workers endured miserable conditions, afflicted by the desert's stifling heat, razor-sharp yucca, scorpions, and tarantulas; allowed only cold showers, they hunted antelope for food. In July 1945, around the same time that a real bomb was being loaded onto a Pacific ship, an experimental device was winched to the top of Ground Zero's cast-iron tower, which stretched 100 feet up and was cemented 20 feet down into the earth. Busloads of visitors arrived to watch the early-morning detonation, but they were unprepared for the extent of the devastation. The photographic and verbal images of blazing suns and mushrooming clouds are familiar. Less so, are some of the statistics—exploded rabbits at 800 yards, temperatures of 750°F at 1500 yards, temporary blindness at nine miles. Afterwards, the scientist and the soldier contemplated the tower's vaporized remains (Figure 50). Oppenheimer recalled Vishnu's cry in the Hindu holy writings—'Now I am become Death, the destroyer of worlds'—but was said to strut around in his hat like a *High Noon* cowboy; Groves commented that the war would only be over when two bombs had been dropped on Japan.

Groves and Oppenheimer both backed the bombing of Hiroshima and Nagasaki the following month, and their Los Alamos colleagues were mostly overjoyed when they realized that those years of dedication had paid off, that their project had proved a success. At least, they were initially. After the photographs were published, and the casualty figures were announced, and radiation sickness appeared, they were less sure. As one German émigré put it, 'it seemed rather ghoulish to celebrate the sudden death of a hundred thousand people even if they were enemies.'[4] The self-congratulation might seem unthinkingly callous, but patriotic militants believed (and still do) that dropping the bombs was the right decision.

Suddenly, physicists had become national heroes. A few of them successfully demanded funding to develop more efficient nuclear weapons, ones that would kill people without damaging buildings. Many accommodated their consciences by working on university research projects that, although sponsored by military organizations, were not immediately directed towards warfare. But others wanted nothing more to do with death and radiation. They turned instead to studying life.

Their inspiration was Erwin Schrödinger, an Austrian pioneer of quantum mechanics who had fled to Dublin during the war. In Figure 49, the photograph of the 1927 Solvay conference, Schrödinger is standing in the back row directly behind Einstein (no coincidence that Schrödinger's suit appears different from the rest—his lifelong habit of wearing hiking clothes meant that he was often turned away from official functions). Like Einstein, although he was responsible for important mathematical equations describing waves and particles, Schrödinger never accepted that probabilities could represent the ultimate answers. In 1945, in a small but hugely influential book called *What is Life?*, he urged scientists to search for the biological equivalent of quantum laws, to formulate physical descriptions of growth, inheritance, and other inexplicable phenomena. In a war-wrecked world, only the USA, Britain, and France were in any position to fund research, and that was where physicists migrated—towards the money, and towards a future in biology, the new manifestation of Big Science.

2 Heredity

For sweetest things turn sourest by their deeds;
Lilies that fester smell far worse than weeds.

—William Shakespeare, 'Sonnet 94'

The newspapers of 1953 had many momentous events to report. That year, Presidents Tito and Eisenhower assumed power, but Josef Stalin lost it; lung cancer was linked with smoking; the Soviet Union exploded a hydrogen bomb; and two men reached the top of Mount Everest. But fifty years later, those stories that had once dominated the headlines no longer seemed so gripping. Instead, anniversary festivities revolved around a short report originally tucked away inside *Nature*, a British academic journal. Written by two unknown scientists from Cambridge, this article's deliberately understated conclusion had at the time been ignored by journalists in search of an exciting lead. 'It has not escaped our notice', the two researchers remarked laconically, 'that the specific pairing we have postulated immediately suggests a possible copying mechanism for the genetic material.'[5] To translate from that dry scientific language—Francis Crick and James Watson claimed that by unravelling the structure of complex molecules lying inside genes, they had revealed the secrets of inheritance. Their low-key announcement in *Nature* now symbolizes a new age of molecular biology.

Since then, the double helix of deoxyribose nucleic acid (DNA) has been transformed into a cultural icon. The early clumsy structures of clamps and hand-made plates (Figure 52) have been stylized by countless artists into elegant twin spirals reproduced not only in biology textbooks, but also as sculptures, perfume bottles, and bracelets (unfortunately, the DNA door-handles at London's Royal Society were initially installed upside down). Like a modern-day caduceus (the ancient twined-snakes symbol of medicine), this molecular model has been abstracted into a double helix that represents the entire scientific enterprise—instantly recognizable, even when not understood. But it has also become a Frankenstein emblem, featuring prominently in propaganda directed against contentious research projects such as cloning, genetically modified crops, and biological weapons.

Converting Crick and Watson's crude model into a universal symbol involved hard publicity work and also self-promotion. As Crick himself liked to emphasize, it was not the scientific couple who made the model, but the model that made them. From the contact prints of Figure 52, Watson picked out the second frame, which has become an icon of scientific discovery, apparently capturing the Cambridge couple at their moment of triumph. The crude Meccano-like structure and the bare laboratory with its old-fashioned sink imply that post-war austerity has been overcome by the youthful discipline of molecular biology. Gesturing with his slide-rule, leather patches on his elbows, Crick assumes the role of scruffy intellectual hero, while Watson gazes up, an American boy genius awed by his marvellous molecule. Although clearly posed, this photograph promises a privileged glimpse into the workshop of scientific knowledge.

The camera never lies, but, but, but . . . this was a demonstration model not used in making the discovery, Crick's slide-rule was an irrelevant stage prop, and the photographs were taken months afterwards (even the date is disputed). Snapshot 2 only became iconic fifteen years later, when Watson included it in *The Double Helix*, his racy scientific thriller about the pursuit of DNA. Much criticized, this bestselling book glorified his own role, but downplayed the contributions of a London research team headed by Maurice Wilkins and Rosalind Franklin, who had published their findings in the same issue of *Nature*. Figure 53 shows a London X-ray photograph, taken by Franklin but leaked by Wilkins, that provided the vital clue for Watson. In his words, 'The instant I saw the picture my mouth fell open and my pulse began to race.'[6]

It is this photograph, not the one of the Cambridge pair, that deserves to be celebrated, because it provided clinching evidence for the structure of DNA. Although not an expert, Watson immediately recognized that the prominent X-shape revealed a helix; later, he realized that the bars and diamonds reveal a double rather than a single spiral, carrying repetitive atomic pattern down its length. Analysing the complexities of this photograph involves careful measurements and long calculations. Nevertheless, Watson's sneak preview inspired him to head off in a new direction.

Watson romanticized science as an exciting, ruthless race. Intelligent yet impulsive and vain, he cast himself in *The Double Helix* as a cavalier American bemused by Cambridge quaintness and obsessed with sex and tennis. According to his own heroic account, Watson defied his boss's instructions to get on with his own work, and instead engaged in clandestine meetings with Crick as they struggled to solve science's biggest puzzle. Although they came from different intellectual backgrounds—Crick a physicist, Watson a biologist—they shared an interest in genetics and their areas of expertise complemented each other. To fill in the gaps, they gleaned information by skimming articles and quizzing the prestigious visitors who passed through Cambridge.

FIG. 52 Photographs of James Watson (dark jacket) and Francis Crick (light jacket) with a DNA model, Cavendish Laboratory, Cambridge (May 1953). Contact prints of the photographs by Antony Barrington-Brown.

In Watson's rule book, all means were valid for reaching his desired goal of being the first to get the right answer—even if it meant appropriating the results of Franklin, denigrated by him as a badly dressed woman who refused to wear lipstick and had foolishly intruded into a man's world. Like Watson, Franklin perceived herself as an outsider, ill-at-ease in the culture of a British laboratory after enjoying a research spell in Paris. Led to believe that she was in charge of her own project, Franklin resented interference and protected herself against discrimination by working alone. In comparison with Watson's trial-and-error approach, she proceeded methodically, systematically investigating the molecules she isolated. Whereas Crick and Watson built models on a trial basis as tools of investigation, Franklin gave them a secondary role of visualizing structures already deduced analytically.

Chance brought together Crick, a long-term itinerant Ph.D. student in his mid-thirties, with Watson, an ambitious and much younger postdoctoral researcher in search of a topic. By then, scientists around the world had already come to the conclusion that genetic information is carried not by proteins, as had been believed until fairly recently, but by nucleic acids, intricate chains of molecules linked into even more complex structures. Drawn by the excitement of the chase, Crick and Watson decided to focus on just one of these acids—DNA. That turned out to be a lucky choice, since it was still unclear that DNA was the key player. In contrast with inert substances, in living cells strings of chemical units are arranged in a definite order. Crucially, this order is determined genetically, so that—somewhere, somehow—there must be a code, a set of instructions, determining how the units are arranged. In retrospect, realizing the need for a code sounds like a sudden flash of inspiration. In reality, like many scientific concepts, it emerged from countless meticulous research projects.

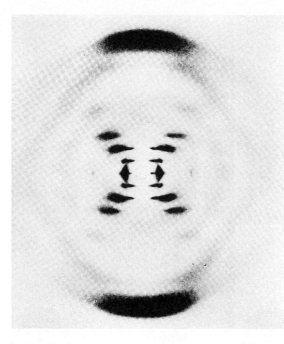

FIG. 53 X-ray diffraction photograph of DNA taken by Rosalind Franklin and Ray Gosling (2 May 1952).

Taking advantage of discoveries emerging from other laboratories, Crick and Watson melded together three different approaches that already existed. Some research groups concentrated on exploring the physical structure of complex molecules, while others examined them from a chemical perspective. In addition, inspired by Schrödinger's *What is Life?*, some scientists were campaigning for a radically different approach to the enigma of life. They believed that just as quantum mechanics had been developed to cope with the uncertainties of the subatomic world, so too a similar leap of intellectual imagination was required to explain the mysteries of inheritance. For them, the key to understanding heredity was information. How is it that living cells pass on characteristics from one generation to another?

Choosing the right organism to study is crucial in biology. Early in the twentieth century, scientists had investigated the genes of fruit-flies (*Drosophila*, Figure 46), but the next generation worked with far simpler organisms—phages, small viruses consisting of a protein coat wrapped around a nucleic acid. Easy to grow, reproducing themselves in around half an hour and consisting of only two molecules, phage viruses proved the ideal subjects for deciding whether proteins or acids are responsible for inheritance. The year after Crick and Watson met at Cambridge, they learnt that a recent phage experiment had conclusively favoured DNA. They set about combining this focus on information with more traditional investigations into the mechanical structure and chemical behaviour of molecules.

Watson was a flagging phage geneticist, bored with chemistry and with little experience in exploring the architecture of large molecules, a type of research especially important in Britain. The major technique was X-ray crystallography, a speciality that included an unusually high number of women in top positions, such as the crystallographer Dorothy Hodgkin (Figure 48), head of Oxford's laboratory and a Nobel Prize-winner. In principle, the technique is simple—by sending beams of X-rays through a crystal, researchers produce patterns of dots on a screen, from which they work out a molecule's internal structure. Actuality is very different. As Franklin's Figure 53 indicates, huge amounts of skill and patience are needed even to obtain a clear image, let alone build up a three-dimensional structure from a series of two-dimensional pictures. X-ray photography demands careful chemistry, precise measurement, and experienced interpretation.

For the expert Franklin, this photograph was just one of many pieces of evidence that she was assessing systematically, taking enough time as she went along to master fully all the necessary techniques—a methodical approach endorsed by Hodgkin. In contrast, Watson described how he lurched from one faulty hypothesis to the next, homing in on the double helix through flashes of intuition and snippets of information borrowed from specialists. As Watson and Crick grappled with their three-dimensional jigsaw puzzle, they garnered only the pieces of information they required to help them juggle their cut-out shapes into a structure compatible with all the data. After many blind alleys and lucky flukes, they eventually hit on a version that made sense and took account of everything. For years afterwards, hordes of molecular biologists dedicated themselves to working out the details, explaining how DNA molecules can unravel into their two separate strands before combining with new partners to wrap themselves into a unique pattern.

Molecular genetics brought together two separate strands of biology—the electrochemical activities inside cells, and Darwinian theories of evolution by natural selection. To trace lines of evolutionary descent, scientists had previously concentrated on examining visible characteristics, such as animals' skeletons or plants' reproductive organs. Once the internal structure of genes had been exposed, they had a new tool for establishing evolutionary relationships, which provided a fresh type of evidence for confirming Darwin's conclusions. Even so, it did little to convince the unconverted—on the contrary, as scientific support for evolution by natural selection piled up in the second half of the twentieth century, opposition got stronger. Fundamentalist Christians retreated to the security of the Bible, while other enthusiasts replaced the traditional God by an Intelligent Designer, neglecting to explain what sort of intelligence is displayed by a designer who plans people with strain-prone backs and large-headed babies.

Deciphering DNA was celebrated as a great triumph, but the enigma of life itself remained unresolved. To cut through that problem, reductionism came back into scientific fashion. In this twentieth-century version, genes acquired a new reputation as the fundamental components of life and society, determining what an organism looks like and how it behaves. Research teams around the world embarked on an ambitious international programme to map the human genome, to find the arrangement of the chemical subgroups making up every single gene. The life sciences had formerly been regarded as a soft option, the province of women and amateurs, but governments starting pouring funds into genetic research, the new rival of physics and space flight. Like landing on the Moon, mapping the human genome provided propaganda material not only for science but also for individual countries. Scientists took advantage of political tensions to solicit state support; for example, in France they stressed the need to prevent American dominance, while their British counterparts highlighted the dangers of the brain drain across the Atlantic.

Genetic research also moved outside laboratories to analyse society. A new scientific discipline emerged in the 1970s—sociobiology, headed by Edward Wilson (usually referred to as E. O. Wilson), an American researcher who originally studied ants, but then leapt to a general theory of human beings. Two basic stages were involved. First, sociobiologists examined their own and other societies to find which elements are common to all; next, they made a theoretical jump, stating that these characteristics are universal because they are coded in people's genes. On that logic, since responsibilities are almost universally carved up between the sexes, men are genetically programmed to work and women to stay at home. Opponents accused sociobiologists of giving scientific validity to political repression—change is fruitless, the argument runs, because people are doomed by genes that have survived three billion years of evolutionary struggle.

A paradox remained at the heart of evolution. If life is a battle for survival, then why are people nice to one another, why do they behave altruistically with no clear benefit to themselves? One of Wilson's disciples, the British zoologist Richard Dawkins, introduced a new term into the English language—'the selfish gene', a metaphor that soon solidified into reality. When Darwin put forward his theory of aggressive survival, he incorporated the competitive ethos of Victorian capitalism; in Dawkins' version, self-interest is encoded in our molecules. Dawkins maintained ruthlessness in the natural world by claiming that individual genes, not entire organisms, are ceaselessly trying to eliminate their molecular competitors. From his sociobiological perspective, although acts of human generosity may appear to be altruistic, they conceal fights being waged deep inside our cells, where the genes are selfishly influencing our behaviour to ensure their own future. Dawkins provided a memorable way of explaining chemical interactions, but in reality—as his critics point out—genes can't think, and they can't have motives, selfish or otherwise. Yet despite its limitations, this verbal model extended its grip.

By the 1980s, any idealistic notion that genetic research was directed solely towards uncovering the truths of nature had evaporated. Molecular biology had been supplanted by biotechnology. Genes were no longer discovered, but were artefacts engineered in the laboratory—which meant they could be patented. The ideology of scientific detachment took yet another knock as commercial companies moved in to market the basic components of life. Universities started to resemble industrial concerns, employing researchers bound by rules of secrecy and aiming to generate profitable inventions owned by the institution, not the individual.

Early hopes that the secrets of life could be found by unwrapping helices proved to be illusory. Real molecules turned out to be far messier than laboratory models, full of mistakes and repetitions. Far from being neatly packed with information, a molecule of DNA contains relatively few effective genes, which lie scattered amongst chemical detritus. Still more problematically, it started to become clear that genes are not responsible for everything—humans and chimpanzees share almost 99% of their DNA, which doesn't leave much over for explaining the differences between them. The old nature/nurture debate reappeared in a new guise, with environmental influences extended to include genes' chemical surroundings inside cells.

Although the human genome project promised great medical benefits, few have been realized because genetic interactions proved to be extremely complicated. There are no single genes for heart disease or cancer, let alone for slimness, sexual proclivity, or intelligence. In any case, new ethical problems have arisen. Tinkering around with cells to be passed onto future generations is an alarming prospect, because things can so easily go wrong. And who gets to decide when a difference becomes a defect? Although many people would feel happy about eradicating Huntington's disease (devastating, progressive, incurable), other inherited conditions seem more equivocal. Tidying up the human race to eliminate supposedly undesirable features sounds too close to Nazi schemes of purification. Like eugenics, gene therapy is a medical science launched with good intentions, but laden with political potential.

3 Cosmology

Two roads diverged in a wood and I—
I took the one less travelled by,
And that has made all the difference

—Robert Frost., 'The Road Not
Taken' (1916)

James Watson turned his status as a non-specialist into a positive advantage through portraying himself as a scientific bricoleur, an intellectual adventurer who had decoded the secrets of inheritance by patching together snippets purloined from various disciplines. But other pioneers who displayed such enterprising panache were pilloried for venturing into areas outside their own expertise. When Alfred Wegener, a German meteorologist, died on an Arctic ice sheet in 1930, he knew that his novel suggestions about the Earth's structure had been rejected by professional geologists. More than thirty years went by before he became a posthumous hero of the Earth sciences, his notion of continental drift finally vindicated in the 1960s.

Like Crick and Watson, Wegener had decided to tackle one of science's great outstanding challenges; also like them, he operated by backing intuitive insights with information borrowed from other specialities, welding them together into a fresh solution. Traditionally, geology had been about dating rocks and identifying fossils, but Wegener studied the Earth as a whole object—like a cosmologist, he tried to understand how our planet has developed since its creation to take on the form it has now. Unfortunately, although he produced a model that was appealingly simple, Wegener had no mechanism to explain how it might work. Orthodox geologists (especially American ones) panned the theories of this German enthusiast, castigating him for gleaning knowledge from books in a library instead of venturing out into the field to gain practical experience.

Wegener's first inspiration came in 1910, when he noticed that the edges of Africa and South America fit together like pieces in a jigsaw. Unsurprisingly, other people had also spotted this match, but Wegener was the first to build it into an entire theory—admittedly, one that was easy to knock down, although it did attempt to resolve long-standing conflicts between different sects of geologists. Old-fashioned disciples of Charles Lyell, the geologist who had influenced Charles Darwin, argued that the Earth is in a steady state of gradual change, slowly being transformed over aeons at the same uniform rate. Ranged against Lyell's supporters were modern catastrophists, who insisted that upheavals were far more dramatic in the past than now; backed up by physicists, they explained that the Earth has been cooling down, shrinking as it goes to create rumpled mountain ranges like the wrinkled skin on an ageing apple.

By the early twentieth century, that picture no longer seemed right either. The sums showed that contraction through cooling was not enough to account for the elephantine folds on the Earth's surface. Another complication was the varied composition of the Earth's crust—it had become clear that the continents are not made of the same material as the ocean floors, but resemble light rafts resting on a harder bed. And to make matters worse, after radioactivity was discovered, physicists claimed that the globe has maintained a steady temperature, centrally heated by nuclear decay deep within its core. Confronted by diverse groups, each with its own preoccupations, Wegener attempted to reconcile their jarring perspectives by picking out those elements that supported his jigsaw view of the world.

Wegener rescued the idea that there had once been a super continent, which he named Pangaea. The top diagram in Figure 54 shows the Earth roughly 300 million years ago, with most of the land concentrated in Pangaea. Very gradually, Wegener explained, this single mass drifted apart into recognizable continents, crumpling up to form mountain ranges. His bottom map shows their positions about 2 million years ago, at the beginning of the present geological period. To support his case, Wegener marshalled plenty of ancillary evidence. An expert in ancient climates, he pointed out how well his theory explained the historical patterns of glaciation far away from the poles. He also summoned up confirmation from fossil records and geological formations, arguing that they continue on either side of oceans like lines of print on a torn newspaper.

Wegener's geological opponents remained unconvinced. All very well for an amateur outsider to draw pretty diagrams, they sneered—but where was the hard evidence? Of all the problems that Wegener had decided to ignore and tidy up later, the most serious was the question of how all this worked. Why and how did continents drift? Neither Wegener nor his followers could come up with a reasonable answer. His notion of continental drift was put on hold until after World War II, when attitudes towards studying the Earth had changed.

By the time of the post-war Cold War, the challenge of deciphering the world's long-distant past was no longer the preserve of traditional geologists examining fossils and strata. Instead, the new umbrella discipline of Earth sciences now also

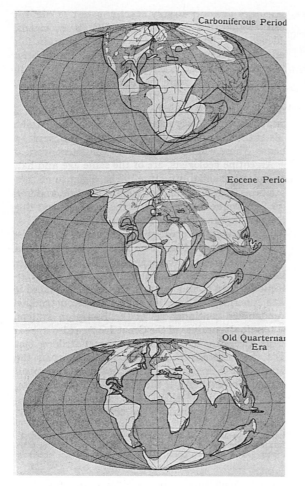

Carboniferous Period

Eocene Period

Old Quarternary Era

FIG. 54 Alfred Wegener's maps showing three stages in the Earth's development.
Alfred Wegener, *The Origin of Continents and Oceans* (1924).

embraced seismologists, meteorologists, oceanographers—specialists who were trained in mathematical physics and regarded the entire Earth as one single unit. As well as examining its surface, they investigated its internal structures, its oceans, and its atmosphere; they also studied the effects of its cosmological environment, such as the space weather of the Sun's magnetic storms. Unlike geology, Earth science was Big Science, attracting massive levels of funding not only from industrial sponsors searching for minerals, but also from states searching for status. The USA was competing against Soviet Russia to conquer space, but it also launched Project Mohole, a colossally expensive plan (later abandoned) to send probes deep, deep into the Earth's interior.

Some Earth scientists worked for the army, mapping sea floors so that enemy submarines could be tracked more easily. They came up with surprising results.

Instead of being thick, old, and even, the ocean beds turned out to be thin, recent (well, in geological terms, that is), and crossed by even younger north–south ridges lying beneath the waters. Stranger still, the rocks on either side of these mid-ocean mountain ranges had striped magnetic bands trapped within them, permanent records of the Earth's past. Meanwhile, up on solid land, geophysicists were uncovering evidence that the Earth's magnetism has changed several times during its history. And while these (and many other) results were accumulating, in 1962 a book appeared that changed how scientists thought about their own activities—Thomas Kuhn's *The Structure of Scientific Revolutions*.

Public disillusion was growing rapidly, fuelled by campaigns to abandon nuclear weapons, ban pesticides, and reappraise the research priorities of a male-dominated society. Science, it was becoming clear, did not necessarily guarantee progress. Kuhn was a catastrophist rather than a uniformitarian. According to him, science's history has been punctuated by a series of revolutions, each precipitated when the evidence against prevailing opinion piles up to an unbearable weight. For instance, before Copernicus, astronomers devised what now seem extraordinary schemes to shore up an Earth-centred system, clinging to complicated epicycles even though their predictions failed to match observations. Eventually, said Kuhn, a crisis point is reached. The old model is jettisoned, and the next generation's efforts are devoted to normal science—refining the new version, testing it against observations, and establishing a fresh paradigm governing how people think about the world. Until, that is, discrepancies start accumulating…and another revolution occurs.

Inspired by the gratifying prospect of being hailed as revolutionaries, Earth scientists self-consciously presented themselves as Kuhnian paradigm shifters. Their Eureka! moment, the geological equivalent of Newton's falling apple or Watt's boiling kettle, came in 1965, when a hypothetical pattern of magnetic stripes meshed perfectly with the observations of an ocean-trawling team. Although much tidying-up was needed, this match symbolized the birth of plate tectonics. Conveniently, a ready-made hero was available—Alfred Wegener, whose ideas shared some features of the new theory. Wegener had imagined the continents drifting around the world, but now they were envisaged as being carried along on top of giant plates kept in constant motion while rock circulates beneath the oceans, welling up at ridges and plunging down again into trenches.

It seemed like an ideal Kuhnian revolution. Introduced rapidly, plate tectonics overturned older models and dramatically resolved intellectual tensions that had been building up since the beginning of the century. Earth scientists settled down to normal science, reconciling this vision of slow perpetual change with Lyell's uniformitarianism. But soon there was another upheaval—this tranquil period of consolidation was disrupted by suggestions that asteroids arrive periodically from outer space, jolting the Earth abruptly from one geological period into another. Reflecting contemporary fears of nuclear winters, disaster scenario scientists envisaged meteoric bombardments producing clouds of debris that blocked off the light, dooming dinosaurs and other species into extinction. Terrestrial uniformitarianism was once again being challenged, this time by cosmological catastrophism.

Geology had previously been distinct from astronomy, but they developed in tandem during the twentieth century. Affected similarly by wars, funding, and mathematization, they became subsumed within the new Big Sciences of Earth science and cosmology. Even those boundaries were blurring. Earth scientists were taking space environments into account, and cosmologists required geological expertise for analysing other-world rocks for traces of life. As the asteroid theory illustrates, they were also addressing the same fundamental question—is change gradual or violent? Earth scientists' debates about continental drift were accompanied by cosmological arguments about the entire Universe—is it eternally stable, or did it originate explosively?

Deciding whether the Universe is uniform or variable was affected by personal convictions as well as by hard evidence. Despite the massive instruments, complex mathematics, and industrial-scale projects characterizing Big Science, the scientists themselves were real people, not rational automata. Take Einstein. In what he later admitted was his greatest blunder, he allowed his own conviction that the Universe is stable to override the prediction of his own General Relativity theory that it is expanding. Although he held out against his critics for years, eventually he heard about some startling results implying that he was wrong. Determined to find out for himself, while Wegener was travelling to his death in Greenland, Einstein was making plans to visit California and meet the astronomer Edwin Hubble.

Hubble is famous for the space telescope named after him, but he was nicknamed 'The Major' for his aristocratic behaviour and English affectations (such as 'come a cropper'). Soon after serving as an officer in World War I, Hubble had become embroiled in American astronomy's biggest controversy. Is there one single massive galaxy, or many smaller ones spread out through space as island universes? Although scientists supposedly make decisions by analysing data, in this case they could not agree on what the data meant. Both sides claimed they had convincing proof, but the same observations can be used to back up different theories—watching a dawn together, Ptolemy would see a rising Sun, but Galileo would see a falling Earth. Gradually, the single-galaxy lobby gained favour, not so much because they had better facts, but more because their leaders were better at arguing.

Astronomers needed a ruler for measuring the Universe, and it was provided by a human computer, Henrietta Leavitt, one of countless women employed during the past three hundred years as scientific drudges. Intelligent enough to undertake calculations, yet sufficiently desperate for work to tolerate long hours and low wages, they include not only pre-electronic mathematicians generating tables of figures, but also 1960s housewives recruited to decipher the photographic tracks of subatomic particles. At Harvard in the early twentieth century, the computers' task was to trawl through photographic plates, gauging the brightness of each star with a standardized palette. Although Leavitt, a reclusive semi-invalid, was paid to toil, not to think, she manipulated her employer into giving her work on her own terms, and she identified a special type of flashing star. When she plotted each star's brightness against its pulse-rate, she came out with a straight-line graph, which Hubble later used to work out the distance of a new star he had discovered. His answer was so vast that the single-galaxy advocates caved in, and admitted that multiple universes were more likely. By then, Leavitt was dead, her personal details forgotten, her glory appropriated by the observatory's director.

Hubble went on to draw his own straight-line graph, using Leavitt's flashing stars to work out the distances of island galaxies and then plot them against speed. It was this result that made Einstein think again, because Hubble showed that the further away a galaxy is, the faster it is flying away from the Earth. As Einstein explained at Oxford on his return (see the blackboard of Figure 40), Hubble's diagram confirmed the consequence of Relativity Theory that Einstein himself had been so reluctant to accept—the Universe started out as a small dense cluster and has been expanding ever since. As Hubble's local newspaper put it, 'Youth who Left Ozark Mountains to Study Stars Causes Einstein to Change His Mind'.[7]

Although Einstein was converted to cosmic expansion, some of the scientists who followed him disagreed. Their reservations were theological as well as scientific. Battle lines were drawn up around the middle of the century, symbolically led by two Cambridge astronomers, Martin Ryle and Fred Hoyle. On one side were ranked Ryle's Big Bang theorists, who claimed that the Universe has exploded outwards from a minute yet massive centre. For them, this was the only way to explain Hubble's expansion and Einstein's relativity. As an added advantage, it was compatible with the Bible's first sentence, 'In the beginning God created the heaven and the earth.' Other scientists—especially Fred Hoyle, a professed atheist—abhorred this intrusion of religious views into science. Rejecting the biblical notion that God's cosmos has a trajectory through time, they insisted that the Universe is in a steady state, gradually expanding as matter is constantly created, but always appearing the same, from wherever you look at it.

The Big Bang advocates were convinced they had won in the early 1960s, when Ryle gained two experimental vindications. One came from the Bell Telephone laboratories in America, where astronomers announced that they had at last worked out the cause of some radio noise that had been disturbing their telescopes (they even cleaned off pigeon-droppings to eliminate every possible source). To them, it seemed clear that they were picking up a low-temperature radiation permeating the whole of space—thus confirming an earlier prediction of what to expect if the cosmos had cooled off after an initial explosion. The other significant discovery was a new type of radio stars, dubbed quasars, found only at huge distances from the Earth and apparently racing rapidly away.

Rather than admit defeat, Hoyle and his supporters behaved like Kuhnian reactionaries, propping up their increasingly untenable theory in the face of mounting counter-evidence. Yet although steady-state ideas faded away, Hoyle had a great effect on cosmology because he popularized it. When Hoyle promoted his own views on radio and in magazines, his Big Bang rivals accused him of dirty tactics. Nevertheless, by bringing abstruse scientific arguments out of rarified journals and into the daily media, Hoyle engaged public support, essential for encouraging governments to increase funding. The astronomers who ridiculed Hoyle ultimately benefited because space research became fashionable, making it easier for scientific pressure groups to solicit state backing.

Although Einstein eventually agreed that the Universe is expanding, he never accepted the ultimate validity of quantum mechanics, despite his own importance in establishing the subatomic rules of probability. For him, they were merely mathematical tools, not descriptions of reality, and he spent decades searching in vain for comprehensive formulae to encapsulate the cosmos. In contrast, during the central third of the twentieth century, most theoretical physicists focused on the quantum realm, providing conceptual tools which could be used by laboratory researchers to investigate minute particles. Einstein's curved space-time became an intellectual backwater, occupied by a small faction of lone mathematicians.

Unexpectedly, General Relativity bounced back into fashion in the 1960s, after Einstein himself had died. This theory without an application suddenly came into its own, because the celestial bodies being revealed by powerful radio telescopes were large and fast enough to be affected by its equations. To describe these relativistic wonders, exotic names started to accumulate. Quasars were rapidly followed by pulsars (which also emit radiation, but appear to flash), detected as rare minute blips on a Cambridge printout by Jocelyn Bell (who indignantly warded off suggestions from her disbelieving supervisor that they were due to interference from the BBC, and has since become a major activist for women's rights in science).

The most renowned astronomical wonders were black holes, the name given in an inspired PR move to theoretical points that Einstein had disdained as irrelevant mathematical curiosities. Like the grin of a Cheshire cat, a black hole is the core of a star that has faded from view to become detectable only from its gravitational pull. By the 1980s, black holes had been joined by wormholes, strings, dark matter, and gravity waves, making high-energy astrophysics a high-profile science amongst people with little idea of what the words meant. It even boasted its own media star, Stephen Hawking, epitome of a disembodied genius, whose *Brief History of Time* is probably the least-read bestseller of all time.

This alien cosmology dazzles its admirers; yet despite its novelty, some objections are familiar. Einstein's distinction between mathematical equations that describe and philosophical models that explain is fundamental to science. Although Einstein recognized that quantum mechanics is useful for describing bizarre phenomena, he believed that its basis in probability was provisional, an interim solution disguising human ignorance of the superior plan devised by a God who does not play dice with the Universe. Similarly, cosmologists acknowledge that while black holes, strings, and their esoteric companions are valuable concepts that work mathematically, their physical existence may make little sense.

By the end of the twentieth century, atheists were boasting aggressively that science had eliminated the need for religion. Yet although cosmologists were peering out to the edge of the Universe and reaching back in time to its origins, they had arrived no nearer to disproving the existence of God. Tracking the history of the Universe from the instant after the Big Bang is a stellar achievement— but it leaves unanswered the fundamental question of how the Big Bang started in the first place. As so often in science, how you interpret the evidence depends on how you want to see it.

4 _Information_

Where is the Life we have lost in living?
Where is the wisdom we have lost in knowledge?
Where is the knowledge we have lost in information?

—T.S. Eliot, _The Rock_ (1934)

Ironically, the history of information processing is shrouded in secrecy. Google now provides instant access to more than anyone needs to know, but computers were developed under a need-to-know policy that restricted information flow. Large electronic calculators originated as military inventions designed to crack enemy codes or compute missile paths, and were protected by stringent security precautions to ensure that no details leaked out. It was only in 2000 that the British Government declassified its official account (whimsically disguised as 'General Report on Tunny') of the wartime equipment developed at Bletchley Park, a camouflaged military base. Electronic information may flash freely round the Internet, but it is also enmeshed in a worldwide web of secrecy.

As science became militarized in the mid-twentieth century, two ideologies clashed. Scientists believed (well, in principle, anyway) in exchanging information freely so that progress could take place as rapidly as possible. In contrast, intelligence personnel compartmentalized activities into small cells, each with limited knowledge. These opposing approaches, both taken for granted by their advocates, came into head-on conflict when military commanders started taking over wartime projects involving scientific researchers—atomic bombs, electronic computers. Instead of sharing their results at international conferences, scientists were obliged to respect the constraints imposed by national security.

This close-guarded culture continued to permeate computer science during the Cold War, when research remained under wraps, directed towards developing clandestine defence systems. In its eagerness to gain electronic superiority over the Russians, the US government poured money not only into the armed forces, but also into universities and private companies producing computers for businesses. As the cartoon of Figure 55 illustrates, military, academic, and commercial

FIG. 55 Machine intelligence. Front cover of *Time* magazine, 23 Jan. 1950.

interests were intertwined. This computer, the Harvard Mark III, was produced in a university laboratory, sponsored by IBM, and—as indicated by the embroidered hat and sleeves—was designed for the Navy.

In this symbiotic relationship, it seemed that everybody stood to win. Commercial enterprises gained state funding to see them through rocky times, and also benefited from a large guaranteed market; at the same time, military experts had immediate access to the latest results. But there were hidden downsides. Scientific researchers not on the government's payroll found it extremely hard to obtain computing facilities—and those who did accept funding could no longer claim allegiance to an ethic of openness, but were bound to secrecy by their employers.

Operating secretly and separately, military inventors in three countries—Britain, Germany, and the USA—worked on computers during World War II. Civilians first gained some inkling of this research in 1946, when the American Army called a press conference to unveil its Electronic Numerical Integrator and Calculator (ENIAC), constructed within a university department but controlled by uniformed

FIG. 56 Electronic Numerical Integrator and Calculator (ENIAC), 1945.

men and women (Figure 56). To enhance its visual impact for the launch, special display panels had been hastily constructed with light bulbs flickering behind halved table tennis balls. The gigantic banks of electrical equipment filled a large room, yet were nowhere near as fast and powerful as a small modern laptop. Even so, the newspaper reports were ecstatic about this artificial machine, whose internal devices operated hundreds of times quicker than neurones inside the brain—an exhilarating if scary concept.

Although the media celebrated ENIAC as the world's first electronic computer, it did suffer from drawbacks. It relied on 18,000 valves (tubes), electronic on–off devices resembling light-bulbs which often blew out and had to be replaced by human operators. As the valves flashed on and off, they generated immense amounts of heat, so that keeping early computers cool was a major problem. There were also insects to contend with, since invasive moths and flies could wreak havoc with the internal connections: the programming term 'to debug' had literal origins in early electronic machines. Still more seriously, there were inherent limitations to the machine's capabilities. ENIAC had originally been designed to generate tables of shell trajectories, but asking it to do anything else—weather forecasting, for instance, or calculating how shock waves move—required several days of manual rewiring, mostly carried out by women. Rather than being a proto-type computer, ENIAC was a giant calculator—there was no way of telling it to carry out a different type of calculation without physically rebuilding it.

Unknown to all but a handful of British scientists, more versatile machines were already operating at Bletchley Park, again with a military purpose—to pene-trate the German intelligence network by deciphering their coded communica-tions. Secrecy was of paramount importance, since the project's success depended on preventing the Germans from realizing that, despite altering the key every day, their diplomatic messages were being understood and acted upon. Speed was also essential, because the code had to be cracked before it changed if a U-boat attack or an air-raid were to be forestalled. To maintain security, the staff at Bletchley worked in small groups, aware only of their own immediate task. Many thousands of these men and women died without breaking their oaths of confidentiality, never revealing that the British, not the Americans, should have been credited with the world's first digital computer.

By the end of the war, ten electronic Colossus machines were churning through intercepted texts, rapidly comparing them with vast numbers of letter patterns until a similarity happened to crop up that suggested a route into the day's concealed code. This task was only feasible because each Colossus could make choices. Instead of mindlessly trawling through every single possibility, it eliminated countless cul-de-sacs at one fell swoop, either by following preset instructions or by pausing to ask a human operator for help. Although far less adaptable than modern computers, Colossus was different from ENIAC because it made decisions. The entire base functioned like a giant information processing machine, taking in garbled messages and generating intelligible details of German plans. Inside, its human, mechanical, and electronic components interacted with each other by following instructions.

Unknown to many of the people there, the world's mathematical expert on decision-making—Alan Turing—worked at Bletchley Park. Nowadays, Turing is celebrated as the founder of our modern information society, in which power and money rely on controlling global communication; yet his own life and reputation were shadowed by the need for silence. In addition to the secrecy surrounding much of his work, he behaved covertly to conceal his homosexual activities, then still illegal. After being forced to confess during a public trial, Turing was sub-jected to a year's experimental hormone therapy, and died in 1954 from eating a poisoned apple. Regarded in his lifetime as obscure, eccentric, and a potential traitor vulnerable to blackmail, Turing has now been transformed into a tragic gay icon, an information guru who patriotically confounded German security.

Pledged to secrecy, Turing and his colleagues remained quiet about their war-time activities, and the earliest programmable computers were invented in igno-rance of their research. Nevertheless, Turing was enormously influential because he thought not only about the technology that makes computers work, but also about their significance. After the war, military and commercial organizations focused on building computers that were larger, faster, and more powerful, and—crucially—could be programmed to switch rapidly from one task to another. Turing posed fundamental questions about machine intelligence, drawing human minds and electronic circuitry ever closer. Ever since his closest friend died when he was a teenager, Turing had doubted the conventional Christian notion of a soul, an ethical conviction that he carried through into his machine philosophy. Resembling the biological determinists who were searching for life in compli-cated molecules, Turing believed that computers can think, even though they are made up of electronic circuitry. Thinking is, he agreed, hard to define—but whatever it is, Turing was convinced that computers and people both do it.

Adopting Turing's position meant establishing new visions of human beings as well as of machines. Computers were modelled on brains—or was it the other way round? Figure 55 asks 'Can man build a superman?', and indicates how electronic accessories might stand in for arms or eyes. The comparisons worked in two direc-tions. Whereas scientists initially enthused that circuits resembled super-fast neurons, soon they were saying that living nervous systems function just like electronic ones. In their visions of the human psyche, people reach decisions after signals have pulsed through a series of zigzag branching points, which act like electronic switches choos-ing between two paths. Typists were a favourite example: a secretary's ears pick up sound waves from her boss's dictation, and her body/brain decodes them into simple electronic signals which activate her fingers (computer wits extended the parallel by adding that she could reach into her memory store for cleaning up the grammar).

In Turing's visionary projects, real and conceptual experiences blended together. In the 1930s, before electronic computers became physically possible, he had invented an imaginary machine which received instructions by reading a long paper tape carrying sequences of marks and blank spaces; in principle, claimed Turing, his machine behaved as if it were a human being. By 1950, it seemed possible that this mathematical inspiration might coalesce into actuality (see Figure 55), and Turing took his analogy still further. Reflecting his familiarity with both code-breaking and sexual subterfuge, Turing first asked how someone could determine from printed answers whether an invisible respondent is male or female. And then he took another step in his imagined universe—could a ques-tioner distinguish between a person and a machine?

This blurring of human–machine boundaries continued during the Cold War, when research into artificial intelligence was driven by government funding. At the same time as engineers were trying to make computers behave like people, psychologists were describing human brains as if they were electronic circuits. In an extension of the symbiotic techno-relationships developed at Bletchley Park, social structures were made compatible with computer systems, which were themselves designed as interactive extensions of human beings. To help military men dispatch orders to the electronic machinery enlisted under their command, programming languages became more and more like human ones; to improve the effectiveness of weaponry, computers were built that could analyse events at the same time as they were actually happening. As computer technology advanced at an explosive rate, these military applications were soon adopted for civilian use—running payroll systems, simulating delivery schedules, reducing production costs. And in the 1970s, when microcircuitry continued to shrink, manufacturers created yet another lucrative market by moving into private homes.

Like many computer addicts, Turing was wildly optimistic about the future. By the end of the century, he predicted, the notion that machines think would be commonplace. Expert after expert made similar rash promises, lured by the stunningly rapid changes in electronic technology which were making computers smaller, faster, and cheaper. But although the Deep Blue computer beat the world chess champion in 1997, it adopted different tactics from any human opponent. Rather than computers becoming like human beings, humans were adapting to fit computer requirements, and physical reality was becoming less significant. Real-life wartime activities—firing missiles, supplying provisions, testing tactics—were simulated on massive computer systems, while people entertained themselves by waging pretend battles on their home screens, or watching films such as *2001* and *Blade Runner*, visionary tales of a near future in which computers reign or are indistinguishable from humans. By the end of the twentieth century, military training was taking place in the virtual realm, as pilots flew simulated bombers rather than risking their lives in real ones, and soldiers learnt safely online the techniques of hand-to-hand combat; conversely, civilian hackers could indulge in the pleasures of warfare by sending out viruses. Little surprise that for the *Star Wars* generation, Second Life can seem more familiar than everyday real life.

Turing was not the only computer expert with utopian visions. During the 1960s, thirty years before the World Wide Web came into being, a Canadian media expert called Marshall McLuhan coined his neat aphorism that electronic technology would recreate the world as a global village. While governments developed computers as military prostheses, Californian gurus were campaigning for information to be freely shared amongst a virtual community based on equal access. In their electronic equivalent of 'Peace, not War', computers would be dedicated to democratic government and universal education. First came the personal computing industry, when calculating power was redistributed away from massive mainframes and into individual desktops. And next came the Internet, which caught on not because it was centrally planned, but because it was under everybody's and nobody's control.

The dreams of Turing and the other information utopians rest unfulfilled. The Web stretches around the globe, but the need to tap in electrically has widened rather than narrowed the gulfs between poor and rich. Information may be freely distributed, but much of it is worthless or even dangerous—online anonymity provides access to child pornography and terrorism manuals. And when everything is coded electronically, privacy disappears: in today's computerized version of McLuhan's global village, concealed cameras record people's daily activities as effectively as inquisitive gossips hiding behind net curtains. Secrecy and warfare still pervade computing.

5 Rivalry

Space isn't remote at all. It's only an hour away if your car could go straight upwards.

—Fred Hoyle, *Observer* (1979)

By the end of World War II, many scientists believed that life must exist elsewhere—after all, there are 100 billion stars in our galaxy alone, so why should planet Earth be unique? For Enrico Fermi, the Italian nuclear physicist who had helped to develop America's atomic bomb, this argument contained a fatal flaw. How come, he asked, that we have found no evidence of any extraterrestrial beings? The obvious answer to Fermi's question is that there are no such aliens, but in the aftermath of Hiroshima, a more sinister solution emerged: could the evolution of intelligence entail a built-in tendency towards self-destruction? During the Cold War, as nuclear reactors proliferated and international tensions rose, worldwide devastation seemed only too likely. Fermi's paradox came to symbolize terrestrial geopolitics.

This fear of global annihilation was reinforced by emphasizing that two superpowers were locked in head-on confrontation. The film *Star Wars* (1977) cast this period as a battle between good and evil, a struggle for survival between light and darkness, as if the entire world had been divided into two rival factions. Just as fiction and fact intermingled in utopian visions of artificial intelligence, so too, celluloid versions of cosmic conflict started to interact with earthly reality. When Ronald Reagan, America's first Hollywood president, proposed building a gigantic missile shield out in space, its nickname became 'Star Wars'.

Never before had science been so blatantly enmeshed with politics. During the Cold War, research programmes that seemed scientific were also being driven by struggles for power. All around the globe, governments jockeyed for position, investing huge portions of their annual budgets into two key areas—space flight and nuclear energy. In particular, the USA and the USSR used their scientific successes to win allies and consolidate their influence. Although they never did engage in nuclear conflict, they both conducted massive propaganda wars to ram

FIG. 57 'In Tune with the Times. Africa!'
Yuri Gagarin salutes Africa from space. *BOCTOK* [*Vostok*] means THE EAST.
The Morning of the Cosmic Era (1961).

home the political implications of their projects. As just one example, the Soviet cartoon of Figure 57 attacks the USA by not only advertising Russia's technological supremacy, but also appealing to developing nations. In 1961, when Yuri Gagarin became the first man to fly into space, his ship carried a potent name— *Vostok*, which means *East*. For Soviet citizens, spacecraft *East* symbolized their continued ascendancy over the West, a message designed to win support in Africa, Asia, and South America. They had already scored an early victory in the Cold War by launching Sputnik, the first artificial satellite to orbit the Earth. Gagarin's flight seemed to confirm Russian claims that only communism could guarantee the progress needed for liberating the developing world from old-fashioned imperial oppression.

Ironically, Sputnik was launched during a project intended to foster scientific cooperation, the International Geophysical Year (IGY) of 1957–8. The IGY was a global enterprise on an unprecedented scale. Sixty-seven countries took part, and around 60,000 scientists spent billions of dollars investigating the Earth in its entirety—not only its surface features, but also its atmosphere and its oceans, its weather and its volcanoes, its shrouds of solar magnetism and cosmic radiation. To guarantee collaboration in the future, outer space and Antarctica were declared to be international laboratories. Originally conceived as the ultimate demonstration that scientific teamwork transcends political differences, by its end the IGY was celebrated for the vast additions that had been made to human knowledge.

Governments pumped money into the IGY because they recognized that other interests were at stake. The geophysical research did increase scientific understanding, but it also carried major commercial and military implications. The same international networks that monitored earthquakes could also detect underground bomb tests. Mapping large mineral deposits was scientifically valuable; financially, it was invaluable. Studying the poles yielded new biological and geological data; strategically, both the northern and southern polar regions are important for defence systems. Naval ships enabled oceanographers to create underwater maps, but their sonar equipment was vital for flushing out enemy submarines. Even predicting the weather suggested a new form of weaponry— seeding clouds to ruin crops, or fomenting storms to destroy cities.

Most exciting of all was the prospect of venturing into space, made feasible by military research into rockets during World War II. Whereas scientists enthused about possibilities for exploring the Earth's upper atmosphere, governments focused on political opportunities. But those aims became tangled together. The boundaries between scientific and military communities had already blurred, and during the IGY it became still harder to distinguish between their activities. For instance, space physicists announced that it would be an excellent idea to investigate the Earth's outer belts of radiation by detonating hydrogen bombs hundreds of miles above the ground. In retrospect, it seems naive to have thought that this global experiment could escape becoming a military operation. Code-named project Argus, it was taken over by the US Army, who clamped down on international discussions of the strange auroral lights being generated over the Pacific by their tests. When the findings were finally declassified, American scientists wriggled out of their obligations to exchange information by maintaining that 'Argus was not an IGY program; it was a Department of Defense effort.'[8]

Sputnik and the other early satellites could be promoted as scientific instruments, but they were also Cold War inventions with the potential to carry out military observations. By the start of the space race, the old ideology of pure science had become untenable. Scientists may have persuaded themselves that they were accepting government funding in order to carry out their own research, but science had become militarized, and military politics had become scientific. Government policy was being directed by scientific possibilities; conversely, the type of knowledge that scientists produced was affected by political requirements.

This is not to say that the information produced by militarized science is wrong, just that it is different from what might otherwise have been produced. For example, international rivalry channelled funds towards surveillance techniques—Cold War politics demanded reconnaissance satellites and high-precision cameras. During the 1960s, the Earth was for the first time viewed from an external vantage point, and images of the Earth as a sphere hanging in space came to dominate geophysical research, changing forever the way in which human beings regard their planetary home.

As the propaganda campaigns intensified, Soviet scientists won the first round by sending Sputnik around the Earth. Soviet politicians also scored a diplomatic coup, aggressively choosing an IGY reception at the Russian embassy in Washington for the announcement of their success. Sputnik's American rival was still on the ground, and although its technological sophistication may have consoled scientists, coming in second did little for the nation's reputation. One of the greatest impacts of the Sputnik project was to persuade the US government that it should pour money into its own programmes of education, defence, and scientific research, all geared towards the ultimate goal of reaching the Moon. To win over resisters who protested about the expense as well as the escalation of hostilities, government leaders emphasized the potential spin-offs from space research. Many of these were impossible to predict in advance, but they turned out to include not only robots and micro-electronic equipment, but also dried food, non-stick frying pans and no-fog ski goggles. US policy makers were set on a neck-to-neck race to the Moon that would boost national pride and deflect attention away from less attractive policies, such as the Vietnam War.

For chauvinistic Americans, the main point of aiming for the Moon was to get there before anyone else. From their perspective, the race started badly. Under starter's orders from President Khruschev, scientists in the Soviet Union aspired to maintain their lead, and they scored several more firsts in close succession—their unmanned probe reached the Moon, Russian dogs beat American chimpanzees into space, and Gagarin went into orbit, soon followed by a Soviet woman. But then their pace slowed down. After several launches went disastrously wrong, Soviet rulers became reluctant to gamble so much money on a contest they might well lose. They had to weigh up the advantages of convincing poorer nations that communism was dedicated to technological improvement against the neglect suffered elsewhere as limited resources were funnelled into space flight.

In the USA, critics were also appalled by the massive expenditure that took funds away from social programmes, but their calls for collaboration rather than competition were quashed. When two American astronauts reached the Moon in 1969, the government did its best to reap the maximum amount of publicity. Like Gagarin's flight, the landing presented a welcome propaganda opportunity.

FIG. 58 The USA Moon landing, 20 July 1969: Neil A. Armstrong and Edwin F. Aldin.

Photographs such as Figure 58 were beamed round the world, showing the Moon's gravelly surface strangely shadowed and disturbed by footprints as the astronauts ventured away from their futuristic ship. The first lunar sentence was carefully crafted in advance to present their achievement as a human rather than an American one—'That's one small step for man, one giant leap for mankind.' Nevertheless, the stars and stripes often shown apparently fluttering in the breeze had been pre-manufactured in rigid material to compensate for the absence of an atmosphere. Similarly, although the plaque left behind by the lunar astronauts proclaimed that 'We came in peace for all mankind', it was written only in English, and was generated by rivalry. Conducted in direct conflict with the USSR, the Moon project yielded military hardware such as spy satellites, communications networks, and defence systems.

Despite the rhetoric, world peace seemed no nearer after that iconic landing. Competition not only continued, but also expanded, so that before the end of the century, several countries had launched their own satellites and were laying plans to reach out into the Solar System, a hitherto untapped zone for global rivalry. In these international battles for prestige, even smaller nations were willing to spend extravagantly for the sake of advertising their independence and their modernity. Individual governments embarked on their own nuclear programmes, nudging the globe nearer and nearer to destruction. As the French minister of defence put it in 1963, 'one is nuclear or one is negligible'.[9]

Like France, countries around the world began buying nuclear power in order to acquire political power. Previously, the devastation wreaked by the American bombs in Japan had brought a temporary near-halt to nuclear research. Many physicists had been so horrified by the outcome of their wartime work that they banded together into pressure groups, determined to spread information about the dangers of nuclear warfare and to disentangle themselves from military control. Nevertheless, others continued with defence work, fascinated by their atomic discoveries and convinced that building better bombs was essential for maintaining peace.

In America, Edward Teller—a Jewish Hungarian émigré who had worked with Fermi during World War II—ignored his colleagues' reservations, insisting that the explosive power of nuclear actions could be more effectively tapped by replicating some of the Sun's activities. In the fission bombs dropped over Japan, energy had been released by splitting large atoms apart. Teller proposed making a more powerful weapon through fusion, by forcing very small atoms to bind together and release energy. After American surveillance revealed that Russia had embarked on making its own bomb, the government gave the go-ahead for Teller's hydrogen super-bomb.

US military scientists converted the South Pacific into a testing ground, detonating nuclear explosives whose effects on local islanders and Japanese fishermen were even more horrendous than anticipated. Determined to maintain their head-start and prevent other countries from building bombs, America imposed such tight restrictions on the distribution of radioactive materials that scientists abroad were unable to carry out experiments or develop medical therapies. Partly in response to this aggressive behaviour, other countries started setting up their own nuclear programmes, and warfare seemed ever more likely. To emphasize the risk of annihilating the entire human race, American physicists invented the Doomsday Clock, a symbolic timepiece whose face has no numbers but is set close to midnight. In response to political crises, the Clock's minute hand nudges towards and away from its final vertical position. It was only two minutes short in 1953 when the super-powers tested thermonuclear devices, retreated to twelve when they signed a Test Ban Treaty in 1963, and then lurched back up to three in the 1980s during the American Star Wars project. The safety margin stretched to its widest in the 1990s as the Cold War ended, but then narrowed again as other countries started testing their own weapons.

Global destruction seemed imminent—so why did it not happen? One answer is that deterrence, not attack, was the major goal. Displaying the ability to retaliate was needed to prevent anyone else from launching the first missile, so that in these battles of bluffs and counter-bluffs, nations assumed power by releasing enough information to make sure that they were seen to be testing. Other diplomatic strategies were also at play. Some nuclear power stations were acquired for show as much as for use. In India, for instance, nuclear reactors served similar political functions to hydroelectric dams and steel manufacturing plants—essential high-status technological installations intended to impress the nation's citizens and celebrate their recent independence from British control. In addition, nuclear power was being promoted as an agent for peace rather than war. At the same time as American statesmen were sanctioning bomb trials over Bikini, they were also boasting how advances in nuclear physics would revolutionize agriculture, medicine, and industry. Atomic energy would, they promised, power the world.

Yet even the peaceful applications of atomic energy were fraught with political machinations, and there was no uniform plan of development. Although the USA did relax its stringent controls and start disseminating nuclear products, this policy stemmed from tactical self-interest rather than scientific altruism. By dispensing atomic expertise, America could appear generous but also make more profit for its own industries and consolidate its global strength through building up blocs of allies. International power networks shifted as African and Asian countries pursued their own political objectives, forcing their way into inter-national discussions by trading on their nuclear capabilities and their uranium reserves. European nations also followed diverse agendas. For instance, Britain started to build nuclear power stations with great enthusiasm, but ground to a halt through a mixture of administrative muddle and growing awareness of the long-term dangers. Conversely, France followed Britain's lead, but soon raced ahead to generate three-quarters of its electricity atomically.

The power struggles of the Cold War made science itself into a political instrument. In diplomatic manoeuvres and commercial negotiations, scientific know-how provided a powerful lever for nations struggling to carve out their independence. As the Indian Prime Minister Jawaharlal Nehru declared, the bomb had demonstrated that economic and military strength 'stem from science and if India is to progress and become a strong nation, second to none, we must build up our science.'[10] Some poorer regions took advantage of their geographical location, converting themselves into indispensable sources of scientific data by building observatories high up unpolluted mountains or at unique equatorial sites. Wealthier ex-colonies—Australia, Canada—concentrated on building high-tech research stations that operated without interference from either Britain or the USA. A more devious tactic was to boycott international projects—scientists could exert political pressure by refusing to collaborate with researchers from specific countries. When the Cold War drew to a close, the head-on rivalry between the USA and the USSR faded away, but science was still generating the power that fuelled world politics.

6 Environment

Praise without end the go-ahead zeal
of whoever it was invented the wheel;
but never a word for the poor soul's sake
that thought ahead, and invented the brake.

—Howard Nemerov, 'To the Congress
of the United States, Entering Its
Third Century' (1989)

When the French explorer Louis de Bougainville wandered round Tahiti in 1768, he enthused that 'I was transported into the Garden of Eden; we crossed a turf, covered by fine fruit trees, and intersected by little rivulets...everywhere we found hospitality, ease, innocent joy, and every appearance of happiness.'[11] Although European visitors soon corrupted this earthly paradise with technological gimmicks and sexually transmitted diseases, they continued to regard the Pacific region as an idyllic Arcadia. Current concerns about the survival of the planet have refuelled such romantic visions of a vanished golden age when natural harmony reigned, unthreatened by ozone holes or vanishing species.

However, preserving the environmental purity of the past is not as straightforward as it sounds. For one thing, much of nature is unnatural: scenes that seem eternal are man-made products. Britain, for example, was originally covered in dense woodland, and bore little resemblance to idealized versions such as that shown in Figure 59, a poster from World War II. This apparently timeless vista of large open fields was fashioned only in the eighteenth century, when wealthy landowners decided to make their farms more profitable by obliterating the small strips of land allocated to individual families. Far from being conservationists, these agricultural reformers overrode protests about decimating traditional village life in order to create the pastures that now characterize picture-book Britain.

Safeguarding the environment might seem a universal ideal, but it is a political issue that has been adopted by very different factions. While this wartime propaganda was urging loyal Britons to fight for the sake of their imagined sun-bathed countryside, their enemies across the Channel (here just glimpsed in the distance)

FIG. 59 'Your Britain...Fight for it Now'. World War II poster by Frank Newbould.

were summoning up nature to back the Nazi cause. Adolf Hitler was a vegetarian whose regime reforested arable land, dispensed organic herbal medicines, and instigated research programmes into natural therapies. His right-hand aide Hermann Goering is now deplored for founding the Gestapo and the concentration camps, but he was also a pioneer environmentalist. After Poland's original woodlands were devastated by the German Occupation, Goering restocked its newly created parks with the animals that had once been native, including a herd of superb bison (a potent Teutonic emblem) bred according to the most recent techniques of post-Darwinian eugenicists. Despite the genocidal campaigns he launched against human beings, Goering insisted that this primeval forest was a sacred grove whose animal inhabitants should remain untouched.

Nature often looks better when it is artificial. That is why the gardener Capability Brown fabricated tranquil English landscapes by digging lakes, planting trees, and moving whole villages—including their inhabitants. The naturalist John Muir was enraptured by the serene Californian meadowlands he visited, but he chose to ignore the influence of Native American farmers who had been fire-clearing the original forest for centuries. The artist James Audubon made a small fortune by selling exotic pictures of birds, carefully crafting them in his studio to symbolize American values of strength and freedom as they soared against painted backdrops of remote mountains. Yet Audubon was no conservationist, but a keen huntsman who obsessively tracked down the rarities he needed to complete his collection, not caring about the risks of extinguishing threatened species.

The appeal of wild nature is relatively recent. For millennia, wilderness was something to push back and overcome as people struggled to carve out a comfortable existence from their hostile surroundings. Survival depended on taming nature, so that barren mountains and dense forests were regarded as suitable for social outcasts, for sinners banished from God's Garden of Eden. Such harsh domains started to become fashionable only a couple of hundred years ago. As the products of civilization started to seem less attractive, Romantic travellers described how they reached states of near-religious ecstasy through contemplating the sublime beauty of precipitous gorges or gloomy cathedral-like groves. After venturing overseas to other continents, they recounted that they had voyaged back in time, encountering primitive societies where life was easier and purer.

This double yearning for the sublime and the primitive manifested itself particularly strongly in the USA. During the nineteenth century, Romantic writers envisaged pioneers steadily pushing back the frontier between the wild and the civilized as they moved ever westwards. This triumphant vision was marred by nostalgic regrets that Americans were losing contact with their immigrant origins as progress obliterated the authentic experiences of the first rugged settlers. To resolve this sentimental dilemma, enterprising naturalists established national parks with a double purpose—to provide sanctuaries for refugees from successful capitalism, and to stand as monuments to America's pioneering spirit.

Most famously, Muir—originally a farmer born in Scotland, nowadays celebrated as the founding father of environmentalism—set about converting Yosemite into a man-made wilderness zone. He intended his national park to appear primal, even though it had never before existed as he designed it. Apparently oblivious to the inherent ironies of their mission, Muir and his contemporaries worked with biblical zeal, aiming not to simulate the grim realities of frontier survival, but instead to resurrect the original Garden of Eden. But manufacturing uninhabited glades of harmony meant forcibly clearing out the indigenous residents, many of whom were slaughtered or subjected to misery in reservations. To guarantee safe access and stop nature from ruining the carefully selected views, conservationists constructed discreetly camouflaged tracks and embarked on continuous maintenance programmes.

Restoring an imagined natural past has always been an expensive business. It also entails interference and oppression—ejecting American Indians from Yosemite, ripping out family strip-farms to make enclosures, relocating villages. Nowadays, privileged eco-tourists who live in cities campaign to preserve endangered species and keep vast tracts of untamed nature as retreats from urban pressures. Maintaining biodiversity may seem a more worthwhile and more scientific ideal than Muir's bid for an original terrestrial paradise. Nevertheless, just as in Yosemite, establishing uninhabited wilderness has entailed evicting the local inhabitants. In the interests of preservation, many victims of involuntary resettlement—Thai, Kenyans, Amazonian Indians—have become conservation refugees confined to shabby squatter camps.

The central paradox is that people are themselves part of nature. In 1964, American conservation law defined wilderness as a place 'where man himself is a visitor who does not remain'—but if people are excluded from nature, then it is intrinsically artificial. In Figure 59, the man blends into the countryside, as much a part of England's natural inheritance as the trees and the animals. Striding through the gentle scenery of rounded hills, this lone shepherd marshals his flock of sheep, an image replete with Christian symbolism. In the Bible, God gave human beings a dual responsibility—to be the world's custodians, but also to exploit it for their own benefit. This mixed message still dogs environmental concerns.

To express this conflict scientifically, the drive to conserve is at odds with the competitive struggle for survival entailed in human evolution. This Darwinian inheritance was interpreted in different ways during the second half of the nineteenth century. Wealthy capitalists justified their cut-throat tactics through invoking the mantra coined by Darwin's disciples—'survival of the fittest'. However, the very success of this ruthless formula prompted critics to focus on its flip-side of exploitation, and in Germany, a very different Darwinian champion appeared— Ernst Haeckel. While American environmentalists were attempting to resurrect uncorrupted paradise, Haeckel was initiating a less oppressive, more holistic approach to biology that strongly influenced later environmentalist movements.

The science of ecology was founded by Haeckel, who invented the word in 1866. Although it has now acquired a moral spin—ecological washing powder is virtuous as well as expensive—ecology started out as the study of the relationships between living creatures and their surroundings. Like 'economy', it comes from the Greek word for a family household, and Haeckel suggested that all the Earth's organisms coexist as a single integrated unit, competing against each other but also offering mutual aid. According to Haeckel's version of Darwinian evolution, if people are to flourish, then they should respect the laws of this universal system instead of trying to dominate it. This non-exploitative approach appealed to Haeckel's disciples, especially in Germany, where their mystical *philosophies* tried to restore a spiritual dimension to the physical universe.

Physicists were also becoming increasingly concerned about the Earth's future. As they tried to make factory equipment more efficient (and hence more profitable), they formulated the laws of thermodynamics, which state mathematically that unless there is some input from outside, the total amount of energy available for use must inevitably diminish. When they started to think about the Universe itself as one massive self-contained machine, scientists realized with alarm that it could grind to a halt. In the worst-case scenario, everything would be uniformly cold, and information would cease to flow—put more technically, the disorder of entropy would have reached a maximum. To slow down the pace of deterioration and safeguard the future, physicists campaigned for waste to be reduced and for non-renewable resources to be conserved.

During the first half of the twentieth century, ecologists synthesized these bio-logical and physical approaches to create a revised vision of nature as a grand eco-nomic machine. Borrowing from the language of industry, they invented a new ecological vocabulary. They started talking about food chains with roots in the humblest of Earth's factory workers—bacteria and plants—that lead up through networks of animal producers to reach human beings, the highest level consum-ers. They reconceived energy as an agent of exchange, the natural equivalent of the currency driving human economies, and they replaced collaborative com-munities of living organisms with ecosystems, in which plants buy in solar energy and repackage it to be stored up for later availability. Such concepts have become everyday terms in global politics, but they originated in ecologists' micro-studies of Thames-side woods and Illinois cornfields.

Once the world had been visualized as a machine, then it seemed right—nat-ural even—for human beings to intervene and make it run more effectively. One approach was to increase nature's productivity through technological manipu-lation. Engineers constructed dams and irrigation schemes, while agricultural experts turned to the chemical industry for help, producing pesticides that enabled farmers to boost their bank balances through increasing their yields. Yet before long, ecologically minded scientists were carrying out research projects which highlighted the knock-on effects of wiping out crop-eating insects, or the risks of flooding some areas to leave others deprived of water. They used these results to emphasize the dangers of short-term policies that converted nature into a set of high-performance components racing at maximum output.

These early protests appeared mainly in academic articles, but they achieved a far greater impact when objectors went public. The environment became a major issue not only because scientific attitudes were altering, but also because the media were expanding—especially television, the brand-new publicity opportunity which became virtually universal during the last third of the twen-tieth century. Scientists in many disciplines grasped the opportunity to promote themselves to wide attentive audiences, and esoteric topics—black holes, genetic decoding, chaos theory—became familiar (at a superficial level, anyway) through documentaries and magazine articles. Yet the influence worked both ways. This public exposure made scientists more vulnerable to criticism, so that their plans and aspirations became subjected to new constraints. Researchers realized that in order to gain financial backing, they needed not only to prove a project's scientific validity, but also to demonstrate its political, commercial, or ethical significance. Gradually, they learnt how to manipulate the media and secure funding through making overblown pronouncements of revolutionary breakthroughs or impending catastrophes.

The early debates about running planet Earth were sparked off by Rachel Carson, a marine biologist who worked for the US government. In 1962, she published *Silent Spring*, a title designed to evoke a potentially near future when all the world's birds have been killed off by toxic chemicals. Carson wrote poetically, but she had compiled her facts thoroughly and she made the scientific arguments easy to follow. She also appreciated the power of bluntness. 'For the first time in the history of the world,' Carson warned her readers, 'every human being is now subjected to contact with dangerous chemicals, from the moment of conception until death'.[12]

Silent Spring made a huge impact. As well as its horror stories about carcinogenic sprays, poisoned reservoirs, and falling birth-rates, the book's critiques resonated with wider Cold War concerns about the twinned power of science and the state. In line with other protest movements of the 1960s, Carson called for citizens to exert more control over their own destiny. An American civil servant, she wrote with an insider's authority when she attacked politicians' failure to challenge the conclusions of self-interested scientists. As Carson explained, the governments that allowed the atmosphere to be polluted with DDT were also sanctioning nuclear programmes generating invisible radiation. Unsurprisingly, political and industrial leaders joined forces to disparage this presumptuous woman who had dared to criticize the establishment by presenting scientific information in a demystified form that everyone could understand.

Campaigns to protect the environment became linked with opposition to conventional government. In Germany, for instance, the Green Party was a powerful political force in the 1970s, despite embarrassing historical associations with the Nazi Party's support of back-to-nature movements. More generally, disillusionment with state-run science forced researchers to recognize the importance of gaining public approval as well as official backing for their projects. By the 1970s, debates about the environment were being conducted in the press and on television. For example, James Lovelock, a chemist specializing in pollution, gained huge publicity for his Gaia hypothesis, even though it was panned by orthodox scientists. In his interactive model, Lovelock imagined the Earth as a giant self-regulating system, a quasi-organic being who can protect herself against the damage induced by human beings. Evocatively wrapped in spiritual packaging, Lovelock's holistic alternative to materialistic science and its technological products gained enormous popular support.

In contrast, most scientists preferred to pursue the centuries-old mechanical approach that had previously brought them so much success—splitting up the world into smaller components which can be tackled separately. That methodology of breaking down a problem into manageable pieces had worked extremely well inside laboratories, but proved less effective when dealing with global phenomena. Take the weather. Meteorologists realized that analysing the atmosphere as separate orderly units was of little help, because even the slightest perturbation could thwart their logical predictions. Chaos reigned, they concluded, so that—in the media-savvy summary—a butterfly flapping its wings in Brazil might appear to cause a tornado in Texas. To cope with this unwanted noise, climatologists applied brute computing force, an option newly available in the 1970s. Yet although they devised ever more complex programs to simulate the atmosphere's behaviour, bigger did not necessarily become better—the digitized models became over-burdened with refinements and riddled with undetectable errors. As critics pointed out, once their virtual structure approached that of the Earth itself, they would become as cumbersome as reality.

For environmentalists, the most effective way of garnering public support and winning government money was to predict catastrophe. One NASA scientist explained frankly to his television listeners, 'It's easier to get funding if you can show some evidence for impending climate disasters…science benefits from scary scenarios.'[13] During the 1970s, climate experts were adamant that another ice age was looming—they insisted that statistical analyses of historical data proved that the globe would once again freeze over into inaction. Twenty years later, that icy vision of the future had been displaced by global warming. According to the latest interpretation, the effects generated by two centuries of industrialization are over riding natural variations in the Earth's climate. By the end of the twentieth century, the debates about global warming had become vitriolic as accusations of vested interests escalated. Experts argued about the relative merits of different specialized techniques, but at the same time, motives were being weighed up alongside facts. Non-scientists wanted to behave like global citizens who cared about the future, but they had become suspicious of scientific pronouncements laden with conflicting conclusions and bitter denunciations.

Over the past fifty years, media-sensitive scientists have learnt that the best way of attracting public attention and government funding is to deliver apocalyptic prognoses—nuclear devastation, meteoric bombardment, an impending ice age, global warming. Modern scientific forecasters seem to fulfil the same psychological needs as religious prophets who preached that the end of the world represents God's punishment of the sinful. In that sense, global warming is more rewarding than an ice age because blame can be assigned to the human race. In contrast with natural disasters, the greenhouse effect and the thinning of the ozone layer are attributed to the industrial activities that drive modern profit-based capitalism. Following this rhetoric, people may be guilty of destroying the world on which they depend, but scientists are offering them the possibility of redemption through altering their behaviour. By enlisting public cooperation to think green and rescue the environment, scientists convert themselves from agents of destruction into secular saviours.

7 Futures

But at my back I always hear
Time's winged chariot hurrying near:
And yonder all before us lie
Deserts of vast eternity.

—Andrew Marvell, 'To his coy
mistress' (1681)

Predicting the future is a hazardous business. Towards the end of the nineteenth century, Western Union rejected telephones as useless and Lord Kelvin pronounced that heavier-than-air flying machines were impossible. Their misplaced caution was, however, trumped by the chairman of IBM, who in 1943 envisaged a world market for five computers. In contrast, technological prophets have—unsurprisingly—more often been over-optimistic about the possibilities opened up by new inventions. When the poet Percy Bysshe Shelley was an Oxford undergraduate, he enthused that electricity would keep poor people warm all winter while balloons glided silently over Africa to map its interior and annihilate slavery for ever. Like so many utopian visionaries, Shelley had not yet learnt that technical feasibility alone is not enough—political motivation is also essential.

Improving the future has been a scientific ideal for the past three hundred years. Progress first became a key leitmotif during the Enlightenment, when reformers declared that the best way forwards was to encourage science. Ever since then, scientific enthusiasts have repeatedly promised that investing in research would make a country richer and help its citizens live better. And they were right; if anything, they underestimated the extent to which science would transform society and dominate the globe. Although the future is unknowable, many further improvements can be forecast with confidence—new drugs will appear, the Internet will become more versatile, genetic techniques will improve, computers will get even cheaper, smaller, and all-pervasive. For people in the right places, life has become longer and more comfortable than it used to be, and will continue do so.

But in some parts of the world, conditions have deteriorated. Futurologists predict that technology is bound to continue on its upward curve, carried on by its own momentum. Similarly, some downward trends have also got to a point where they are hard to reverse. Natural resources will become even scarcer, infectious diseases will proliferate, and overcrowding in urban slums will soar still higher. The net outcome of scientific innovation has apparently been to widen the division between rich and poor, not reduce it. Now that scientific research is so tightly tied together with political interests, deciding how to tackle disparities is a global issue in which all the world's citizens have a stake.

Time after time, technophiles have predicted that one or another new invention would change human behaviour and revolutionize society. During the past couple of centuries, the world effectively shrunk as first trains and ships, then telephones and radio, and now the Internet have successively made it easier for people to communicate with one another. Yet despite the hopes that readier contact would bring closer understanding, world peace is no nearer. The opposite technological route to global harmony—increasingly powerful weapons for cowing enemies into submission—has proved equally ineffective. Another social benefit that has been repeatedly claimed for technical innovation is equality. According to this line of argument, new inventions liberate oppressed groups. At various times in the past, technological optimists have predicted that textile workers would benefit from factory automation, that women would be emancipated by washing machines and vacuum cleaners, and that racial discrimination would vanish in the age of computers. If only.

One way of predicting the future is to plot the number of inventions that appear each year, and then extrapolate forwards. However, to concentrate exclusively on dates can be a misleading way of mapping technological progress. Another way of thinking about it is to consider how many people are using different pieces of technology: take-up may reveal more than novelty. Looking back over the past, it is clear that even the most successful inventions have not necessarily eliminated older ones, many of which have continued to stay in use or even gained strength. The American horse population went on expanding long after cars were being mass produced, because animals were needed to power agricultural equipment. Although bombs, poison gas, and other outstanding military innovations of the twentieth century were heavily publicized, in terms of effectiveness—numbers killed—guns scored higher. Now, unlike thirty years ago, twice as many bicycles as motor cars are being manufactured every year.

During the twentieth century, governments increasingly encouraged their citizens to pin hopes for the future on science and technology. But when it arrived, the future did not always conform to expectations. DDT boosted agricultural production but decimated the countryside, nuclear reactors conserved coal but accidents unleashed radiation, and the Internet promised democratic access but enabled pornography to flourish. In medicine, the single most acclaimed invention was penicillin, converted into a usable drug during World War II to save the lives of wounded soldiers. Fifty years later, hospitals were plagued by resistant bacteria, and in the USA, half the antibiotics being manufactured were going into animal food as growth promoters.

Nevertheless, science's undoubted successes make it the best prospect for creating a better future. After World War II, when the industrialized nations banded together to remedy world poverty, they recommended science as the obvious cure, and set in place development schemes that aimed to bring the poorer parts of the world up to the same technological level as their own. Although in many ways enormously successful, these projects were not as disinterested as they might seem, but were permeated with political interests. By distributing the benefits of their industrial expertise around the globe, rich and powerful nations reinforced their power. Spreading science reinforced supremacy, so that philanthropic development programmes were also a disguised form of imperialism.

Scientific politics went global during the Cold War, when a new international player appeared on the scene—the Third World, a term invented in 1952. While the western bloc of North America and Europe was pitted against the Soviet-dominated East, there was also a conflict between the 'North' and the 'South'. These inverted commas are a short-hand way of indicating that the globe can be metaphorically divided in two—the wealthy industrialized powers (originally in north-west Europe, but now including Australia) were competing to win the allegiance of the poorer Third World nations, many of which (but not all) lay in the southern hemisphere. As illustrated by Figure 57 (Gagarin's Russian space-ship flying over Africa), technological assistance was one of the major bribes on offer. In these asymmetrical 'North–South' interactions, scientific help came with a political price tag. Although less visible during the Cold War, these 'North South' interactions were in many ways just as important as the East–West hostilities.

'Northern' wealth ensured the dominance of the scientific and technological styles of dealing with the world that had proved so successful for industrialized nations. Such scientific-centrism is insidious because it organizes the entire globe in its own image. The problem is not so much that it ignores other ways of thinking, but that it makes it impossible to think in any other way. The upside-down Australian map at the beginning of this book (Figure 1) was designed to protest against narrow assumptions that the European perspective is the only one possible. More generally, it reveals the arrogance underpinning not only terrestrial geography, but also Northern views of knowledge and beliefs in general.

The very concept of development implies not only that modern science and technology are intrinsically superior, but also that the 'North' knows best. Intrinsically flawed, development strategies for improving the future failed to iron out scientific inequalities. Distributing high-tech scientific apparatus to Third

World nations made the powerful donor look good, but was not necessarily the best solution for poverty. Accepting such gifts entailed political subordination and imposed modernity on people who did not necessarily want it. For example, Colombia and Paraguay agreed to receive a US nuclear reactor not because they needed to generate energy or detonate a bomb, but because they wished to display their political loyalty. Instead of developing, many poor nations became poorer, or even destitute, during the second half of the twentieth century.

Development projects pushed Third World countries into a state of scientific dependency, in which they were obliged to rely on equipment produced by 'Northern' industrialized nations. They were denied the possibility of scientific equality, because the only research projects set up within poorer countries were dedicated to technological applications carrying an immediate social benefit. However generous the donations, they were dedicated to alleviating poverty, not to creating rival research centres. Abstract theoretical investigations, the high-status aspect of science, remained the privilege of rich countries, who objected to directing funds away from their own institutions. Educational projects did encourage children in poorer nations to study science, but if they wanted to pursue a career as a professional scientist, they were forced to emigrate. Scientific research became a luxury that the Third World could not afford.

Scientific development programmes were designed to help the Third World catch up with the industrialized 'North'. However, instead of the world's countries drawing closer together, they headed off in different directions, rather like evolving species who irrevocably diverge. The poorer nations were not simply passive recipients of 'Northern' methods and equipment, but instead adapted what they were given to carve out their own technological routes towards the future. Rather than importing modern cars or motorbikes, people devised their own means of transport to fit local conditions—cycle rickshaws in India and Singapore, boats driven by borrowed irrigation pumps in Bangladesh. The massive shanty towns of Africa and Asia seem uninhabitable when viewed from the outside, but they manage to function because local inventors have produced new versions of existing materials such as corrugated iron, now relatively unusual in 'Northern' nations, and asbestos-cement, banned elsewhere on safety grounds.

Decisions about using science and technology carried great political implications. In several countries, manufacturers retained hand-operated tools which were labour intensive—sewing machines, for instance—in preference to building expensive factories relying on foreign automated equipment. Technological and political power are bonded together. As Mahatma Gandhi put it, he hoped to reject mass production in favour of production by the masses, and the Indian flag now features a spinning wheel to symbolize independence from industrial Britain. Towards the end of the twentieth century, this reliance on individual activity enabled Indian computer programmers to undercut American ones—and because of this skilled labour force, India started to emerge as an international political force to be reckoned with.

The most ambitious scientific development project became known as the Green Revolution. In the mid-1960s, governments and international organizations decided to tackle world poverty by transforming global agriculture. To eliminate starvation and increase food output in heavily populated areas, they started replacing traditional methods with the latest scientific techniques. These economic philanthropists promised that by adopting science-based agriculture, Third World nations would be able to support themselves, or even make a financial profit by exporting tropical fruit and vegetables to 'Northern' countries. In addition to introducing chemical fertilizers and industrial irrigation schemes, within a few years scientists were disseminating seeds that had been genetically engineered by the new specialists of biotechnology.

Genetic engineering initially sounded like a contradiction in terms, because it metaphorically brought together two disciplines that had previously been placed at opposite ends of a spectrum ranging from the hard masculine sciences to the soft feminine ones. Biologists associated themselves with industrialists by adopting mechanical terms such as splicing and cutting to describe how they were exploring the double helix of DNA through manipulating its chemical groups. Genetic modification was not in itself new. The opening chapters of Charles Darwin's book on evolution describe how farmers and pigeon breeders used selective breeding to style cattle and birds for carrying out particular tasks. In contrast, the new biotechnologists alter genes from the inside. And like manufacturing entrepreneurs, they have set up commercial companies to market products derived from the natural world.

At first, scientific remedies for global poverty seemed to be working splendidly. By 1980, India had become self-sufficient in wheat and rice, while several other areas of the world were reporting record harvests. Science was apparently living up to its reputation of providing a miracle recipe to guarantee prosperity. Nevertheless, the Green Revolution was already under heavy attack from disillusioned critics. Many sceptics focused on the environmental damage caused by this unprecedented transfer of plants from one part of the world to another, which generated all sorts of unanticipated consequences. Powerful chemicals were being introduced to improve the barren soil or cope with tropical pests, and their deleterious effects rippled out along food chains. Water diversion projects altered existing drainage systems, so that although some places benefited from bumper yields, others experienced droughts.

Genetic modification came under particularly heavy fire. On the plus side, plants that had been specially engineered to cope with local conditions were converting large tracts of barren land into high-yielding fertile fields. Yet as an unanticipated consequence, the survival of these artificially adapted plants meant that other species were dying out. Opponents emphasized the possibilities of some nightmarish scenarios. If crops are tailor-made to repel insect infestations, then cross-breeding might lead to resistant weeds, which could proliferate out of control. Or as another grim possibility, if one super-crop replaces thousands of different varieties, then it runs the risk of being wiped out if some super-predator should appear in the future. In Europe, although not in the USA, genetically modified (GM) products were denounced as 'Frankenfoods'.

In addition, the Green Revolution had adverse social consequences because its very makeup incorporated political power structures. Instead of importing food, poorer nations were now buying in the expensive chemicals, seeds, and expertise they needed to maintain their altered style of agriculture. Whereas the profits of large landowners with affluent contacts soared, small farmers were squeezed out of business and migrated to swell the urban slums still further. GM organisms were being produced in distant research laboratories, pouring 'southwards' from rich countries to poor ones—notably from North to South America. In contrast, financial benefits flowed in the opposite direction: the manipulated genes originated in local plants and were being drained 'northwards' to boost the profit and prestige of biotechnology companies.

Development implies that poorer countries will be fast-tracked to achieve financial and scientific equality with their privileged helpers. But just as was happening with heavy industrial technology, the Green Revolution resulted in irreversible changes that increased divergence. All over the world, scientific research was being dedicated to modernizing agricultural techniques, but they became far more efficient far more quickly in wealthy countries, where governments could afford to protect their own producers against cheap imports. Whereas idealistic reformers once envisaged a rosy future in which poorer countries would feed the industrialized 'North', the reverse was taking place. For example, the USA started selling wheat to the USSR and exporting raw cotton to China, where newly established factories processed it into clothing suitable for rich Americans.

Finding fault with science is relatively easy—the problem is deciding how to improve its impact. Reactionaries have always protested about declining standards, insisting that the future can only get worse. Romantic technophobes lambast innovation and yearn after an imaginary ideal past, insisting that science has generated new means of destruction, not only by producing powerful bombs but also by becoming a political weapon in its own right. Writing on word-processors in the comfort of their centrally heated homes, they deploy selective vision to ignore the countless ways in which scientific research has resulted in undeniable benefits.

The problem is not that scientific technology is in itself bad, but that it can too easily become a tool for domination and coercion. Futuristic predictions abound of the technological marvels that lie ahead in the twenty-first century, such as nanotransponders (miniature brain implants designed to link human beings into global electronic networks), artificial genes, and fuel cells (tiny perpetual batteries based on chemical interactions). At the same time, environmentalists are issuing urgent warnings that the human race will pollute itself out of existence unless steps are taken now to prevent further global warming. Studying the past makes it clear that choosing a route towards the future is not just a matter of getting the scientific equations right, but also entails making worthwhile political decisions.

Postscript

People have always tried to make sense of the world about them. However, looking at the past makes it clear that there is no single correct way of imposing order, even though earlier schemes can seem strange. Mediaeval Aristotelians saw three colours in the rainbow, whereas Newton split it into seven; before clockmakers started measuring out time in equal units, days and nights were divided into hours of varying length; and while many collectors arranged flowers according to their colours or leaves, Linnaeus classified them numerically by their reproductive organs. The system that you are brought up with is the one that seems obvious—any others feel intuitively wrong, however rationally they may have been constructed.

The Argentinian writer Jorge Luis Borges highlighted this taxonomic dilemma by imagining a Chinese encyclopaedia that sorts animals into the following:

> (a) those that belong to the Emperor, (b) embalmed ones, (c) those that are trained, (d) suckling pigs, (e) mermaids, (f) fabulous ones, (g) stray dogs, (h) those that are included in this classification, (i) those that tremble as if they were mad, (j) innumerable ones, (k) those drawn with a very fine camel's hair brush, (l) others, (m) those that have just broken a flower vase, (n) those that resemble flies from a distance.[14]

Although it might be possible to envisage real-life situations in which an individual category would be useful, this fictional list is rife with ambiguities and overlaps, and mocks the universality aimed at by scientific classifiers.

Borges devised this parody for his enigmatic account of John Wilkins, a seventeenth-century scholar who set out to create an international language by organizing the Universe into forty classes, each subdivided into smaller sets allocated their own identifying symbols. Once versed in his method, Wilkins argued, readers would be able not only to understand what a word meant but also to work out how the object or idea it represented fitted into the grand scheme of things. Which sounds fine—until you realize that, from a modern perspective, Wilkins's categories appear as unsatisfactory as those of Borges's fictional encyclopaedia. For example, he has four types of stone—ordinary, precious, transparent, and insoluble.

Should slate be slotted into ordinary or insoluble? Are sapphires transparent or precious? A system that seemed universally valid to Wilkins would be useless for modern mineralogists.

Far from being an isolated crank, Wilkins was a leading light of the early Royal Society—he even chaired the meeting at which it was founded. Since Wilkins was a major player in formulating the Society's experimental approach, he might be considered one of England's first scientists. On the other hand, looking back, Wilkins refuses to fit any modern classification scheme. For one thing, he was ordained, holding several church positions before eventually being consecrated Bishop of Chester. Furthermore, as well as working on his visionary philosophical language, Wilkins dedicated himself to several projects nowadays not seen as legitimate science—perpetual motion, magical illusions, naval vocabulary, secret codes.

As Borges concluded in his cautionary tale, all human schemes are provisional. Now that science dominates the world, it is hard to believe that only two hundred years ago, the word 'scientist' had not even been invented. Over the past few millennia, many peoples—Babylonians and Chinese, farmers and navigators, colonizers and slaves, miners and monks, Muslims and Christians, astrophysicists and biochemists—have contributed towards building up our current understanding of the cosmos. Like human societies, knowledge is never definitively fixed, but is constantly changing as old categories dissolve and new ones coalesce. There can be no cast-iron guarantee that the cutting-edge science of today will not represent the discredited alchemy of tomorrow. Even so, one thing is certain: science has changed the Universe and its inhabitants forever.

Notes

These notes provide references for quotations only. Please see the section on Special Sources for additional information.

1 Origins

1 SEVENS

1. Quoted in David Brown, *Mesopotamian Planetary Astronomy-Astrology* (Groningen: Styx, 2000), 151, 135 (with slight changes).

2 BABYLON

2. Quoted in Eleanor Robson. More than Metrology: Mathematics Education in an Old Babylonian Scribal School', in John M. Steele and Annette Imhausen (eds), *Under One Sky: Astronomy and Mathematics in the Ancient Near East* (Münster: Ugarit-Verlag, 2002), 325–65, esp. 349–52.

5 LIFE

3. John Locke, *An Essay concerning Human Understanding* (Oxford: Clarendon Press, 1975), 446–7 (book III, ch. 6, section 12).

4. Charles Singer, *Galen: On Anatomical Procedures* (London: Oxford University Press, 1956), 190.

6 MATTER

5. Democritus, Fragment 125.

7 TECHNOLOGY

6. Samuel Johnson, Preface, *A Dictionary of the English Language* (1755), unpaginated.

II *Interactions*

1. Sir Robert Gorden Menzies, quoted in *Sydney Morning Herald*, 27 April 1939.

2. Quoted in Nathan Sivin, 'Science in China's Past', in Leo A. Orleans (ed.), *Science in Contemporary China* (Stanford: Stanford University Press, 1980), 1–29, esp. 6.

3. Quoted in Nathan Sivin, 'Shen Gua', in *Dictionary of Scientific Biography*, ed. Charles C. Gillispie, 16 vols (New York: Scribner and Sons, 1970–80), xii.369–93, esp. 390.

4. Needham quoted in Toby E. Huff, *The Rise of Early Modern Science: Islam, China and the West* (Cambridge: Cambridge University Press, 1993), 314.

5. From *The Rubaiyat of Omar Khayyam*, trans. Edward Fitzgerald (1879).

6. Abu Yūsuf Ya'qūb ibn Ishāq Al-Kindī, quoted in David C. Lindberg, *The Beginnings of Western Science: The European Scientific Tradition in Philosophical, Religious, and Institutional Context, 600 BC to AD 1450* (Chicago/London: University of Chicago Press, 1992), 176.

7. Quoted ibid.

8. Roger Bacon, *Opus Maius*, quoted in David C. Lindberg, *The Beginnings of Western Science*, 226.

9. David C. Lindberg, *Roger Bacon's Philosophy of Nature* (Oxford: Clarendon Press, 1985), 5 (slightly altered from Lindberg's translation).

10. From Albert's commentary on Aristotle's *De Anima* (*On the Soul*), quoted in Edward Grant, *The Foundations of Modern Science in the Middle Ages* (Cambridge: Cambridge University Press, 1996), 164.

11. Sir Andrew Aguecheek and Sir Toby Belch, in William Shakespeare, *Twelfth Night*, I.iii.

12. Roger Bacon, *Excellent Discourse of the Admirable Force and Efficacie of Art and Nature*, opening sentence quoted in Stanton J. Linden, *The Alchemy Reader: From Hermes Trismegistus to Isaac Newton* (Cambridge: Cambridge University Press, 2003), 13.

III *Experiments*

I EXPLORATION

1. Letter of 4 July 1471, quoted in *Dictionary of Scientific Biography*, ed. Charles C. Gillispie, 16 vols (New York: Scribner and Sons, 1970–80), xi.351.

2. Quoted in Paula Findlen, 'Inventing Nature: Commerce, Art, and Science in the Early Modern Cabinet of Curiosities', in Pamela H. Smith and Paula Findlen (eds), *Merchants and Marvels: Commerce, Science and Art in Early Modern Europe* (New York/London: Routledge, 2002), 297–323, 299 (Paris, 3 Feb. 1644).

2 MAGIC

3. John Maynard Keynes, 'Newton, the Man', in *The Royal Society Newton Tercentenary Celebrations* (Cambridge: Cambridge University Press, 1947), 27–34, esp. 27.

4. William Shakespeare, *The Tempest*, I.ii.399–406.

5. William Shakespeare, *A Midsummer Night's Dream*, II.i.166–7.

6. John Dee, 'Preface', in *The Elements of Geometrie of the Most Auncient Philosopher Euclide of Megara*, trans. Henry Billingsley (London, 1570), cuts Aj and Aij.

3 ASTRONOMY

7. Quoted in Michael Hoskin (ed.), *Astronomy* (Cambridge: Cambridge University Press, 1997), 119.

8. From the frontispiece of *The Starry Messenger* (1610), quoted in Mario Biagioli, *Galileo, Courtier: The Practice of Science in the Culture of Absolutism* (Chicago/London: Chicago University Press, 1993), 103.

4 BODIES

9. William Harvey, *The Circulation of the Blood and Other Writings*, trans. Kenneth Franklin (London: Everyman, 1990), 46.

10. Ibid. 3.

11. John Aubrey, quoted in Andrew Weir's introduction to ibid. p. xxv.

5 MACHINES

12. Quoted from *L'Homme* in Stephen Gaukroger, *Descartes's System of Natural Philosophy* (Cambridge: Cambridge University Press, 2002), 180.

13. Descartes's response to Frans Burman, quoted in John Cottingham, *Descartes* (Oxford: Basil Blackwell, 1986), 120–1.

14. Robert Boyle, *Notion of Nature*, quoted in William B. Ashworth, 'Christianity and the Mechanistic Universe', in David C. Lindberg and Ronald L. Numbers (eds), *When Science and Christianity Meet* (Chicago/London: University of Chicago Press, 1993), 61–84, esp. 79.

6 INSTRUMENTS

15. Francis Bacon, *The New Organon*, ed. Lisa Jardine and Michael Silverthorne (Cambridge: Cambridge University Press, 2000), 69 (Book I, Aphorism LXXXIV).

16. Robert Hooke, *Micrographia* (London, 1665), p. 4 of unpaginated Preface.

17. Ibid. 210–11.

18. Isaac Newton to Edmond Halley, letter, 20 June 1686, *The Correspondence of Isaac Newton*, ed. H. W. Turnbull et al., 7 vols (Cambridge: Cambridge University Press, 1959–77), ii.437.

7 GRAVITY

19. George Byron, *Don Juan* (Harmondsworth: Penguin, 1973), 375 (Canto X, stanzas 1–2).

20. William Stukeley, *Memoirs of Sir Isaac Newton's Life, being some Account of his Family and Chiefly of the Junior Part of his Life*, ed. A. Hastings White (London: Taylor and Francis, 1936), 20.

21. Isaac Newton to Robert Hooke, letter, 5 Feb. 1676, *Correspondence*, i.416.

22. Letter to Willian Derham, quoted in Stephen D. Snobelen, 'On Reading Isaac Newton's *Principia* in the 18th Century', *Endeavours*, 22 (1998), 159–63, esp. 159.

23. Letter to Caroline of Ansbach, Nov. 1715, quoted in H. G. Alexander, *The Leibniz-Clarke Correspondence* (Manchester: Manchester University Press, 1956), 11.

24. François-Marie Arouet Voltaire, *Letters on England*, trans. L. Tancock (Harmondsworth: Penguin, 1980), 68.

25. Stephen Hales, *Vegetable Staticks*, ed. M. A. Hoskin (London : Oldbourne, 1969), 147.

26. John Theophilus Desaguliers, *The Newtonian System of the World, the Best Model of Government: An Allegorical Poem* (London, 1728), 22–4.

27. Xavier Bichat quoted in Thomas S. Hall, 'On Biological Analogs of Newtonian Paradigms', *Philosophy of Science*, 35 (1968), 6–27, esp. 6.

IV *Institutions*

I SOCIETIES

1. John Beale, quoted in Michael Hunter, *Science and Society in Restoration England* (Cambridge: Cambridge University Press, 1981), 195.

2. Quoted in J. E. McClellan, *Science Reorganised: Scientific Societies in the Eighteenth Century* (New York: Columbia University Press, 1985), 212.

3. Robert Walton, in Mary Shelley, *Frankenstein or The Modern Prometheus: The 1818 Text* (Oxford/New York: Oxford University Press, 1993), 7.

4. Quoted in Richard Drayton, *Nature's Government: Science, Imperial Britain, and the 'Improvement of the World'* (New Haven/London: Yale University Press, 2000), 104.

2 SYSTEMS

5. Both examples from Richard Yeo, *Encyclopaedic Visions: Scientific Dictionaries and Enlightenment Culture* (Cambridge: Cambridge University Press, 2001), 31.

6. Quoted in L. Schiebinger, *Nature's Body: Gender in the Making of Modern Science* (Boston: Beacon Press, 1993), 22–3.

3 CAREERS

7. Lord Camden, quoted in William Cobbett (ed.), *The Parliamentary History of England from the Earliest Period to the Year 1803*, vols 13–36 (London: Longman, 1812–20), xvii.999–1000 (1774).

8. Benjamin Martin, *The Young Gentleman and Lady's Philosophy*, 2 vols (London: 1759–63), i.319.

9. Donald F. Bond, *The Spectator*, 5 vols (Oxford: Clarendon Press, 1965), i.44 (12 March 1711).

10. Humphry Davy, *The Collected Works of Sir Humphry Davy*, ed. John Davy, 9 vols (London: Smith, Elder, 1839–40), ii.319 (1802 lecture on chemistry).

4 INDUSTRIES

11. David Miller, '"Puffing Jamie": The Commercial and Ideological Importance of Being a "Philosopher" in the Case of the Reputation of James Watt (1736–1819)', *History of Science*, 38 (2000), 1–24, esp. 2.

12. Arthur Young, quoted from *Annals of Agriculture* (1785) in Francis D. Klingender, *Art and the Industrial Revolution* (London: Paladin, 1968), 77.

13. Joseph Priestley, *Experiments and Observations on Different Kinds of Air* (London: J. Johnson, 1774–7), vol. i, p. xiv.

14. Letter to Thomas Bentley, 1769, quoted in Neil McKendrick, 'Josiah Wedgwood and Factory Discipline', *Historical Journal*, 4 (1961), 30–55, esp. 34.

15. James Boswell, quoted in Jenny Uglow, *The Lunar Men: The Friends Who Made the Future, 1730–1810* (London: Faber and Faber, 2002), p. xi.

16. Quoted in Jan Golinski, *Science as Public Culture: Chemistry and Enlightenment in Britain, 1760–1820* (Cambridge: Cambridge University Press, 1992), 147.

17. Erasmus Darwin, *Loves of the Plants* (London: J. Johnson, 1794), canto II, ll. 99–104.

18. Friedrich Engels, quoted in Francis Wheen, *Karl Marx* (London: Fourth Estate, 1999), 81.

5 REVOLUTIONS

19. Le Turc, 1794, quoted in Margaret Jacob, *Scientific Culture and the Making of the Industrial West* (New York/Oxford: Oxford University Press, 1997), 165.

20. Davy, *Collected Works*, viii.282 (1808 lecture on electro-chemical science).

21. Quoted from Max Planck, *A Scientific Autobiography* (1949), in Gerard Holton, *Thematic Origins of Scientific Thought: Kepler to Einstein* (Cambridge, MA: Harvard University Press, 1973), 394.

6 RATIONALITY

22. Probably Augustus de Morgan, quoted in Charles Couston Gillispie, *Pierre-Simon Laplace, 1749–1827: A Life in Exact Science* (Princeton: Princeton University Press, 1997), 272.

7 DISCIPLINES

23. Jane Austen, *Pride and Prejudice* (1813; Ware: Wordsworth, 1992), 22.

24. Adam Sedgwick, quoted in James A. Secord, *Victorian Sensation: The Extraordinary Publication, Reception, and Secret Authorship of* Vestiges of the Natural History of Creation (Chicago/London: University of Chicago Press, 2000), 405.

V *Laws*

2 GLOBALIZATION

1. From *Personal Narrative,* quoted in Mary Louise Pratt, *Imperial Eyes: Travel Writing and Transculturation* (London/New York: Routledge, 1992), 130.

2. William Thomson (1883), quoted in Crosbie Smith and M. Norton Wise, *Energy and Empire: A Biographical Study of Lord Kelvin* (Cambridge: Cambridge University Press, 1989), 455.

3 OBJECTIVITY

3. *Times Literary Supplement* (17 March 1927), 167.

4. Two British doctors of 1867, quoted in Thomas L. Hankins and Robert Silverman, *Instruments of the Imagination* (Princeton, NJ: Princeton University Press, 1995), 138.

5. Gertrude M. Prescott, 'Faraday: Image of the Man and the Collector', in David Gooding and Frank James (eds), *Faraday Rediscovered: Essays on the Life and Work of Michael Faraday, 1791–1867* (New York: Macmillan, 1985), 15–32, esp. 17.

6. William Farr, quoted in G. Gigerenzer et al, *The Empire of Chance: How Probability Changed Science and Everyday Life* (Cambridge: Cambridge University Press, 1989), 38.

4 GOD

7. Quoted in Frank M. Turner, *Contesting Cultural Authority: Essays in Victorian Intellectual Life* (Cambridge: Cambridge University Press, 1993), 192.

8. Robert Chambers, quoted in Theodore Porter, *The Rise of Statistical Thinking, 1820–1900* (Princeton, NJ: Princeton University Press, 1986), 57.

9. James Hutton, quoted in David Goodman and Colin A. Russell, *The Rise of Scientific Europe 1500–1800* (Kent: Hodder and Stoughton, 1991), 291, 293.

10. Thomas Henry Huxley, 'On a Piece of Chalk' (1868), reproduced in Alan P. Barr, *The Major Prose of Thomas Henry Huxley* (Athens, GA/London: University of Georgia Press, 1997), 154–73, esp. 156.

11. Alfred Tennyson, *In Memoriam*, in *Poems*, ed. Christopher Ricks (London: Longmans, 1969), 909, 973 (sect. 54, l. 5; sect. 123, ll. 1–4).

5 EVOLUTION

12. Letter, Charles Darwin to Charles Lyell, quoted in James A. Secord, *Victorian Sensation: The Extraordinary Publication, Reception, and Secret Authorship of* Vestiges of the Natural History of Creation (Chicago/London: University of Chicago Press, 2000), 431.

13. Charles Darwin, *On The Origin of Species* (Oxford: Oxford University Press, 1996), 396.

14. From Charles Darwin, *The Descent of Man* (1871), 119, quoted in Evelyn Richards, 'Redrawing the Boundaries: Darwinian Science and Victorian Women Intellectuals', in Bernard Lightman (ed.), *Victorian Science in Context* (Chicago/London: University of Chicago Press, 1987), 119–42.

6 POWER

15. H. G. Wells, *The Time Machine* (1895; London: Pan, 1953), 94.

16. Pierre Duhem, quoted in Iwan Rhys Morus, *When Physics Became King* (Chicago: Chicago University Press, 2005), 85.

17. Hermann von Helmholtz, 'The Interaction of Natural Forces', in *Science and Culture: Popular and Physical Essays*, ed. David Cahan (Chicago: University of Chicago Press, 1990), 20.

18. Quoted in Zaheer Baber, *The Science of Empire: Scientific Knowledge, Civilisation, and Colonial Rule in India* (Albany: State University of New York Press, 1996), 254.

7 TIME

19. Quoted in Simon Schaffer, 'Accurate Measurement is an English Science', in M. Norton Wise (ed.), *The Values of Precision* (Princeton, NJ: Princeton University Press, 1995), 135–72, esp. 136.

20. John Scott Haldane, quoted in Ronald Clark, *Einstein: The Life and Times* (London: Hodder and Stoughton, 1973), 412.

VI *Invisibles*

I LIFE

1. Mary Shelley, *Frankenstein or The Modern Prometheus: The 1818 Text*, ed. Marilyn Butler (Oxford/New York: Oxford University Press, 1993), 38–9.

2. Quoted from *Anthropogenie* in Nick Hopwood, 'Pictures of Evolution and Charges of Fraud: Ernst Haeckel's Embryological Illustrations', *Isis*, 97 (2006), 260–301, esp. 291.

2 GERMS

3. Quoted in Fiona Haslam, *From Hogarth to Rowlandson: Medicine in Art in Eighteenth-Century Britain* (Liverpool: Liverpool University Press, 1996), 236.

4. James Young-Simpson, quoted in John Waller, *Fabulous Science: Fact and Fiction in the History of Scientific Discovery* (Oxford: Oxford University Press, 2002), 163.

5. Quoted in Susan Sontag, *Illness as Metaphor* (London: Allen Lane, 1979), 7.

6. Mrs Alving in Act 2, in Henrik Ibsen, *Ghosts*, trans. Christopher Hampton (New York: Samuel French, 1983), 47.

3 RAYS

7. William Crookes, 'Spiritualism Viewed by the Light of Modern Science', *Quarterly Journal of Science*, 7 (1870), 316–21, reprinted in Noel G. Coley and Vance M. D. Hall (eds), *Darwin to Einstein: Primary Sources on Science and Belief* (Harlow: Longman/ Open University, 1980), 60–3, esp. 61.

8. Quoted in Iwan Rhys Morus, *When Physics Became King* (Chicago: Chicago University Press, 2005), 186.

9. Quoted in John Waller, *Fabulous Science: Fact and Fiction in the History of Scientific Discovery* (Oxford: Oxford University Press, 2002), 43.

10. Quoted in Abraham Pais, *Inward Bound: Of Matter and Forces in the Physical World* (Oxford/New York: Oxford University Press, 1986), 189.

11. Ernest Rutherford, *The Newer Alchemy* (Cambridge: Cambridge University Press, 1937), 65.

4 PARTICLES

12. Stanford and Berkeley logbook of 1974, quoted in Peter Galison, *How Experiments End* (Chicago/London: University of Chicago Press, 1987), 1.

5 GENES

13. Charles Darwin, *The Descent of Man*, quoted in Tim Lewens, *Darwin* (London/ New York: Routledge, 2007), 216.

14. James Barr, 'Some Eugenic Ideals', in *King Albert's Book: A Tribute to the Belgian King and People from Representative Men and Women throughout the World*, quoted in Nicholas Humphrey, 'History and Human Nature', *Prospect* (Sept. 2006), 126.

15. A. N. Studitskii, 'Fly-Lovers—Man-Haters', *Ogonek* (13 Mar. 1949), 14–16. I am very grateful to Simon Franklin for translating this article for me.

6 CHEMICALS

16. Karl Vogt, quoted in Roy Porter, *The Greatest Benefit to Mankind: A Medical History of Humanity from Antiquity to the Present* (London: HarperCollins, 1997), 329.

17. Anonymous claim of 1933, quoted in Nelly Oudshoorn, *Beyond the Natural Body: An Archaeology of Sex Hormones* (London/New York: Routledge, 1994), 93.

7 UNCERTAINTIES

18. Quoted from *Why War?* in Ronald Clark, *Einstein: The Life and Times* (London: Hodder and Stoughton, 1973), 348.

VII *Decisions*

1 WARFARE

1. Bertrand Russell, 'Philosophy and Politics', in *Unpopular Essays* (London: Allen and Unwin, 1950), 9–34, esp. 18.

2. Quoted in Richard Rhodes, *The Making of the Atomic Bomb* (London: Penguin, 1986), 89.

3. Laura Fermi, *Atoms in the Family: My Life with Enrico Fermi* (Chicago: Chicago University Press, 1954), 173.

4. Otto Frisch, quoted in G. I. Brown, *Invisible Rays: The History of Radioactivity* (Stroud: Sutton, 2002), 125.

2 HEREDITY

5. J. D. Watson and F. H. C. Crick, 'A Structure for Deoxy Ribose Nucleic Acid', *Nature*, 171 (25 Apr. 1953), 737–8.

6. James D. Watson, *The Double Helix* (London: Penguin, 1997), 132.

3 COSMOLOGY

7. George Johnson, *Miss Leavitt's Stars: The Untold Story of the Woman Who Discovered How to Measure the Universe* (New York: Norton, 2005), 108.

5 RIVALRY

8. Richard Porter, 'Introductory Remarks', *Journal of Geophysical Research*, 64 (1959), 865–7.

9. Quoted in the editors' introduction to John Krige and Kai-Henrik Barth (eds), *Global Power Knowledge: Science and Technology in International Affairs* (Chicago: Chicago Univesity Press, 2006), 5 (*Osiris*, vol. 21: 'Historical Perspectives on Science, Technology, and International Affairs').

10. Quoted in Itty Abraham, 'The Ambivalence of Nuclear Histories', in Krige and Barth (eds), *Global Power Knowledge*, 49–65, esp. 62.

6 ENVIRONMENT

11. Louis de Bougainville, quoted in Bernard Smith, *European Vision and the South Pacific* (Melbourne: Oxford University Press, 1989), 42.

12. Rachel Carson, *Silent Spring* (London: Penguin, 1999), 31.

13. Roy Spencer, quoted in D. Jones, *The Greenhouse Conspiracy* (London: Channel 4, 1990), 24.

POSTCRIPT

14. Jorge Luis Borges, 'The Analytical Language of John Wilkins', in *Other Inquisitions 1937–52*, trans. Ruth Simms (New York: Washington Square Press, 1966), 108.

Photographic Acknowledgements
(by figure number)

akg-images: **42**; Musée Condé, Chantilly/akg-images: **2**; Argonne National Laboratory, USA: **51**; © S. McArthur/Artarom: **1**; The Art Archive: **7, 9, 13, 35**; The British Library/ The Art Archive: **10**; Ironbridge Gorge Museum/The Art Archive: **27**; Courtesy of the Warden & Scholars of New College, Oxford/The Bridgeman Art Library: **21**; The Science Museum/The Bridgeman Art Library: **32**; Yale Center for British Art, Paul Mellon Collection, USA/The Bridgeman Art Library: **20**; The British Museum: **26**; Cambridge University Library: **24, 33, 34, 38**; Bettmann/Corbis: **50**; Derby Museums and Art Gallery: **22**; Mary Evans Picture Library: **31, 36, 43**; Courtesy of Roger Gaskell Rare Books: **11, 16**; Time Life Pictures/Getty Images: **55**; Sonia Halliday Photographs: **8**; Oxford Science Archive/Heritage Image Partnership: **54**; Nick Hopwood: **41**; Imperial War Museum: **59**; Institut International de Physique Solvay, Brussels: **49**; © 1942 The Kosciuszko Foundation: **14**; Leiden University Library: **6**; The Metropolitan Museum of Art, purchase, Mr. and Mrs. Charles Wrightsman Gift, in honour of Everett Fahy, 1977 (1977.10): **28**; Museum of the History of Science, Oxford: **40**; courtesy of NASA: **58**; National Portrait Gallery, London: **48**; The Royal Observatory, Edinburgh: **5**; The Royal Society: **23, 29**; The National Gallery/photo © 2005 Ann Ronan/HIP/Scala, Florence: **12**; Science Photo Library: **53**; Carl Anderson/Science Photo Library: **45**; Antony Barrington-Brown/ Science Photo Library: **52**; George Bernard/Science Photo Library: **44**; University College London Library: **3**; US Army Photo: **56**; Wellcome Library, London: **30**; Whipple Museum of the History of Science, University of Cambridge: **4**

Special Sources

As *Science: A Four Thousand Year History* is intended to provide an introductory overview of science's past, I have not included the full academic apparatus of footnotes, although I have specified the origin of all my direct quotations. I am indebted to the work of many, many scholars, and a full reading list would be far too long. However, I express my special gratitude to the authors of the following books and articles, on which I leant particularly heavily.

Introduction

I first saw the Australian map of the world in Jeremy Black, *Maps and Politics* (London: Reaktion, 1997).

1 *Origins*

1 SEVENS

I took several examples of special sevens from Annemarie Schimmel, *The Mystery of Numbers* (New York/Oxford: Oxford University Press, 1993), 127–55.

2 BABYLON

I am indebted to Eleanor Robson for advice about ancient Babylon, and I relied greatly on her groundbreaking paper, 'More than Metrology: Mathematics Education in an Old Babylonian Scribal School', in John M. Steele and Annette

Imhausen (eds), *Under One Sky: Astronomy and Mathematics in the Ancient Near East* (Münster: Ugarit-Verlag, 2002), 325–65. My other major specialized sources were David Brown, *Mesopotamian Planetary Astronomy-Astrology* (Groningen: Styx, 2000) and Francesca Rochberg, *The Heavenly Writing: Divination, Horoscope, and Astronomy in Mesopotamian Culture* (Cambridge: Cambridge University Press, 2004).

3–7 HEROES TO TECHNOLOGY

My major sources were the two classic books by Geoffrey E. R. Lloyd: *Early Greek Science: Thales to Aristotle* (London: Chatto and Windus, 1970) and *Greek Science after Aristotle* (London: Chatto and Windus, 1973). I also used material from Andrew Gregory, *Eureka! The Birth of Science* (Duxford: Icon, 2001) and Serafina Cuomo, *Pappus of Alexandria and the Mathematics of Late Antiquity* (Cambridge: Cambridge University Press, 2000).

II *Interactions*

I EUROCENTRISM

I relied on three main sources: Zachary Lockman, *Contending Visions of the Middle East: The History and Politics of Orientalism* (Cambridge: Cambridge University Press, 2004), esp. 8–65; John M. Hobson, *The Eastern Origins of Western Civilisation* (Cambridge: Cambridge University Press, 2004), esp. chs. 1 and 5; and Julia M. H. Smith, *Europe after Rome: A New Cultural History* (Oxford: Oxford University Press, 2005), esp. Introduction and ch. 8.

2 CHINA

I took my summary of Needham's life and impact from the *Oxford Dictionary of National Biography* article by Gregory Blue, and also Francesca Bray, 'Eloge of Joseph Needham', *Isis*, 87 (1996), 312–17. My major guide around Joseph Needham's *Science and Civilisation in China* was vol. 7.2, edited by Kenneth Robinson and with a valuable introduction by Mark Elvin. I relied heavily on two overviews by Nathan Sivin: 'Science in China's Past', in Leo A. Orleans (ed.), *Science in Contemporary China* (Stanford: Stanford University Press, 1980), 1–29; and 'Editor's Introduction', in Joseph Needham (ed.), *Science and Civilisation in China*, Vol. 6.6 (Cambridge: Cambridge University Press, 2000), 1–37 (this has a particular emphasis on medicine). For general analyses, I relied on Toby E. Huff, *The Rise of Early Modern Science: Islam, China and the West* (Cambridge: Cambridge University Press, 1993), 237–320; and John M. Hobson, *The Eastern Origins of Western Civilisation* (Cambridge: Cambridge University Press, 2004), ch. 3. My account of Shen Gua is based mainly on Nathan Sivin, 'Shen Gua', in *Dictionary of Scientific Biography*, xii.369–93, which also provides an excellent survey of Chinese civilization around the eleventh century. I learned of Wang Ho from William H. McNeill, *The Pursuit of Power: Technology, Armed Force, and Society since AD 1000* (Oxford: Basil Blackwell, 1983), 41.

3–4 ISLAM AND SCHOLARSHIP

My guiding text for representing an Islamic perspective was Seyyed Hossein Nasr, *Science and Civilisation in Islam* (Cambridge, MA: Harvard University Press, 1968). I also turned to Michael Hoskin (ed.), *Astronomy* (Cambridge: Cambridge University Press, 1997) and Toby E. Huff, *The Rise of Early Modern Science: Islam, China and the West* (Cambridge: Cambridge University Press, 1993). For the Baghdad translation project, I used Dimitri Gutas, *Greek Thought, Arabic Culture: The Graeco-Arabic Translation Movement in Baghdad and Early 'Abbāsid Society* (London: Routledge, 1998).

5 6 EUROPE AND ARISTOTLE

My major standard sources were David C. Lindberg, *The Beginnings of Western Science: The European Scientific Tradition in Philosophical, Religious, and Institutional Context, 600 BC to AD 1450* (Chicago/London: University of Chicago Press, 1992); and Edward Grant, *The Foundations of Modern Science in the Middle Ages* (Cambridge: Cambridge University Press, 1996). My examples of agricultural change came from John M. Hobson, *The Eastern Origins of Western Civilisation* (Cambridge: Cambridge University Press, 2004), ch. 9. For the trade windows at Chartres, I relied on Jane Welch Williams, *Bread, Wine, and Money: The Windows of the Trades at Chartres Cathedral* (Chicago/London: University of Chicago Press, 1993). My understanding of mediaeval optics is taken mainly from chs. 3 and 4 of Dalibor Vesely, *Architecture in the Age of Divided Representation: The Question of Creativity in the Shadow of Production* (Cambridge, MA: MIT Press, 2004). My major sources for time and clocks were Jo Ellen Barnett, *Time's Pendulum: The Quest to Capture Time—From Sundials to Atomic Clocks* (New York/London: Plenum, 1998); David S. Landes, *Revolution in Time: Clocks and the Making of the Modern World* (Cambridge, MA/London: Harvard University Press, 1983); and Samuel Macey, *Clocks and the Cosmos: Time in Western Life and Thought* (Hamden, CT: Archon Books, 1980).

7 ALCHEMY

For analyses, I relied mainly on two recent guides—Bruce T. Moran, *Distilling Knowledge: Alchemy, Chemistry, and the Scientific Revolution* (Cambridge, MA/London: Harvard University Press, 2005), and the short introduction in Stanton J. Linden, *The Alchemy Reader: From Hermes Trismegistus to Isaac Newton* (Cambridge: Cambridge University Press, 2003). I also used William Eamon, *Science and the Secrets of Nature: Books of Secrets in Medieval and Early Modern Culture* (Princeton: Princeton University

Press, 1994), esp. 15–90, and W. F. Ryan and Charles B. Schmitt (eds), *Pseudo-Aristotle The 'Secret of Secrets': Sources and Influences* (London: Warburg Institute, 1982).

III *Experiments*

1 EXPLORATION

My major sources for printing, commodification, and communication were Lisa Jardine, *Worldly Goods: A New History of the Renaissance* (London: Macmillan, 1996); and Jessica Wolfe, *Humanism, Machinery, and Renaissance Literature* (Cambridge: Cambridge University Press, 2004), esp. 96–103 for Holbein, covered in detail in Susan Foister, Ashok Roy, and Martin Wyld, *Holbein's Ambassadors* (London: National Gallery Publications, 1997), esp. 30–43. I relied heavily on the following collections, which include some marvellous essays: Pamela H. Smith and Paula Findlen (eds), *Merchants and Marvels: Commerce, Science and Art in Early Modern Europe* (New York/London: Routledge, 2002), esp. the editors' Introduction, ch. 1 (Larry Silver and Pamela Smith, 'The Powers of Nature and Art in the Age of Dürer'), ch. 2 (Pamela Long, 'Objects of Art/Objects of Nature'), and ch. 12 (Paula Findlen's 'Inventing Nature' on cabinets of curiosity); N. Jardine, J. A. Secord, and E. C. Spary (eds), *Cultures of Natural History* (Cambridge: Cambridge University Press, 1996), esp. ch. 2 (William Ashworth, 'Emblematic Natural History of the Renaissance', the source of my comments on Gesner's fox) and ch. 4 (Paula Findlen, 'Courting Nature', on court natural history); Londa Schiebinger and Claudia Swan (eds), *Colonial Botany: Science, Commerce, and Politics in the Early Modern World* (Philadelphia: University of Pennsylvania Press, 2005), esp. the editors' Introduction, ch. 5 (Daniela Bleichmar, 'Books, Bodies and Fields' on New World medicines in Europe), and ch. 12 (Judith Carney, 'Out of Africa' on rice and other exports). I also incorporated insights from Brian W. Ogilvie, *The Science of Describing: Natural History in Renaissance Europe* (Chicago/London: University of Chicago Press, 2006).

2 MAGIC

For *The Tempest*, I used Frank Kermode's notes in the 1954 Arden edition, and also Frances A. Yates, *Theatre of the World* (London: Routledge and Kegan Paul, 1987); and Charles Nicholl, *The Chemical Theatre* (London: Routledge and Kegan Paul, 1980). In addition to the seminal texts on Agrippan magic—Frances Yates, *Giordano Bruno and the Hermetic Tradition* (London: Routledge and Kegan Paul, 1964) and *The Occult Philosophy in the Elizabethan Age* (London: Routledge and Kegan Paul, 1979)—I also relied on the essays by Brian Copenhaver and William Eamon in David C. Lindberg and Robert S. Westman, *Reappraisals of the Scientific Revolution* (Cambridge: Cambridge University Press, 1990). For John Dee, I turned to Peter

J. French, *John Dee: The World of an Elizabethan Magus* (London: Routledge and Kegan Paul, 1972); and Nicholas H. Clulee, *John Dee's Natural Philosophy: Between Science and Religion* (London/New York: Routledge, 1988); I took the notion of his experimental lifestyle from the splendid article by Deborah E. Harkness, 'Managing an Experimental Household: The Dees of Mortlake and the Practice of Natural Philosophy', *Isis*, 88 (1997), 247–62. For alchemy and Paracelsus, I mainly used Bruce T. Moran, *Distilling Knowledge: Alchemy, Chemistry, and the Scientific Revolution* (Cambridge, MA/London: Harvard University Press, 2005).

3 ASTRONOMY

For Copernicus's strategies and influence, I referred mainly to Owen Gingerich, 'The Copernican Quinquecentennial and its Predecessors: Historical Insights and National Agendas', *Osiris*, 14 (1999), 37–60; and Robert Westman, 'Proof, Poetics, and Patronage: Copernicus's Preface to *De Revolutionibus*', in D. C. Lindberg and R. S. Westman (eds), *Reappraisals of the Scientific Revolution* (Cambridge: Cambridge University Press, 1990), 167–205. For the status of early modern astronomy, I used Westman's 'The Astronomer's Role in the Sixteenth Century: A Preliminary Study', *History of Science*, 18 (1980), 105–47; and Nicholas Jardine, 'The Places of Astronomy in Early-Modern Culture', *Journal for the History of Astronomy*, 29 (1998), 49–62. The best discussion of Tycho Brahe's iconography is in John Robert Christianson, *On Tycho's Island: Tycho Brahe and His Assistants, 1570–1601* (Cambridge: Cambridge University Press, 2000). For Galileo, I turned to Mario Biagioli, *Galileo, Courtier: The Practice of Science in the Culture of Absolutism* (Chicago/London: Chicago University Press, 1993); and David Lindberg, 'Galileo, the Church, and the Cosmos', in David C. Lindberg and Ronald L. Numbers, *When Science and Christianity Meet* (Chicago/London: University of Chicago Press, 1993).

4 BODIES

My major source for Vesalius and Fabricius was Andrew Cunningham, *The Anatomical Renaissance: The Resurrection of the Anatomical Projects of the Ancients* (Aldershot: Scolar Press, 1997). For Vesalius's artistic imagery, I turned to Pamela Long's essay, 'Objects of Art/Objects of Nature', in Pamela H. Smith and Paula Findlen (eds), *Merchants and Marvels: Commerce, Science and Art in Early Modern Europe* (New York/London: Routledge, 2002); and also to ch. 5 of Katharine Park, *Secrets of Women: Gender, Generation, and the Origins of Human Dissection* (New York: Zone Books, 2006).

5 MACHINES

I took comments on time from Harold Cook, 'Time's Bodies', in Pamela H. Smith and Paula Findlen (eds), *Merchants and Marvels: Commerce, Science and Art in Early Modern Europe* (New York/London: Routledge, 2002); and Rob Iliffe, 'The Masculine Birth of Time: Temporal Frameworks of Early Modern Natural

Philosophy', *British Journal for the History of Science*, 33 (2000), 427–53. For scientific aspects of Descartes's thought, I relied on Stephen Gaukroger, *Descartes's System of Natural Philosophy* (Cambridge: Cambridge University Press, 2002); and William B. Ashworth, 'Christianity and the Mechanistic Universe', in David C. Lindberg and Ronald L. Numbers (eds), *When Science and Christianity Meet* (Chicago/London: University of Chicago Press, 1993), 61–84.

6 INSTRUMENTS

I based my analysis of instruments on Jim Bennett's classic paper, 'The Mechanics: Philosophy and the Mechanical Philosophy', *History of Science*, 24 (1986), 1–28. For Hooke, I drew particularly on Michael Dennis, 'Graphic Understanding: Instruments and Interpretation in Robert Hooke's *Micrographia*', *Science in Context*, 3 (1989), 309–64. For instruments as demonstration devices, I used Thomas L. Hankins and Robert Silverman, *Instruments of the Imagination* (Princeton, NJ: Princeton University Press, 1995), and for Newton's prism experiment, Simon Schaffer, 'Glass Works', in David Gooding, Trevor Pinch, and Simon Schaffer (eds), *The Uses of Experiment: Studies in the Natural Sciences* (Cambridge: Cambridge University Press, 1989).

7 GRAVITY

All my sources are listed in the bibliography of my own *Newton: The Making of Genius* (London: Picador, 2002).

IV *Institutions*

I SOCIETIES

For the early Royal Society, I referred to Michael Hunter, *Science and Society in Restoration England* (Cambridge: Cambridge University Press, 1981), and for the Transit of Venus expeditions, to J. E. McClellan, *Science Reorganised: Scientific Societies in the Eighteenth Century* (New York: Columbia University Press, 1985). I based my analyses of Banks and imperialism on John Gascoigne, *Joseph Banks and the English Enlightenment* (Cambridge: Cambridge University Press, 1994), and *Science in the Service of Empire* (Cambridge: Cambridge University Press, 1998); and Richard Drayton, *Nature's Government: Science, Imperial Britain, and the 'Improvement of the World'* (New Haven, CT/London: Yale University Press, 2000).

2 SYSTEMS

I took my account of John Ray from Anna Pavord, *The Naming of Names: The Search for Order in the World of Plants* (London: Bloomsbury, 2005), 372–94. My major source for Linnaeus was Lisbet Koerner, *Linnaeus: Nature and Nation*

(Cambridge, MA/London: Harvard University Press, 1999). For the history of globalization, I was most influenced by C. A. Bayly, *The Birth of the Modern World 1780–1914: Global Connections and Comparisons* (Oxford: Blackwell, 2004), esp. 1–83. I took the nutmeg debate from E. C. Spary, 'Of Nutmegs and Botanists', in Linda Schiebinger and Claudia Swan (eds), *Colonial Botany: Science, Commerce, and Politics in the Early Modern World* (Philadelphia: University of Pennsylvania Press, 2005), and the case of the hermaphrodite monkey from Anna Maerker, 'The Tale of the Hermaphrodite Monkey: Classification, State Interests and Natural Historical Expertise between Museum and Court, 1791–4', *British Journal for the History of Science*, 39 (2006), 29–47. For the arguments about race, I used David Bindman, *Ape to Apollo: Aesthetics and the Idea of Race in the Eighteenth Century* (London: Reaktion, 2002).

3 CAREERS

I took the concept of science's intellectual class system from Bernice A. Carroll, 'The Politics of "Originality": Women and the Class System of the Intellect', *Journal of Women's History*, 2 (1990), 136–63. My analysis of Davy comes from Jan Golinski, *Science as Public Culture: Chemistry and Enlightenment in Britain, 1760–1820* (Cambridge: Cambridge University Press, 1992), and my ideas about *Frankenstein* and the new men of science originated with Ludmilla Jordanova, 'Melancholy Reflection: Constructing an Identity for Unveilers of Nature', in Stephen Bann (ed.), *Frankenstein Creation and Monstrosity* (London: Reaktion Books, 1994), 60–76.

4 INDUSTRIES

I took my ideas on Watt from Christine MacLeod, 'James Watt, Heroic Invention and the Idea of the Industrial Revolution', in Maxine Berg and Kristine Bruland (eds), *Technological Revolutions in Europe: Historical Perspectives* (Cheltenham/Northampton, MA: Edward Elgar, 1998), 96–116; and from David Philip Miller, '"Puffing Jamie": The Commercial and Ideological Importance of Being a "Philosopher" in the Case of the Reputation of James Watt (1736–1819)', *History of Science*, 38 (2000), 1–24. My comments on Africa's importance are based on Joseph E. Inikori, *Africans and the Industrial Revolution in England: A Study in International Trade and Economic Development* (Cambridge: Cambridge University Pres, 2002). The best book on the Lunar Society is Jenny Uglow, *The Lunar Men: The Friends Who Made the Future, 1730–1810* (London: Faber and Faber, 2002). Darwin's exclusion of workers and women from his poetry is analysed by Maureen McNeil, 'The Scientific Muse: The Poetry of Erasmus Darwin', in Ludmilla Jordanova (ed.), *Languages of Nature: Critical Essays on Science and Literature* (London: Free Association Books, 1986), 159–203; and by Janet Browne, 'Botany for Gentlemen:

Erasmus Darwin and *The loves of the plants*', *Isis*, 80 (1989), 593–620; I also used Deborah Valenze, *The First Industrial Woman* (New York/Oxford: Oxford University Press, 1995).

5 REVOLUTIONS

For industrial chemistry, I relied mainly on Colin Russell, *Science and Social Change 1700–1900* (London: Macmillan, 1983), 96–135, but the best account of Enlightenment British chemistry, including Humphry Davy, is Jan Golinski, *Science as Public Culture: Chemistry and Enlightenment in Britain, 1760–1820* (Cambridge: Cambridge University Press, 1992). As a biographical source for Lavoisier, I used Jean-Pierre Poirier, *Lavoisier: Chemist, Biologist, Economist* (Philadelphia: University of Pennsylvania Press, 1993); for his career and iconography, the best analyses are by Marco Beretta, in his 'Chemical Imagery and the Enlightenment of Matter', in William R. Shea (ed.), *Science and the Visual Image in the Enlightenment* (Canton, MA: Science History Publications, 2000), 57–88, and *Imaging a Career in Science: The Iconography of Antoine Laurent Lavoisier* (Canton, MA: Science History Publications, 2001). My account of William Lewis's laboratory (Figure 29) is mostly from F. W. Gibbs, 'William Lewis, MB, FRS (1708–1781)', *Annals of Science*, 8 (1952), 122–51.

6 RATIONALITY

My account of Laplace is based mainly on Robert Fox's essay, 'Laplacian Physics', from *Historical Studies in the Physical Sciences*, 4 (1974), 89–136, reprinted as ch. 18 of R. C. Olby et al (eds), *Companion to the History of Modern Science* (London/New York: Routledge, 1990). The classic account of metrology is Kula Witold, *Measures and Men* (Princeton: Princeton University Press), and I am also indebted to Ken Alder's works, especially *The Measure of All Things: The Seven-Year Odyssey That Transformed the World* (London: Little, Brown, 2002), and his article on French engineering in William Clark et al (eds), *The Sciences in Enlightened Europe* (Chicago/London: University of Chicago Press, 1999).

7 DISCIPLINES

My original introduction to these ideas was Andrew Cunningham, 'Getting the Game Right: Some Plain Words on the Identity and Invention of Science', *Studies in the History and Philosophy of Science*, 19 (1988), 365–89. I took the controversy over 'scientist' from Sydney Ross, '*Scientist*: The Story of a Word', *Annals of Science*, 18 (1962), 65–85, and also from Paul White, *Thomas Huxley: Making the 'Man of Science'* (Cambridge: Cambridge University Press, 2003), esp. the Introduction and Conclusion. Two other major sources for this chapter were James A. Secord, *Victorian Sensation: The Extraordinary Publication, Reception, and Secret Authorship of* Vestiges of the Natural History of Creation (Chicago/London: University of

Chicago Press, 2000); and Martin Rudwick, *The New Science of Geology* (Ashgate: Variorum, 2004).

v *Laws*

I PROGRESS

My major source for publishing and science in the nineteenth century was James A. Secord, *Victorian Sensation: The Extraordinary Publication, Reception, and Secret Authorship of* Vestiges of the Natural History of Creation (Chicago/London: University of Chicago Press, 2000), esp. 41–56, 515–32. My comments on the BAAS and method were based on Richard Yeo, 'Scientific Method and the Image of Science 1831–1891', in Roy MacLeod and Peter Collins (eds), *The Parliament of Science: The British Association for the Advancement of Science 1831–1981* (Northwood, Middx: Science Reviews, 1981), 65–88. I originally came across the intellectual class system in Bernice A. Carroll, 'The Politics of "Originality": Women and the Class System of the Intellect', *Journal of Women's History*, 2 (1990), 136–63. The Manchester weavers are marvellously discussed in Anne Secord, 'Science in the Pub: Artisan Botanists in Early Nineteenth Century Lancashire', *History of Science*, 32 (1994), 269–315; for Mary Anning, see Hugh Torrens, 'Mary Anning (1799–1847) of Lyme: The Greatest Fossilist the World Ever Knew', *British Journal for the History of Science*, 28 (1996), 257–84. The best biography of Mary Somerville is Kathryn A. Neeley, *Mary Somerville: Science, Illumination, and the Female Mind* (Cambridge: Cambridge University Press, 2001); the best accounts of her writing are in *Collected Works of Mary Somerville*, 9 vols, ed. and intro. James A. Secord (London: Thoemmes Continuum, 2004).

2 GLOBALIZATION

My major sources for Humboldt's influential visions of the New World were Mary Louise Pratt, *Imperial Eyes: Travel Writing and Transculturation* (London/New York: Rouledge, 1992), esp. 105–97; and Michael Dettelbach, 'Humboldtian Science', in N. Jardine, J. A. Secord, and E. C. Spary (eds), *Cultures of Natural History* (Cambridge: Cambridge University Press, 1996), 287–304; I also relied on Nancy Leys Stepan, *Picturing Tropical Nature* (London: Reaktion, 2001), 31–56. The innovatory and now classic study of visual languages is Martin Rudwick, 'The Emergence of a Visual Language for Geological Science 1760–1840', *History of Science*, 14 (1976), 149–95. An excellent review of British colonial science in the nineteenth century, with full references, is Mark Harrison, 'Science and the British Empire', *Isis*, 96 (2005), 56–63. The seminal study of early Victorian magnetism is John Cawood, 'The Magnetic Crusade: Science and Politics in Early Victorian Britain', *Isis*, 70 (1979), 493–518.

For telegraphy from a British perspective, I used Iwan Rhys Morus, *Frankenstein's Children: Electricity, Exhibition, and Experiment in Early-Nineteenth-Century London* (Princeton, NJ: Princeton University Press, 1998), 194–230; and for its imperial aspects, the excellent account in Bruce Hunt, 'Doing Science in a Global Empire', in Bernard Lightman (ed.), *Victorian Science in Context* (Chicago/London: Chicago University Press, 1997), 312–33. The best source for Thomson and telegraphy is Crosbie Smith and M. Norton Wise, *Energy and Empire: A Biographical Study of Lord Kelvin* (Cambridge: Cambridge University Press, 1989), 445–94, 649–83.

3 OBJECTIVITY

The classic article on objective representation in the nineteenth century is Lorraine Daston and Peter Galison, 'The Image of Objectivity', *Representations*, 40 (1992), 81–128. For instruments and problems of deciphering, I relied on Thomas L. Hankins and Robert Silverman, *Instruments of the Imagination* (Princeton, NJ: Princeton University Press, 1995), 113–47; and Peter Galison, 'Judgement Against Objectivity', in C. Jones and P. Galison (eds), *Picturing Science Producing Art* (London: Routledge, 1998), 327–59. My major general sources for my discussions of photography were John Tagg, *The Burden of Representation: Essays on Photographies and Histories* (London: Palgrave, 1988); Peter Hamilton and Roger Hargreaves, *The Beautiful and the Damned: The Creation of Identity in Nineteenth Century Photography* (London: Lund Humphries, 2001); Jennifer Tucker, *Nature Exposed: Photography as Eyewitness in Victorian Science* (Baltimore: Johns Hopkins University Press, 2005); in addition, I took many ideas from Alex Soojung-Kim Pang. '"Stars should henceforth register themselves"', *British Journal for the History of Science*, 30 (1997), 177–202; and Holly Rothermel, 'Images of the Sun: Warren De la Rue, George Biddell Airy and Celestial Photography', *British Journal for the History of Science*, 26 (1993), 137–69.

4 GOD

I based my discussion of the Prayer Gauge debate and the Victorian struggles for scientific authority on Frank M. Turner, *Contesting Cultural Authority: Essays in Victorian Intellectual Life* (Cambridge: Cambridge University Press, 1993), 151–200. For social sciences and statistics, I relied on Theodore Porter, *The Rise of Statistical Thinking, 1820–1900* (Princeton, NJ: Princeton University Press, 1986). The best writer on nineteenth-century geology is Martin Rudwick: I referred to the essays collected in his *The New Science of Geology: Studies in the Earth Sciences in the Age of Revolution* (Aldershot: Ashgate, 2004).

5 EVOLUTION

For Robert Chambers, I relied on James A. Secord, *Victorian Sensation: The Extraordinary Publication, Reception, and Secret Authorship of* Vestiges of the Natural

History of Creation (Chicago/London: University of Chicago Press, 2000), and the facsimile edition by University of Chicago Press, 1994, which has an excellent introduction by J. A. Secord. For caricatures and for Darwin's attitudes towards women, I used two essays in Bernard Lightman (ed.), *Victorian Science in Context* (Chicago/London: University of Chicago Press, 1987)—James Paradis, 'Satire and Science in Victorian Culture', 143–75, and Evelyn Richards, 'Redrawing the Boundaries: Darwinian Science and Victorian Women Intellectuals', 119–42.

6 POWER

For basic information, I turned to Iwan Rhys Morus, *When Physics Became King* (Chicago: Chicago University Press, 2005). I took my comparisons of Britain, France, America, and Germany from Terry Shinn, 'The Industry, Research, and Education Nexus', in Mary Jo Nye (ed.), *The Cambridge History of Science*, vol. 5 (Cambridge: Cambridge University Press, 2003), 133–53; and from Joseph Ben-David, *The Scientist's Role in Society: A Comparative Study* (Englewood Cliffs, NJ: Prentice-Hall, 1971). For global modernization, I relied on C. A. Bayly, *The Birth of the Modern World 1780–1914* (Oxford: Blackwell, 2004), esp. 284–324. My other main sources were James R. Bartholomew, *The Formation of Science in Japan: Building a Research Tradition* (New Haven, CT/London: Yale University Press, 1989); Nathan Sivin, 'Science in China's Past', in Leo A. Orleans (ed.), *Science in Contemporary China* (Stanford: Stanford University Press, 1980), 1–29; Zaheer Baber, *The Science of Empire: Scientific Knowledge, Civilisation, and Colonial Rule in India* (Albany: State University of New York Press, 1996); and Satpal Sangwan, 'Indian Response to European Science and Technology 1757–1857', *British Journal for the History of Science*, 21 (1988), 211–32.

7 TIME

I took my account of the Parisian pneumatic clocks from Peter Galison, *Einstein's Clocks, Poincaré's Maps* (London: Hodder and Stoughton, 2003), 92–8; this marvellous book is also my major source for Einstein and relativity. For telegraphy and precision, I relied on Simon Schaffer, 'Metrology, Metrication, and Victorian Values', in Bernard Lightman (ed.), *Victorian Science in Context* (Chicago/London: University of Chicago Press, 1987), 438–74; and Simon Schaffer, 'Accurate Measurement is an English Science', in M. Norton Wise (ed.), *The Values of Precision* (Princeton, NJ: Princeton University Press, 1995), 135–72. I took Einstein as a Germanic hero from Richard Staley, 'On the Histories of Relativity: The Propagation and Elaboration of Relativity Theory in Participant Histories in Germany, 1905–11', *Isis*, 89 (1998), 263–99; and my account of Eddington is closely based on John Earman and Clark Glymour, 'Relativity and Eclipses: The British Eclipse Expeditions of 1919 and their Predecessors', *Historical Studies in Physical Science*, 11 (1980), 49–85.

VI *Invisibles*

I LIFE

My sources for basic information were Peter Bowler, *Evolution: The History of an Idea* (Berkeley: University of California Press, 1984); and William Coleman, *Biology in the Nineteenth Century: Problems of Form, Function, and Transformation* (Cambridge: Cambridge University Press, 1997). I based my comments about Mary Shelley's science on Marilyn Butler's introduction in Mary Shelley, *Frankenstein or The Modern Prometheus: The 1818 Text* (Oxford/New York: Oxford University Press, 1993). The relationships between natural history and biology in the nineteenth century are discussed in Lynn K. Nyhart, 'Natural History and the "New" Biology', in N. Jardine, J. A. Secord, and E. C. Spary (eds), *Cultures of Natural History* (Cambridge: Cambridge University Press, 1996), 426–43. For the Pasteur-Pouchet debate, I relied on Gerald L. Geison, *The Private Science of Louis Pasteur* (Princeton, NJ: Princeton University Press, 1995), 110–42; and the shorter account in John Waller, *Fabulous Science: Fact and Fiction in the History of Scientific Discovery* (Oxford: Oxford University Press, 2002), 15–46. For my account of Haeckel's diagrams, I am indebted to Nick Hopwood, Pictures of Evolution and Charges of Fraud: Ernst Haeckel's Embryological Illustrations', *Isis*, 97 (2006), 260–301.

2 DISEASE

The basic texts I used were Mark Harrison, *Disease and the Modern World: 1500 to the Present Day* (Cambridge: Polity Press, 2004); and William F. Bynum, *Science and the Practice of Medicine in the Nineteenth Century* (Cambridge: Cambridge University Press, 1994). For my analyses of Snow and Lister, I relied mainly on John Waller, *Fabulous Science: Fact and Fiction in the History of Scientific Discovery* (Oxford: Oxford University Press, 2002), 114–31, 160–75. For images of illness, I used Susan Sontag, *Illness as Metaphor* (London: Allen Lane, 1979); and Sander L. Gilman, *Disease and Representation: Images of Illness from Madness to AIDS* (Ithaca, NY/London: Cornell University Press, 1988), 245–72.

3 RAYS

As a standard history of radioactivity, I used G. I. Brown, *Invisible Rays: The History of Radioactivity* (Stroud: Sutton, 2002). For discussions of spiritualism and photography, I relied on Iwan Rhys Morus, *When Physics Became King* (Chicago: Chicago University Press, 2005) and Jennifer Tucker, *Nature Exposed: Photography as Eyewitness in Victorian Science* (Baltimore: Johns Hopkins University Press, 2005). I took my account of N-rays from Mary Jo Nye, *Science in the Provinces: Scientific Communities and Provincial Leadership in France, 1860–1930* (Berkeley/London: University of California Press, 1986), 53–77.

4 PARTICLES

My biographical information about Mendeleev came mainly from B. M. Kedrov's article in the *Dictionary of Scientific Biography*. For cloud chambers, I relied on Peter Galison and Alexi Assmus, 'Artificial Clouds, Real Particles', in David Gooding, Trevor Pinch, and Simon Schaffer (eds), *The Uses of Experiment: Studies in the Natural Sciences* (Cambridge: Cambridge University Press, 1989), 225–74; and Peter Galison, *How Experiments End* (Chicago/London: University of Chicago Press, 1987). For quarks, I turned to Michael Riordan, *The Hunting of the Quark: A True Story of Modern Physics* (New York: Simon and Schuster, 1987); and for mass, to Gordon Kane, 'The Mysteries of Mass', *Scientific American* (July 2005).

5 GENES

For basic information, I turned to Garland Allen, *Life Science in the Twentieth Century* (Cambridge: Cambridge University Pres, 1978); and also to two of Peter Bowler's books, *Evolution: The History of an Idea* (Berkeley: University of California Press, 1984), and *The Fontana History of the Environmental Sciences* (London: Fontana Press, 1992). The strong links between American and German eugenics are exposed in Stefan Kühl, *The Nazi Question: Eugenics, American Racism, and German National Socialism* (New York/Oxford: Oxford University Press, 1994). Mendel's predecessors and life are closely examined in Robert Olby, *Origins of Mendelism* (London: Constable, 1966).

6 CHEMICALS

I took background material on medicine from Roy Porter, *The Greatest Benefit to Mankind: A Medical History of Humanity from Antiquity to the Present* (London: HarperCollins, 1997). To discuss the various anaemias, I used Keith Wailoo, *Drawing Blood: Technology and Disease Identity in Twentieth Century America* (Baltimore/London: Johns Hopkins University Press, 1997), and *Dying in the City of the Blues: Sickle Cell Anemia and the Politics of Race and Health* (Chapel Hill/London: University of North Carolina Press, 2001). For my discussion of Alexander Fleming, I relied on Robert Bud, 'Penicillin and the New Elizabethans', *British Journal for the History of Science*, 31 (1998), 305–33; and on John Waller, *Fabulous Science: Fact and Fiction in the History of Scientific Discovery* (Oxford: Oxford University Press, 2002), 246–67 (and for insulin, 222–45). My accounts of sexual hormones and the pill are based on Nelly Oudshoorn, *Beyond the Natural Body: An Archaeology of Sex Hormones* (London/New York; Routledge, 1994); and Suzanne White Junod and Lara Marks, 'Women's Trials: The Approval of the First Contraceptive Pill', *Journal of the History of Medicine*, 57 (2002), 117–60; I based my comments on Viagra on Malcolm Potts, 'Two Pills, Two Paths: A Tale of Gender Bias', *Endeavour*, 27 (2003), 127–30.

7 UNCERTAINTIES

I took my opening discussion of Einstein and Freud mainly from Ronald Clark, *Einstein: The Life and Times* (London: Hodder and Stoughton, 1973), 297–355. My main sources for discussing Freud's photograph was J. C. Spector, *The Aesthetics of Freud: A Study in Psychoanalysis and Art* (Westport, CT: Praeger, 1972); and for his life, I turned to Peter Gay, *Freud: A Life for our Time* (London: Dent, 1988), and also James Strachey's brief but excellent introduction in Sigmund Freud, *Two Short Accounts of Psycho-analysis* (London: Penguin, 1991). For military psychiatry, I relied on Elaine Showalter, *The Female Malady: Women, Madness, and English Culture, 1830–1980* (London: Virago, 1987), 167–219; and Hans Pols, 'Waking up to Shell Shock: Psychiatry in the US Military during World War II', *Endeavour*, 30 (2006), 144–9.

VII *Decisions*

1 WARFARE

I took my basic account of British science and war from the classic text, Hilary Rose and Steven Rose, *Science and Society* (Harmondsworth: Penguin, 1969). For Big Science and the Manhattan Project, I relied heavily on Jeff Hughes, *The Manhattan Project: Big Science and the Atom Bomb* (Duxford: Icon, 2002); and Richard Rhodes, *The Making of the Atomic Bomb* (London: Penguin, 1986). For a powerful reappraisal of science-technology relationships, I read David Edgerton, *The Shock of the Old: Technology and Global History since 1900* (London: Profile Books, 2007).

2 HEREDITY

For examples of helix iconography, I used Soraya de Chadarevian and Harmke Kamminga, *Representations of the Double Helix* (Cambridge: Whipple Museum, 2002). I took the story of the photograph from Soraya de Chadarevian, 'Portrait of a Discovery: Watson, Crick, and the Double Helix', *Isis*, 94 (2003), 90–105. I based my account of DNA not only on James D. Watson, *The Double Helix* (London: Penguin, 1997), with an introduction by Steve Jones, but also on Garland Allen, *Life Science in the Twentieth Century* (Cambridge: Cambridge University Press, 1978); Horace Freeland Judson, *The Eighth Day of Creation: Makers of the Revolution in Biology* (London: Jonathan Cape, 1979); and Brenda Maddox, *Rosalind Franklin: The Dark Lady of DNA* (London: HarperCollins, 2002). My main sources for the political implications of genes were R. C. Lewontin, *Biology as Ideology: The Doctrine of an Idea* (New York: HarperPerennial, 1991); and Jean-Paul Gaudillière,

'Globalization and Regulation in the Biotech World: The Transatlantic Debates over Cancer Genes and Genetically Modified Crops', *Osiris*, 21 (2006), 251–72.

3 COSMOLOGY

My main sources for Wegener and the relationships between geology and the Earth sciences were Robert Muir Wood, *The Dark Side of the Earth* (London: Allen and Unwin, 1985); David Oldroyd, *Thinking about the Earth: A History of Ideas in Geology* (London: Athlone, 1996), chs. 11 to 13; and Peter Bowler, *The Environmental Sciences* (London: Fontana, 1992), ch. 9. I also benefited from a lecture by Jon Agar, which helped me think anew about the 1960s. I took my account of Leavitt from George Johnson, *Miss Leavitt's Stars: The Untold Story of the Woman Who Discovered How to Measure the Universe* (New York: Norton, 2005). For the changing fortunes of General Relativity, I relied on Jean Eisenstaedt, *The Curious History of Relativity: How Einstein's Theory of Gravity Was Lost and Found Again* (Princeton, NJ: Princeton University Press, 2006). For ways of thinking about science, I was inspired by Peter Dear, *The Intelligibility of Nature: How Science Makes Sense of the World* (Chicago: University of Chicago Press, 2006).

4 INFORMATION

For Alan Turing himself, I referred to the major biography, Andrew Hodges, *Alan Turing: The Enigma* (London: Burnett Books, 1983); for his significance, I relied heavily on Jon Agar, *Turing and the Universal Machine: The Making of the Modern Computer* (Duxford: Icon, 2001). My emphasis on secrecy and the Cold War was inspired by Michael Aaron Dennis, 'Secrecy and Science Revisited: From Politics to Historical Practice and Back', in Ronald E. Doel and Thomas Söderqvist (eds), *The Historiography of Contemporary Science, Technology and Medicine: Writing Recent Science* (London/New York: Routledge, 2006), 172–84; and by Paul N. Edwards, *The Closed World: Computers and the Politics of Discourse in Cold War America* (Cambridge, MA: MIT Press, 1996). The original article accompanying Figure 55, 'The Thinking Machine', is in *Time* (23 Jan. 1950) (available on the Internet).

5 RIVALRY

My main source for the space race was Walter A. McDougall, *The Heavens and the Earth: A Political History of the Space Age* (New York: Basic Books, 1985). For the political implications of nuclear power, I relied heavily on the essays by John Krige and Kai-Henrik Barth (eds), *Global Power Knowledge: Science and Technology in International Affairs* (*Osiris*, vol. 21); in particular, I took the *Star Wars* connection from Sheila Jasanoff, 'Biotechnology and Empire: The Global Power of Seeds and Science', 273–92. I took several comments about the military/scientific production of knowledge from Michael Aaron Dennis, 'Earthly Matters: On the Cold War and the Earth Sciences', *Social Studies of Science*, 33 (2003), 809–19.

6 ENVIRONMENT

I based my discussion of landscape and wilderness on William Cronon, 'The Trouble with Wilderness: Or, Getting Back to the Wrong Nature', in William Cronon (ed.), *Uncommon Ground: Toward Reinventing Nature* (New York/London: Norton, 1995), 23–90; Mark Dowie, 'Conservation Refugees', *Orion* (Nov/Dec 2005); and Simon Schama, *Landscape and Memory* (London: HarperCollins, 1995). For ecology and environment, I relied on Peter J. Bowler, *The Fontana History of the Environmental Sciences* (London: Fontana Press, 1992, esp. 503–53; and Donald Worster, *Nature's Economy: A History of Ecological Ideas* (Cambridge: Cambridge University press, 1977). For Rachel Carson's impact, I used the 'Afterword' by Linda Lear in *Silent Spring* (London: Penguin, 1999). I took my comments on meteorological computer models from Mott T. Greene, 'Looking for a General for Some Modern Major Models', *Endeavour*, 30 (2006), 55–9.

7 FUTURES

My comments on the politicization of development programmes relied on Alexis de Grieff and Mauricio Nieto Olarte, 'What We Still Do Not Know about South-North Technoscientific Exchange: North-Centrism, Scientific Diffusion, and the Social Studies of Science', in Ronald E. Doel and Thomas Söderqvist (eds), *The Historiography of Contemporary Science, Technology and Medicine: Writing Recent Science* (London/New York: Routledge, 2006), 239–59; and on Sheila Jasanoff, 'Biotechnology and Empire: The Global Power of Seeds and Science', in John Krige and Kai-Henrik Barth (eds), *Global Power Knowledge: Science and Technology in International Affairs* (*Osiris*, vol. 21), 273–92. My use-based approach towards technological innovation was taken from David Edgerton, *The Shock of the Old: Technology and Global History since 1900* (London: Profile Books, 2007).

Index

Page numbers in **bold** refer to illustrations.